T0213692

Lecture Notes in Computer Science 9998

Commenced Publication in 1973
Founding and Former Series Editors:
Gerhard Goos, Juris Hartmanis, and Jan van Leeuwen

Shaoxu Song · Yongxin Tong (Eds.)

Web-Age Information Management

WAIM 2016 International Workshops
MWDA, SDMMW, and SemiBDMA
Nanchang, China, June 3–5, 2016
Revised Selected Papers

 Springer

Editors
Shaoxu Song
Tsinghua University
Beijing
China

Yongxin Tong
Beihang University
Beijing
China

ISSN 0302-9743   ISSN 1611-3349   (electronic)
Lecture Notes in Computer Science
ISBN 978-3-319-47120-4   ISBN 978-3-319-47121-1   (eBook)
DOI 10.1007/978-3-319-47121-1

Library of Congress Control Number: 2016940123

LNCS Sublibrary: SL3 – Information Systems and Applications, incl. Internet/Web, and HCI

Printed on acid-free paper

This Springer imprint is published by Springer Nature
The registered company is Springer International Publishing AG
The registered company address is: Gewerbestrasse 11, 6330 Cham, Switzerland

Preface

Web-Age Information Management (WAIM) is a leading international conference for researchers, practitioners, developers, and users to share and exchange their cutting-edge ideas, results, experiences, techniques, and tools in connection with all aspects of Web data management. The conference invites original research papers on the theory, design, and implementation of Web-based information systems. As the 17th event in the increasingly popular series, WAIM 2016 was held in Nanchang, China, during June 3–5, 2016, and it attracted more than 400 participants from all over the world.

Along with the main conference, WAIM workshops intend to provide international forum for researchers to discuss and share research results. This WAIM 2016 workshop volume contains the papers accepted for the following three workshops that were held in conjunction with WAIM 2016. These three workshops were selected after a public call for proposals process, each of which focuses on a specific area that contributes to the main themes of the WAIM conference. The three workshops were as follows:

- The International Workshop on Spatiotemporal Data Management and Mining for the Web (SDMMW 2016)
- The International Workshop on Semi-Structured Big Data Management and Applications (SemiBDMA 2016).
- The International Workshop on Mobile Web Data Analytics (MWDA 2016)

All the organizers of the previous WAIM conferences and workshops have made WAIM a valuable trademark, and we are proud to continue their work. We would like express our thanks to all the workshop organizers and Program Committee members for their great effort in making the WAIM 2016 workshops a success. In total, 27 papers were accepted for the workshops. In particular, we are grateful to the main conference organizers for their generous support and help.

July 2016

Shaoxu Song
Yongxin Tong

Organization

SDMMW 2016

Workshop Chairs

Di Jiang	Beihang University, China
Deqing Wang	Beihang University, China
Hui Zhang	Beihang University, China

Program Committee

Chen Cao	Hong Kong Financial Data Technology, Ltd., SAR China
Yunfan Chen	The Hong Kong University of Science and Technology, SAR China
Yurong Cheng	Northeastern University, China
Xiaonan Guo	Stevens Institute of Technology, USA
Kuiyang Liang	Beihang University, China
Mengxiang Lin	Beihang University, China
Rui Liu	Beihang University, China
Rui Meng	The Hong Kong University of Science and Technology, SAR China
Jieying She	The Hong Kong University of Science and Technology, SAR China
Fabrizio Silverstri	Yahoo Research, UK
Chi Su	Peking University, China
Zhiyang Su	Microsoft, China
Li Zhao	IBM, China

SemiBDMA 2016

Workshop Chairs

Baoyan Song	Liaoning University, China
Linlin Ding	Liaoning University, China
Ye Yuan	Northeastern University, China

Program Committee

Xiangmin Zhou	RMIT University, Australia
Jianxin Li	Swinburne University of Technology, Australia

Bo Ning	Dalian Maritime University, China
Yongjiao Sun	Northeastern University, China
Guohui Ding	Shenyang Aerospace University, China
Bo Lu	Dalian Nationalities University, China
Yulei Fan	Zhejiang University of Technology, China

MWDA 2016

Workshop Chairs

Xiangliang Zhang	King Abdullah University of Science and Technology, Saudi Arabia
Li Li	Southwest University, China
Li Liu	Chongqing University, China

Program Committee

Jiong Jin	Swinburne University of Technology, Australia
Ming Liu	Southwest University, China
Guoxin Su	National University of Singapore, Singapore
Min Gao	Chongqing University, China
Shiping Chen	CSIRO, Australia
Rong Xie	Wuhang University, China
Huawen Liu	Zhejiang Normal University, China
Lifei Chen	Fujian Normal University, China
Basma Alharbi	King Abdullah University of Science and Technology, Saudi Arabia
Ling Ou	Southwest University, China
Zehui Qu	Southwest University, China
Xianchuan Yu	Beijing Normal University, China
Yufang Zhang	Chongqing University, China
Yonggang Lu	Lanzhou University, China

Contents

MWDA 2016

Modeling User Preference from Rating Data Based on the Bayesian Network with a Latent Variable

Renshang Gao[1], Kun Yue[1(✉)], Hao Wu[1], Binbin Zhang[1],
and Xiaodong Fu[2]

[1] Department of Computer Science and Engineering,
School of Information Science and Engineering,
Yunnan University, Kunming, China
kyue@ynu.edu.cn
[2] Faculty of Information Engineering and Automation,
Kunming University of Science and Technology, Kunming, China

Abstract. Modeling user behavior and latent preference implied in rating data are the basis of personalized information services. In this paper, we adopt a latent variable to describe user preference and Bayesian network (BN) with a latent variable as the framework for representing the relationships among the observed and the latent variables, and define user preference BN (abbreviated as UPBN). To construct UPBN effectively, we first give the property and initial structure constraint that enable conditional probability distributions (CPDs) related to the latent variable to fit the given data set by the Expectation-Maximization (EM) algorithm. Then, we give the EM-based algorithm for constraint-based maximum likelihood estimation of parameters to learn UPBN's CPDs from the incomplete data w.r.t. the latent variable. Following, we give the algorithm to learn the UPBN's graphical structure by applying the structural EM (SEM) algorithm and the Bayesian Information Criteria (BIC). Experimental results show the effectiveness and efficiency of our method.

Keywords: Rating data · User preference · Latent variable · Bayesian network · Structural EM algorithm · Bayesian information criteria

1 Introduction

With the rapid development of mobile Internet, large volumes of user behavior data are generated and many novel personalized services are generated, such as location-based services and accurate user targeting, etc. Modeling user preference by analyzing user behavior data is the basis and key of these services. Online rating data, an important kind of user behavior data, consists of the descriptive attributes of users themselves, relevant objects (called items) and the scores that users rate on items. For example, MovieLens data set given by GroupLens [2] involves attributes of users and items, as well as the rating scores. The attributes of users include sex, age, occupation, etc., and the attributes of items include type (or genre), epoch, etc. Actually, rating data reflects user preference (e.g., type of items), since a user may rate an item when he is preferred

© Springer International Publishing AG 2016
S. Song and Y. Tong (Eds.): WAIM 2016 Workshops, LNCS 9998, pp. 3–16, 2016.
DOI: 10.1007/978-3-319-47121-1_1

to this item. Moreover, the rating frequency and corresponding scores w.r.t. a specific type of item also indicate the degree of user preference to this type of item.

In recent years, many researchers proposed various methods for modeling user preference by means of matrix factorization or topic model [13, 17–19, 21]. However, these methods were developed upon the given or predefined preference model (e.g., the topic model is based on a fixed structure), which is not suitable for describing arbitrary dependencies among attributes in data. Meanwhile, the inherent uncertainties among the scores, attributes of users and items cannot be well represented by the given model. Thus, it is necessary to construct a preference model from user behavior data to represent arbitrary dependencies and the corresponding uncertainties.

Bayesian network (BN) is an effective framework for representing and inferring uncertain dependencies among random variables [15]. A BN is a directed acyclic graph (DAG), where nodes represent random variables and edges represent dependencies among variables. Each variable in a BN is associated with a table of conditional probability distributions (CPDs), also called conditional probability table (CPT) to give the probability of each state when given the states of its parents. Making use of BN's mechanisms of uncertain dependency representation, we are to model user preference by representing the arbitrary dependencies and the corresponding uncertainties.

However, latent variables for describing user preference implied in rating data cannot be observed directly, i.e., hidden or latent w.r.t. the observed data. Fortunately, BN with latent variables (abbreviated as BNLV) [15] are extensively studied in the paradigm of uncertain artificial intelligence. This makes it possible to model user preference by introducing a latent variable into BN to describe user preference and represent the corresponding uncertain dependencies. For example, we could use the BNLV ignoring CPTs shown in Fig. 1 to model user preference, where U_1, U_2, I, L and R is used to denote user's sex, age, movie genre, user preference and the rating score of users on movies respectively. Based on this model, we could fulfill relevant applications based on BN's inference algorithms.

Fig. 1. A BNLV ignoring CPTs

Particularly, we call the BNLV as Fig. 1 as user preference BN (UPBN). To construct UPBN from rating data is exactly the problem that we will solve in this paper. For this purpose, we should construct the DAG structure and compute the corresponding CPTs, as those for learning general BNs from data [12]. However, the introduction of the latent variable into BNs leads to some challenges. For example, learning the parameters in CPTs cannot be fulfilled by using the maximum likelihood estimation directly, since the data of the latent variable is missing w.r.t. the observed data. Thus, we use the Expectation-Maximization (EM) algorithm [5] to learn the parameters and the Structural EM (SEM) algorithm [7] to learn the structure respectively. In this paper, we extend the classical search & scoring method that concerns.

It is worth noting that the value of the latent variable in a UPBN cannot be observed, which derives strong randomness if we learn the parameters by directly using EM and further makes the learned DAG incredible to a great extent. In addition, running SEM with a bad initialization usually leads a trivial structure. In particular, if we set an empty graph as the initial structure, then the latent variable will not have connections with other variables [12]. Thus, we consider the relation between the latent and observed variables, and discuss the property as constraints that a UPBN should satisfy from the perspective of BNLV's specialties.

Generally speaking, the main contributions can be summarized as follows:

- We propose user preference Bayesian network to represent the dependencies with uncertainties among latent or observed attributes contained in rating data by using a latent variable to describe user preference.
- We give the property and initial structure constraint that make the CPDs related to the latent variable fit the given rating data by EM algorithm.
- We give a constraint-based method to learn UPBN by applying the EM algorithm and SEM algorithm to learn UPBN's CPDs and DAG respectively.
- We implement the proposed algorithms and make preliminary experiments to test the feasibility of our method.

2 Related Work

Preference modeling has been extensively studied from various perspectives. Zhao et al. [21] proposed a behavior factorization model for predicting user's multiple topical interests. Yu et al. [19] proposed a user's context-aware preferences model based on Latent Dirichlet Allocation (LDA) [3]. Tan et al. [17] constructed an interest-based social network model based on Probabilistic Matrix Factorization [16]. Rating data that represents user's opinion upon items has been widely used for modeling user preference. Matrix factorization and topic model are two kinds of popular methods. Koren et al. [13] proposed the timeSVD ++ model for modeling time drifting user preferences by extending the Singular Value Decomposition method. Yin et al. [18] extended LDA and proposed a temporal context-aware model for analyzing user behaviors. These methods focus on parameter learning of the given or predefined model, but the graph model construction has not been concerned and the arbitrary dependencies among concerning attributes cannot be well described. In this paper, we focus on both parameter and structure learning by incorporating the specialties of rating data.

BN has been studied extensively. For example, Yue et al. [20] proposed a parallel and incremental approach for data-intensive learning of BNs. Breese et al. [4] first applied BN, where each node is corresponding to each item in the domain, to model user preference in a collaborative filtering way. Huang et al. [9] adopted expert knowledge of travel domain to construct a BN for estimating travelers' preferences. In the general BN without latent variables, user preference cannot be well represented due to the missing of corresponding values.

Meanwhile, there is a growing study on BNLV in recent years. Huete et al. [10] described user's opinions of one item's every component by latent variables and constructed the BNLV for representing user profile in line with expert knowledge. Kim et al. [11] proposed a method about ranking evaluation of institutions based on BNLV where the latent variable represents ranking scores of institutions. Liu et al. [14] constructed a latent tree model, a tree-structured BNLV, from data for multidimensional clustering. These findings provide basis for our study, but the algorithm for constructing BNLV that reflects the specialties of rating data should be explored.

3 Basis for Learning BN with a Latent Variable

3.1 Preliminaries

BIC scoring metric is to measure the coincidence of BN structure with the given data set. The greater the BIC score, the better the structure. Friedman [6] gave the expected BIC scoring function for the case where data is incomplete, defined as follows:

$$BIC(G|D^*) = \sum_{i=1}^{m} \sum_{X_i} P(X_i|D_i, \theta^*) \log P(X_i, D_i|G, \theta^*) - \frac{d(G)}{2} \log m. \quad (1)$$

where G is a BN, D^* is a complete data obtained by EM algorithm, θ^* is an estimation of model parameter, m is the total number of samples and $d(G)$ is the number of independent parameters required in G. The first term of $BIC(G|D^*)$ is the expected log likelihood, and the second term is penalty of model complexity [12].

As a method to conduct BN's structure learning w.r.t. incomplete data [7], SEM first fixes the current optimal model structure and exerts several optimizations on the model parameter. Then, the optimizations for structure and parameter are carried out simultaneously. The process will be repeated until convergence.

3.2 Properties of BNLV

Let $X_1, X_2, \ldots, X_n$ denote observed variables that have dependencies with the latent variable respectively. Let Y denote the set of observed variables that have no dependency with the latent variable, and L denote the latent variable. There are three possible forms of local structures w.r.t. the latent variable in a BNLV, shown as Fig. 2, where the dependencies between observed variables are ignored.

Property 1. The CPTs related to the latent variable can fit data sets by EM if and only if there is at least one edge where the latent variable points to the observed variable, shown as Fig. 2 (a).

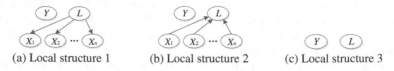

| (a) Local structure 1 | (b) Local structure 2 | (c) Local structure 3 |

Fig. 2. Local structure related to the latent variable

For the situation in Fig. 2 (a), the CPTs related to the latent variable will be changed in the EM iteration, while the CPTs related to the latent variable will be the same as the initial state in the EM iteration by mathematical derivation of EM for the situations in Fig. 2 (b) and (c). For space limitation, the detailed derivation will not be given here. Accordingly, Property 1 implies that a BNLV must contain the substructure shown in Fig. 2 (a) if we are to make the BNLV fully fit the data set.

4 Constraint-Based Learning of User Preference Bayesian Network

Let $U = \{U_1, U_2, ..., U_n\}$ denote the set of user's attributes. Let I denote the type of an item, and $I = c_j$ means that the item is of the jth type c_j. Let latent variable L denote user preference to an item, described as the type of the preferring item (i.e., $L = l_j$ means that a user has preference to the item whose type is c_j). Similarly, let R denote the rating score on items. Following, we first give the definition of UPBN, which is used to represent the dependencies among the latent and observed variables.

Definition 1. A user preference Bayesian network, abbreviated as UPBN, is a pair $S = (G, \theta)$, where

(1) $G = (V, E)$ is the DAG of UPBN, where $V = U \cup \{L\} \cup \{I\} \cup \{R\}$ is the set of nodes in G. E is the directed edge set representing the dependencies among observed attributes and user preference.
(2) θ is the set of UPBN's parameters constituting the CPT of each node.

4.1 Constraint Description

Without loss of generality, we suppose a user only rates the items that he is interested in. The rating frequency and the corresponding scores for a specific type of items indicate the degree of user preference. Accordingly, we give the constraints to improve the effectiveness of model construction, where constraint 1 means that the initial structure of UPBN learning should be the same as the structure shown in Fig. 3 and constraint 2 means that the CPTs corresponding to I and R should satisfy the inequality for random initialization.

Constraint 1. The initial structure of UPBN is shown as Fig. 3. This constraint demonstrates that the type of a rated item is dependent on user preference and the corresponding rating score is dependent on the type of itself and user preference.

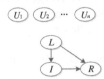

Fig. 3. The initial structure of UPBNs

Constraint 2. Constraint on the initial CPTs:

(1) $P(I = c_i | L = l_i) > P(I = c_j | L = l_i, i \neq j)$, namely the probability of the users rate c_i will be greater than that of they rate c_j if the user preference value takes l_i.

(2) If R takes the rating values such as $R \in \{1, 2, 3, 4, 5\}$, then R_1 and R_2 will take values from $\{4, 5\}$ and $\{1, 2, 3\}$, respectively. This means that the users tend to rate high score (4 or 5) instead of rate low score (1, 2, or 3) when their preferences are consistent with the type of items, represented by the following two inequalities:

$$P(R = R_1 | I = c_i, L = l_i) > P(R = R_2 | I = c_i, L = l_i) \text{ and}$$
$$P(R = R_2 | I = c_i, L = l_j, i \neq j) > P(R = R_1 | I = c_i, L = l_j, i \neq j)$$

4.2   Parameter Learning of UPBN

UPBN's parameter learning starts from an initial parameter θ^0 randomly generated under Constraint 2 in Sect. 4.1 and we apply EM to iteratively optimize the initial parameter until convergence.

Suppose that we have conducted t times of iterations and obtained the estimation value θ^t, then the $(t + 1)$th iteration process will be built as the following E-step and M-step, where there are m samples in data set D, and the cardinality of the variable denoting user preference L is c (i.e., c values of user preference, $l_1, l_2, \ldots, l_c$).

E-step. In light of the current parameter θ^t, we calculate the posterior probability of different user preference value l_j by Eq. (2), $P(L = l_j \mid D_i, \theta^t)$ $(1 \leq j \leq c)$ for every sample D_i $(1 \leq i \leq m)$ in D, making data set D complete as D^t. Then we obtain expected sufficient statistics by Eq. (3).

$$P(L = l_j | D_i, \theta^t) = \frac{P(L = l_j, D_i | \theta^t)}{\sum_{j=1}^{c} P(L = l_j, D_i | \theta^t)}. \tag{2}$$

$$m_{ijk}^t = \sum_{l=1}^{m} P(V_i = k, \pi(V_i) = j | D_l^t). \tag{3}$$

M-step. Based on the expected sufficient statistics, we can get the new greatest possible parameter θ^{t+1} by Eq. (4).

$$\theta_{ijk}^{t+1} = \frac{m_{ijk}^t}{\sum_{k=1}^{r_i} m_{ijk}^t}. \tag{4}$$

To avoid overfitting and ensure the convergence efficiency of the EM iteration, we give a method to measure parameter similarity. The parameter similarity between θ_1 and θ_2 of a UPBN is defined as the follows:

$$sim(\theta_1, \theta_2) = |\log P(D|G, \theta_1) - \log P(D|G, \theta_2)| \tag{5}$$

UPBN's parameter learning will converge if $sim(\theta^{t+1}, \theta^t) < \delta$.

For a UPBN structure G' and data set D, we generate initial parameter randomly under Constraint 2 and make D become the complete data set D^0. We use Eq. (3) to calculate the expected sufficient statistics and obtain parameter estimation θ^1 by Eq. (4). Then, we use θ^1 to make D become the complete data set D^1 again. By repeating the process until convergence or stop condition is met, the optimal parameter θ will be obtained. The above ideas are given in Algorithm 1.

Algorithm 1. EM(G', D, θ^0, δ, T)

Input: G', UPBN structure
            D, rating data
            θ^0, the initial parameter randomly generated under
            the constraint 2
            δ, the threshold of convergence
            T, the number of iteration
Output: θ, the optimal parameter of UPBN
Steps:
```
1:  G←G'
2:  θ←θ⁰
3:  FOR t = 0 TO T DO
4:     E-step: make D become the complete dataset Dᵗ
```
$$5: \quad m_{ijk}^t = \sum_{l=1}^m P(V_i = k, \pi(V_i) = j \mid D_i^t)$$
$$6: \quad \text{M-step: } \theta^{t+1} \leftarrow \theta_{ijk}^{t+1} = \frac{m_{ijk}^t}{\sum_{k=1}^{r_i} m_{ijk}^t}$$
```
7:     IF sim(θ^{t+1}, θ^t) < δ   THEN
8:        RETURN θ^{r+1}
9:     END IF
10: END FOR
```

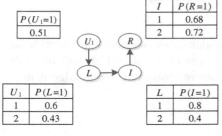

U_1	$P(L=1)$
1	0.6
2	0.43

I	$P(R=1)$
1	0.68
2	0.72

$P(U_1=1)$
0.51

L	$P(I=1)$
1	0.8
2	0.4

Fig. 4. Current UPBN and θ^1

Table 1. Dataset D

Sample	U_1	I	R	L	Count
D_1	1	1	1		271
D_2	1	1	2		69
D_3	1	2	1		99
D_4	1	2	2		67
D_5	2	1	1		139
D_6	2	1	2		125
D_7	2	2	1		186
D_8	2	2	2		44

Example 1. The current UPBN structure and data set D is presented in Fig. 4 and Table 1 respectively, where *Count* is to depict the number of the same sample. By the E-step in Algorithm 1 upon the initial parameter, we make D become the complete data set D^0 and use Eq. (3) to compute expected sufficient statistics. Then, we obtain parameter θ^1 by Eq. (4), shown in Fig. 4.

4.3 Structure Learning of UPBN

UPBN's structure learning starts from the initial structure and CPTs under the constraints given in Sect. 4.1. First, we rank the order of nodes of the UPBN and make the initial model be the current one. Then, we execute Algorithm 1 to conduct parameter learning of the current model and use BIC to score the current model. Following, we modify the current model by edge addition, deletion and reversal to obtain a series of candidate models which should satisfy Property 1 for the purpose that the candidate ones will be fully fit to the data set.

For each candidate structure G' and the complete data set D^{t-1}, we use Eq. (3) to calculate the expected sufficient statistics and obtain maximum likelihood estimation θ of parameter by Eq. (4) for model selection by BIC scoring metric. The maximum likelihood estimation is presented as Algorithm 2.

Algorithm 2. SEM_ML (G', D^{t-1})
Input: G', candidate structure of UPBN
        D^{t-1}, the complete dataset derived from EM algorithm.
Output: θ, the maximum likelihood estimation of the
        current UPBN parameter
Steps:
1: $G \leftarrow G'$
2: $m_{ijk}^{t-1} = \sum_{l=1}^{m} P(V_i = k, \pi(V_i) = j \mid D^{t-1})$
3: $\theta \leftarrow \theta_{ijk}^{t} = \dfrac{m_{ijk}^{t-1}}{\sum_{k=1}^{r_i} m_{ijk}^{t-1}}$
4: RETURN θ

By comparing the current model with candidate ones, we adopted that with the maximum BIC score as the basis for the next time of search, which will be made iteratively until the score is not increased. The above ideas are given in Algorithm 3.

Algorithm 3. Learn_UPBN

Input: G_0, the initial structure of UPBN
  D, rating data
  T, the number of parameter optimization between
  two times of adjacent structure optimization
  δ, the threshold of convergence in EM algorithm.

Output: the optimal UPBN

Steps:

```
1:   G←G₀
2:   θ←θ⁰˒ ⁰      //the initial parameter θ⁰˒ ⁰
3:   θ', Dᵀ⁻¹ ← EM(G, D, θ, δ, T)
4:   oldScore ← BIC(G, θ'|Dᵀ⁻¹)
5:   WHILE true DO
6:      FOR i = 1 TO n DO //The order search from each node
7:         newScore ← -∞
8:         FOR each candidate model G' DO
            //Candidate model G' derived from an edge addition,
            edge deletion, or edge reversal on G
9:            θ'← SEM_ML(G', Dⁱ˒ ᵀ⁻¹)
10:           tempScore ← BIC(G', θ'|Dⁱ˒ ᵀ⁻¹)
11:           IF tempScore > newScore THEN
12:              Gⁱ⁺¹, G'; θⁱ⁺¹˒ ⁰←θ'; newScore←tempScore
13:           END IF
14:        END FOR
15:        IF newScore > oldScore THEN
16:           θⁱ⁺¹˒ ᵀ, Dⁱ⁺¹˒ ᵀ ← EM(Gⁱ⁺¹, D, θⁱ⁺¹˒ ⁰, δ, T)
17:           G←Gⁱ⁺¹; θ←θⁱ⁺¹˒ ᵀ
18:           oldScore ← BIC(G, θ|Dⁱ⁺¹˒ ᵀ)
19:        ELSE
20:           RETURN (G, θ)
21:        END IF
22:     END FOR
23: END WHILE
```

Example 2. For the data set D in Table 1 and initial structure of UPBN in Fig. 5(a), we first conduct parameter learning of the initial structure and compute the corresponding BIC score by Algorithm 1. We then execute the three operators on U_1 and obtain three candidate models, shown in Fig. 5(b). Following, we estimate the parameters of the candidate models by Algorithm 2 and compute the corresponding BIC scores by Eq. (1). Thus, we obtain the optimal model G_3' as the current model G. Executing these three operators on other nodes and repeating the process until convergence, an optimal structure of UPBN can be obtained, shown in Fig. 5(c).

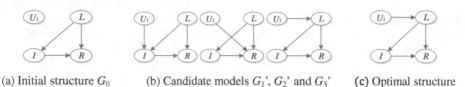

(a) Initial structure G_0 (b) Candidate models G_1', G_2' and G_3' (c) Optimal structure

Fig. 5. UPBN's structure learning

5 Experimental Results

5.1 Experiment Setup

To verify the feasibility of the proposed method, we implemented the algorithms for the parameter learning and structure learning of UPBN. The experiment environment is as follows: Intel Core i3-3240 3.40 GHz CPU, 4 GB main memory, running Windows 10 Professional operating system. All codes were written in C++.

All experiments were established on synthetic data. We manually constructed the UPBN shown as Fig. 1 and sampled a series of different scales of data by means of Netica [1]. As for the situation where UPBN contains more than 5 nodes, we randomly generated the corresponding value of sample data. For ease of the exhibition of experimental results, we made use of some abbreviations to denote different test conditions and adopted sign '+' to combine these conditions, where initial CPTs obtained under constraints, initial CPTs obtained randomly, and Property 1 is abbreviated as CCPT, RCPT, P1 respectively. Moreover, we use 1 k to denote 1000 instances.

5.2 Efficiency of UPBN Construction

First, we tested the efficiency of Algorithm 1 for parameter learning with the increase of data size when UPBN contains 5 nodes, and that of Algorithm 1 with the increase of UPBN nodes on 2 k data under different conditions of the initial CPTs, shown in Fig. 6 (a) and (b) respectively. It can be seen that the execution time of Algorithm 1 is increased linearly with the increase of data size. This shows that the efficiency of Algorithm 1 mainly depends on the data size.

Second, we recorded the execution time of Algorithm 1 with the increase of data size and nodes under the condition of CCPT, shown in Fig. 6(c) and (d) respectively. It can be seen that the execution time is increased linearly with the increase of data size no matter how many nodes there are in a UPBN. This means that the execution time is not sensitive to the scale of UPBN.

Third, we tested the efficiency of Algorithm 3 for structure learning with the increase of data size when UPBN contains 5 nodes, and that of Algorithm 3 with the increase of UPBN nodes on 2 k data under different conditions, shown in Fig. 7(a) and (b) respectively. It can be seen from Fig. 7(a) that the execution time of Algorithm 3 is increased linearly with the increase of data size. Moreover, Constraint 2 is obviously beneficial to reduce the execution time under Property 1 when the data set is larger than

6 k. It can be seen form Fig. 7(b) that the execution time of Algorithm 3 is increased sharply with the increase of nodes, and the execution time under CCPT is larger than that under RCPT.

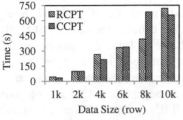
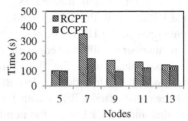

(a) Execution time with the increase of data size when UPBN containing 5 nodes

(b) Execution time with the increase of nodes under the situation where data size is 2k

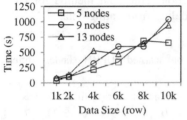
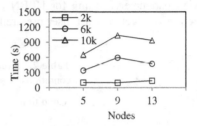

(c) Execution time with the increase of data size under the condition of CCPT

(d) Execution time with the increase of nodes under the condition of CCPT

Fig. 6. Execution time of parameter learning

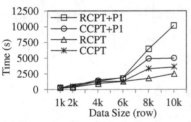
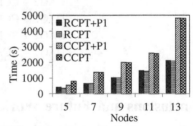

(a) Execution time with the increase of data size when UPBN containing 5 nodes

(b) Execution time with the increase of nodes under the situation where data size is 2k

Fig. 7. Execution time of structure learning

5.3 Effectiveness of UPBN Construction

It is pointed out [6] that a BNLV resulted from SEM makes sense under specific initial structures. According to Property 1, a UPBN should include the constraint "$L \rightarrow X$" at least, where L is the latent variable and X is an observed variable. Thus, we introduced the initial structure in Fig. 8 with the least prior knowledge. We constructed 50 UPBNs under the constraint in Fig. 3, denoted as DAG1, and each combination of different conditions respectively. Meanwhile, we also constructed 50 UPBNs under the constraint in Fig. 8, denoted as DAG2, and each combination of different conditions respectively.

To test the effectiveness of the method for UPBN construction, we constructed the UPBN by the clique-based method [8], shown as Fig. 1. We then compared our constructed UPBNs with this UPBN, and recorded the number of different edges (e.g., no different edges in the UPBN shown in Fig. 1). We counted the number of UPBNs with various number of different edges (0 ~ 8), shown in Table 2. It can be seen that the UPBN constructed upon Fig. 3 is better than that upon Fig. 8 under the same conditions, since the former derives less different edges than the latter. Moreover, the number of the constructed UPBNs with less different edges under CCPT is obviously larger than that under RCPT (e.g., the number of UPBNs with 0 different edges under DAG1 + CCPT is greater than that under DAG1 + RCPT), which means that our constraint-based method is beneficial and better than the traditional method by EM directly in parameter learning for UPBN construction. Thus, our method for UPBN construction is effective w.r.t. user preference modeling from rating data.

Fig. 8. Initial structure with the least constraint

Table 2. Structures of learned UPBN under different conditions

Condition	The number of different edge								
	0	1	2	3	4	5	6	7	8
DAG1 + CCPT + P1	18		27	3	2				
DAG1 + CCPT	18		27	3	2				
DAG1 + RCPT + P1			2	1	47				
DAG1 + RCPT			2	1	47				
DAG2 + CCPT + P1			12		11	5	13	9	
DAG2 + CCPT			5		10	16	15	2	2
DAG2 + RCPT + P1			1		1	4	17	27	
DAG2 + RCPT			1		1	4	17	18	9

6   Conclusions and Future Work

In this paper, we aimed to give a constraint-based method for modeling user preference from rating data to provide underlying techniques for the novel personalized services in the context of mobile Internet like applications. Accordingly, we gave the property that enables CPTs related to the latent variable to fit data sets by EM and constructed UPBN to represent arbitrary dependencies between user preference and explicit attributes in rating data. Experimental results showed the efficiency and effectiveness. However, only test on synthetic data is not enough to verify the feasibility of our method in realistic situations. So, we will make more experiments on real rating data sets further. As well, modeling preference from massive, distributed and dynamic rating data is what we are currently exploring.

Acknowledgements. This paper was supported by the National Natural Science Foundation of China (Nos. 61472345, 61562090, 61462056, 61402398), Natural Science Foundation of Yunnan Province (Nos. 2014FA023, 2013FB009, 2013FB010), Program for Innovative Research Team in Yunnan University (No. XT412011), and Program for Excellent Young Talents of Yunnan University (No. XT412003).

References

1. Netica Application (2016). http://www.norsys.com/netica.html
2. MovieLens Dataset (2016). http://grouplens.org/datasets/movielens/1m
3. Blei, D.M., Ng, A.Y., Jordan, M.I.: Latent Dirichlet Allocation. J. Mach. Learn. Res. **3**, 993–1022 (2003)
4. Breese, J., Heckerman, D., Kadie, C.M.: Empirical analysis of predictive algorithms for collaborative filtering. In: UAI 1998, pp. 43–52. Morgan Kaufmann (1998)
5. Dempster, A., Laird, N., Rubin, D.: Maximum-likelihood from Incomplete Data via the EM algorithm. J. Royal Stat. Soc. **39**(1), 1–38 (1977)
6. Friedman, N.: Learning belief networks in the presence of missing values and hidden variables. In: ICML 1997, pp. 452–459. ACM (1997)
7. Friedman, N.: The Bayesian structural EM algorithm. In: UAI 1998, pp. 129–138. Morgan Kaufmann (1998)
8. Elidan, G., Lotner, N., Friedman, N., Koller, D.: Discovering Hidden variables: a structure-based approach. In: NIPS 2000, pp. 479–485 (2000)
9. Huang, Y., Bian, L.: A bayesian network and analytic hierarchy process based personalized recommendations for tourist attractions over the internet. Expert Syst. Appl. **36**(1), 933–943 (2009)
10. Huete, J., Campos, L., Fernandez-luna, J.M.: Using structural content information for learning user profiles. In: SIGIR 2007, pp. 38–45 (2007)
11. Kim, J., Jun, C.: Ranking evaluation of institutions based on a bayesian network having a latent variable. Knowl. Based Syst. **50**, 87–99 (2013)
12. Koller, D., Friedman, N.: Probabilistic Graphical Models: Principles and Techniques. MIT Press, Cambridge (2009)
13. Koren, Y.: Collaborative filtering with temporal dynamics. Commun. ACM **53**(4), 89–97 (2010)
14. Liu, T., Zhang, N.L., Chen, L., Liu, A.H., Poon, L., Wang, Y.: Greedy learning of latent tree models for multidimensional clustering. Mach. Learn. **98**(1–2), 301–330 (2015)
15. Pearl, J.: Fusion, propagation, and structuring in belief networks. Artif. Intell. **29**(3), 241–288 (1986)
16. Salakhutdinov, R., Mnih, A.: Probabilistic Matrix Factorization. In: NIPS 2007, pp. 1257–1264 (2007)
17. Tan, F., Li, L., Zhang, Z., Guo, Y.: A multi-attribute probabilistic matrix factorization model for personalized recommendation. In: Dong, X.L., Yu, X., Li, J., Sun, Y. (eds.) WAIM 2015. LNCS, vol. 9098, pp. 535–539. Springer, Heidelberg (2015). doi:10.1007/978-3-319-21042-1_57
18. Yin, H., Cui, B., Chen, L., Hu, Z., Huang, Z.: A Temporal context-aware model for user behavior modeling in social media systems. In: SIGMOD 2014, pp. 1543–1554. ACM (2014)

19. Yu, K., Zhang, B., Zhu, H., Cao, H., Tian, J.: Towards personalized context-aware recommendation by mining context logs through topic models. In: Tan, P.-N., Chawla, S., Ho, C.K., Bailey, J. (eds.) PAKDD 2012. LNCS (LNAI), vol. 7301, pp. 431–443. Springer, Heidelberg (2012). doi:10.1007/978-3-642-30217-6_36
20. Yue, K., Fang, Q., Wang, X., Li, J., Liu, W.: A parallel and incremental approach for data-intensive learning of bayesian networks. IEEE Trans. Cybern. **45**(12), 2890–2904 (2015)
21. Zhao, Z., Cheng, Z., Hong, L., Chi, E.H.: Improving user topic interest profiles by behavior factorization. In: WWW 2015, pp. 1406–1416. ACM (2015)

A Hybrid Approach for Sparse Data Classification Based on Topic Model

Guangjing Wang, Jie Zhang, Xiaobin Yang, and Li Li$^{(\boxtimes)}$

Faculty of Computer and Information Science, Southwest University,
Chongqing 400715, China
lily@swu.edu.cn

Abstract. With an increasing number of short text emerging, sparse text classification is becoming crucial in data mining and information retrieval area. Many efforts have been devoted to improve the efficiency of normal text classification. However, it is still immature in terms of high-dimension and sparse data processing. In this paper, we present a new method which fancifully utilizes Biterm Topic Model (BTM) and Support Vector Machine (SVM). By using BTM, though the dimensionality of training data is reduced significantly, it is still able to keep rich semantic information for the sparse data. We then employ SVM on the generated topics or features. Experiments on 20 Newsgroups and Tencent microblog dataset demonstrate that our approach can achieve excellent classifier performance in terms of precision, recall and F1 measure. Furthermore, it is proved that the proposed method has high efficiency compared with the combination of Latent Dirichlet Allocation (LDA) and SVM. Our method enhances the previous work in this field and establishes the foundation for further studies.

1 Introduction

More and more textual data is unfolding before people's eyes in more diverse forms with the rise of web 2.0. For example, multifarious data is generated from queries and questions in Web search, social networks, various internet news and so on. As a consequence, researchers are urged to solve the problem that internet users sometimes get bored because they are subject to a myriad of turbid information and the restraint of limited message coverage [19].

As an essential topic, lots of methods are put forward for the above problem. Text categorization used in information retrieval, news classification, spam mail filtering to acquire better user experience is studied roundly [10]. However, the applicability of classification for high dimensional and sparse data often becomes a short slab in many models. Like a teeter-board, the efficiency of processing sparse data and performance quality are hard to be fairness considered. On one hand, the classification accuracy would be descending if the dimension was cut down at an efficient level. On the other hand, for sparse and high dimensional datasets, the computing efficiency has to be sacrificed since the dimension will get to thousands or even more [13].

S. Song and Y. Tong (Eds.): WAIM 2016 Workshops, LNCS 9998, pp. 17–28, 2016.
DOI: 10.1007/978-3-319-47121-1_2

Researchers usually characterize sparse data by building semantics association or employing external knowledge base to settle the sparse feature problems. For instance, Wikipedia was used in [15] as an external corpus to rich the corpus. Cataldi et al. [2] used semantics relation rules to build relation rules library, so as to rich feature corpus. Xia et al. [20] introduced topics for multi-granularity, and then discriminative features are generated for sparse data classification. Nevertheless, it is hard to introduce external corpora to sparse text due to specific situations, and appropriate semantic association that can enhance the effect of sparse data classification [23]. What's more, the problem of accuracy and efficiency in classification are difficult to get an optimal solution [8].

A novel way to address the above problem is presented in our paper. To classifying sparse text accurately and fleetly, Biterm Topic Model (BTM) algorithm [21] is used for generating features, so that we can utilize topic information in Vector Space Model (VSM). Then the Support Vector Machine (SVM) is acted on it to obtain better classification result. Through the experiments on 20 Newsgroups datasets and dataset from Tencent Microblogs, we found that the combination of BTM and SVM enhances performance much more than other classification models for sparse data. Moreover, the proposed method provides a novel way to process sparse data.

The rest of the paper is organized as follows: the related work is reviewed in Sect. 2. Section 3 discusses our approach using BTM+SVM, and then the implementation is detailed in Sect. 4. Further discussion is presented experimentally in Sect. 5. Finally, Sect. 6 is the conclusion.

2   Related Work

Text classification is an important task for natural language process, and topic model is popular among researchers to process natural language. Liu et al. [9] devised a semi-supervised learning with Universum algorithm based on boosting technique. In their method, they aims to study a collection of nonexamples that do not belong to any class of interest. Luss et al. [11] developed an analytic center cutting plane method to solve the kernel learning problem efficiently, this method exhibits linear convergence but requires very few gradient evaluations. Lai et al. [7] applied a recurrent structure to capture contextual information as far as possible when learning word representations, and it is said that the proposed method shows better results than the state-of-the-art methods on document-level. By contrast, our method uses the generation of word co-occurrence pattern to keep main information while reducing dimensionality. Landeiro et al. [8] estimated the underlying effect of a text variable on the class variable based on Pearls back-door adjustment.

SVM is widely uesed in text classification. Yin et al. [22] used semi-supervised learning and SVM to improve the traditional method and it can classify a large number of short texts to mine the useful massage from the short text, however the efficiency is not satisfactory. Song et al. [18] illustrated Chinese text feature selection method based on category distinction and feature location information, while this method has boundedness that location information is not easy to obtain. Nguyen et al. [14] proposed the improving multi-class text classification method combined the SVM classifier with OAO and DDAG strategies. In Seetha et al. [16], nearest neighbour and SVM classifiers are chosen as text classifiers for their good classification accuracy. Luo et al. [10] presented a method which combines the Latent Dirichlet Allocation (LDA) algorithm and SVM. However, the method is not good at deal with sparse text data according to our experiments. Altinel et al. [1] proposed a novel semantic smoothing kernel for SVM based on a meaning measure.

3   Problem Formalization

Motivated by researches on classification models, this study first formalizes the data collection to meet the prerequisites in algorithms. As usual, we use a vector to represent a document, and the whole text data can be regarded as a matrix. The problem is formalized technically as follows.

Every document and extracted term are supposed to be mapped into a vector [5] to represent text documents as a document-term matrix according to VSM.

$$d_j - (w_{1j}, w_{2j}, \ldots, w_{tj}) \tag{1}$$

Each dimension related to a separate term, where the value corresponds to the term is usually computed by term frequency-inverse document frequency model (TF-IDF). The weight vector for document d is

$$v_d = [w_{1,d}, w_{2,d}, \ldots, w_{N,d}]^T \tag{2}$$

where $w_{t,d} = tf_{t,d} \cdot log \dfrac{|D|}{|\{d\prime \in D | t \in d\prime\}|}$ and tf_{td} is the term frequency of term t in document d, $|D|$ is the total number of documents in the set; $|\{d\prime \in D | t \in d\prime\}|$ is the number of documents containing the term t.

For dimension reduction, there are two general ways to apply. One is feature extraction, large data is transformed into a reduced features vector, so that the desired task can be solved using the reduced representation [13]. The data transformation model can be nonlinear like kernel principal component analysis, linear like latent semantic indexing, linear discriminant analysis and so on. The other one is known as feature selection, such as χ^2 statistic, document frequency and so forth, those are selecting a subset of relevant features for use in model construction.

4   Novel Method for Sparse Data Classification

In this section, we will illustrate our method for sparse data classification carefully. To begin with, an overview of BTM and SVM model is presented.

After that we will elaborate how to employ BTM to generate the document topic matrix, and then explain how to utilize the SVM to classify and predict the category of sparse data.

4.1   Matrix of Topic Distribution

BTM is a probabilistic model that learns topics over short texts by directly using the generation of biterms in the whole corpus [21]. The notation of "biterm" refers to an instance of unordered word pair occurrence, and any two distinct words in a document compose a biterm. The model in graph is showed in Fig. 1. The key point is that two words are more likely to be in the same topic if they co-occur more frequently.

Given a corpus with N_D documents, we can utilize a K-dimensional multinomial distribution $\theta = \{\theta_k\}_{k=1}^{K}$ with $\theta_k = P(z = k)$ and $\sum_{k=1}^{K} \theta_k = 1$ to show the prevalence of topics. Suppose each biterm is drawn from a specific topic independently, the specific generative process of the corpus in BTM can be shown as follows [4]. The notations used in BTM are listed in Table 1.

1. For each topic z, draw a topic-specific word distribution $\phi_z \sim Dir(\beta)$.
2. Extracting a topic distribution $\theta \sim Dir(\alpha)$ for the whole collection.

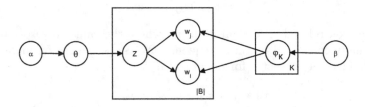

Fig. 1. BTM: a generative graphical model

Table 1. Notations in BTM

N_D	The number of documents
K	The number of latent topics
W	The number of unique words
$|B|$	The number of biterms
$\mathbf{B} = \{b_i\}_{i=1}^{|B|}$	The collection of biterms
$b_i = w_{i,1}, w_{i,2}$	The i-th biterm
$\theta = \{\theta_k\}_{k=1}^{K}$	A K-dimensional multinomial distribution
$\theta_k = P(z = k)$	The prevalence of topic k where $\sum_{k=1}^{K} \theta_k = 1$
Φ	A $K \times W$ matrix
Φ_k	A W-dimensional multinomial distribution in k-th row
α, β	Dirichlet hyperparameters

3. For each biterm b in the biterm set B, draw a topic assignment: $z \sim Multi(\theta)$, and draw two words: $w_i, w_j \sim Multi(\phi_z)$.

The joint probability of a biterm $b = (w_i, w_j)$ over topic z can be written as:

$$P(b) = \sum_z P(w_i|z)P(w_j|z) = \sum_z \theta_z \phi_{i|z} \phi_{j|z} \tag{3}$$

Similar as LDA, Gibbs sampling can be adopted to perform approximate inference. In the process, the topic-word distribution ϕ and global topic distribution θ can be generated as:

$$\phi_{w|z} = \frac{n_{w|z} + \beta}{\sum_w n_{w|z} + M\beta} \tag{4}$$

$$\theta_z = \frac{nz + \alpha}{|B| + K\alpha} \tag{5}$$

where $|B|$ is the aggregated number of biterms. The matrix θ is an essential part of our method as the matrix of topic distribution.

4.2 Support Vector Machine (SVM)

SVM plays an important part in lots of domains, and hyperplanes are constructed when it performs classification tasks in a multidimensional space. It is reported that SVM can generate better results than other learning algorithms in classification [6]. The basic theory of SVM is elaborated next:

When the training dataset of n points in the form of $(\boldsymbol{x_1}, \boldsymbol{y_1}), \ldots, (\boldsymbol{x_n}, \boldsymbol{y_n})$ is known, where y_i is either 1 or -1, the optimization problem is defined as:

$$min \; \frac{1}{2}w^T w + C \sum_{i=1}^{n} \zeta_i \; s.t. \; y(w^T \phi(x_i) + b) \geq 1 - \zeta_i, \zeta_i \geq 0 \tag{6}$$

where function ϕ can map training vectors x_i into a higher dimensional space b. $C > 0$ is the penalty parameter of the error instances, which should be chosen with care to avoid over fitting. SVM supports both regression and classification tasks and can handle multiple continuous and categorical variables. On the basis of Mercer theorem [12], there always exists an equation $K(x_i, x_j) = \phi(x_i)^T \phi(x_j)$ called the kernel function. The problem 6 can be derived as:

$$f(x) = \sum_{i=1}^{l} a_i y_i K(x_i, x_j) + b \tag{7}$$

By solving the optimization, parameters of the maximum-margin hyperplane are derived specifically. Note that the core of SVM which is good at processing high dimensional data is that the number of dimensions can be turned from $\phi(x_i)$ to x_i. What's more, LIBSVM [3] has some attractive training time properties. Each convergence iteration takes linear time to read the training data and the iterations also have a Q-Linear Convergence property, which makes the algorithm extremely fast [17].

4.3   Experimental Procedure for Enhancement

For less complexity and higher performance, our method retrieves optimal set of features, which reflects the original data distribution. The steps in document classification are listed as follows.

Step 1. Making a document-term matrix according to the vector support model.
Step 2. Analysing the topic distribution and building a matrix about topic distribution for documents.
Step 3. Acquiring the weight of vector support model by using the topic distribution values.
Step 4. Testing documents by building the classifier.

We firstly formalize the data collection in order that it can be used in SVM, so a document-term matrix must be built in Step 1. Since Step 2 utilizes matrix θ to indicate the relationship between texts and topics, we need to generate it by BTM estimation with Gibbs sampling first. In Step 4, SVM is used to build upon the characteristics identified in Step 2.

5   Experimental Evaluation

In this section, we conduct several experiments to show the great superiority of our method, results are presented below followed by discussion.

5.1   Data Preparation

We evaluate our method on two popular datasets used in large scale and sparse text classification study. One is Tencent microblogs, which contains 11,285,538 messages from seven different micro-channels posted by users from July 2 to July 14 in 2013 [19] on Tencent microblog platform (http://t.qq.com/). The other dataset

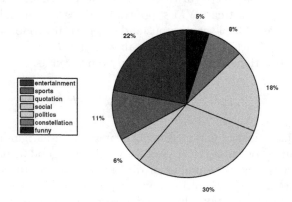

Fig. 2. Category distribution of Tencent messages

Table 2. Data description for 20 Newsgroups

Dataset	Category	Training data	Test data
20 Newsgroups	alt.atheism	480	319
	comp.graphics	584	389
	comp.os.ms-windows.misc	591	394
	comp.sys.ibm.pc.hardware	590	392
	comp.sys.mac.hardware	578	385
	comp.windows.x	593	395
	misc.forsale	585	390
	rec.auto	594	396
	rec.motorcycles	598	398
	rec.sport.baseball	597	397
	rec.sport.hockey	600	399
	sci.cypt	595	396
	sci.electronics	591	393
	sci.med	594	396
	sci.space	593	394
	sci.religion.christian	599	398
	talk.politics.guns	546	364
	talk.politics.mideast	564	376
	talk.politics.misc	465	310
	talk.religion.misc	377	251

is 20 Newsgroups (http://qwone.com/~jason/20Newsgroups/), which has 20 categories and is widely used in text classification.

The raw data of these collections is very noisy. For preprocessing, the terms like the punctuation marks, stop words, links and other non-words in the raw microblogging datasets are removed in data preparation using a punctuation list and a stop words dictionary. Specifically, for the process of word segmentation, the ICTCLAS (http://www.ictclas.org/) is used in this paper.

To further describe the datasets for classification, Fig. 2 is showed for category distribution of Tencent messages, and Table 2 illustrates the classical data proportion on 20 Newsgroups.

5.2 Evaluation Criteria

In our experiment, the $Macro/Micro - precision$, $Macro/Micro - Recall$ and $Macro/Micro - F1$ criteria are employed to evaluate the method. The definitions are showed below.

$$Micro - Precision = \frac{\sum_{i=1}^{m} TP_i}{\sum_{i=1}^{m} TP_i + FP_i} \tag{8}$$

$$Micro - Recall = \frac{\sum_{i=1}^{m} TP_i}{\sum_{i=1}^{m} TP_i + FN_i} \tag{9}$$

$$Micro - F1 = \frac{Micro - Precision \times Micro - Recall \times 2}{Micro - Precision + Micro - Recall} \tag{10}$$

$$Macro - Precision = \frac{1}{m} \sum_{i=1}^{m} P_i \tag{11}$$

$$sMacro - Recall = \frac{1}{m} \sum_{i=1}^{m} R_i \tag{12}$$

$$Micro - F1 = \frac{Macro - Precison \times Macro - Recall \times 2}{Macro - Precision + Macro - Recall} \tag{13}$$

5.3   Results and Analysis

We choose two other methods PCA+SVM and LDA+SVM as baselines to verify the advantage of our approach. Documents used in our experiments are mapped into document-term matrix firstly. Considering topic model as a method of dimensionality reduction firstly, we then trained the document vectors by LIBSVM (http://www.csie.ntu.edu.tw/~cjlin/libsvm/index.html), and we then predicted the categories of new documents. Unlike the PCA method which treats terms as features of document vector, the LDA and BTM methods use the topics as features of documents vectors. In order to obtain document-topic matrix, the widely used LDA tool GibbsLDA++ (http://gibbslda.sourceforge.net/) was employed in our experiments. BTM (http://shortext.org/) is first used to acquire the matrix of topic distribution for documents. The number of Gibbs sampling iterations in the following experiment is set to 1000 to insure the classification accuracy.

We use $Macro - Precision$, $Macro - Recall$, $Macro - F1$ and $Micro - F1$ to evaluate the classifiers PCA+SVM, LDA+SVM and BTM+SVM based on 20 Newsgroups which are depicted in Figs. 3 and 4, respectively. What need to mention is that Micro-Precision and Micro-Recall are the same as Micro-F1 since we suppose each instance has exactly one correct label. From the result, we can see that the values in Fig. 3 reach peak value after the dimensionality is brought down at 400. By contrast, as we can see from Fig. 4, when the number of topics is merely set to 180 for BTM+SVM, the $Macro - Precision$, $Macro - Recall$, $Macro - F1$ and $Micro - F1$ undulate slightly around 0.87,0.86,0.87,0.90, respectively. It can be seen from that the values of those criteria for BTM+SVM are relatively higher than those of PCA+SVM and LDA+SVM, respectively.

Comparison experiments were made in order to verify the high performance of BTM for feature selection, we estimated the number of iterations needed to obtain high accuracy by spending less time on topic-matrix generation.

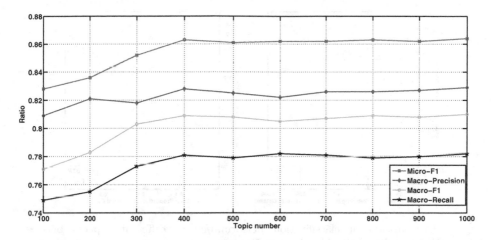

Fig. 3. The values of evaluation criteria under diverse number of features reduced by PCA+SVM method on 20 newsgroups collection

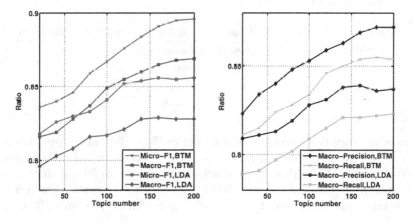

Fig. 4. The values of evaluation criteria under diverse number of features reduced by LDA+SVM, BTM+SVM methods on 20 newsgroups collection

The accuracy on 5-fold cross validation is reported in Fig. 5. It can be seen that 900 iterations is a relatively better choice on Tencent Dataset, and accuracy keeps around 90 % with 60 features generated. From Fig. 5(b), we can see that all the methods work better with training data size grows. It suggests that the LDA+SVM method is not able to overcome the sparsity problem, while BTM+SVM can achieve better performance than LDA+SVM, which also shows the superiority of our method.

BTM+SVM can resolve the over-fitting and feature redundancy problem, and yields better classification results than others. Utilizing the topical model is able to accelerate the process of classification. What's more, for sparsity problem in conventional topical model, BTM is better at capturing the topics by using

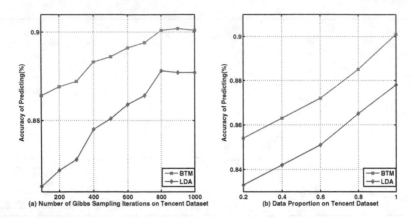

Fig. 5. Comparision of classification performance in different aspects between LDA+SVM and BTM+SVM on Tencent Dataset

Table 3. Time cost for dimensionality generated on 20 Newsgroups by three following methods using 3.0 GHz CPU, 2G memory

Methods	File quantity	Time consumed	Dimensionality generated
PCA+SVM	18846	Roughly 250 min	100
LDA+SVM	18846	Roughly 80 min	100
BTM+SVM	18846	Roughly 50 min	100

word co-occurrence patterns in the whole corpus [21]. The Table 3 presents information about training speed of three provided methods, which also shows the high efficiency of BTM+SVM by comparison. It only takes 50 min to generate a topic matrix by GibbsLDA++ with 100 topics and 1000 iterations, which saves about 30 min than LDA+SVM and is only one fifth of the time PCA+SVM consumed.

6   Conclusion

In this paper, we proposed a hybrid approach called BTM+SVM for sparse data classification. We explored the difference among BTM+SVM, PCA+SVM and LDA+SVM, and the results showed that our method has superiority over accuracy and efficiency when sparse text is processed. We figured out the number of topics to use when approximating the matrix properly. Comparing with traditional methods, we improved the classification accuracy and tested the training speed over the experiments. Overall, our method is able to cope with sparse problem properly, which is promising and can be used extensively in real applications.

Acknowledgments. This work is supported by Natural Science Foundations of China (No. 61170192), National High-tech R&D Program of China (No. 2013AA013801), Fundamental Research Funds for the Central Universities (No. XDJK2016E064).

References

1. Altınel, B., Ganiz, M.C., Diri, B.: A corpus-based semantic kernel for text classification by using meaning values of terms. Eng. Appl. Artif. Intell. **43**, 54–66 (2015)
2. Cataldi, M., Di Caro, L., Schifanella, C.: Emerging topic detection on twitter based on temporal and social terms evaluation. In: Proceedings of the Tenth International Workshop on Multimedia Data Mining, p. 4. ACM (2010)
3. Chang, C.-C., Lin, C.-J.: Libsvm: a library for support vector machines. ACM Trans. Intell. Syst. Technol. (TIST) **2**(3), 27 (2011)
4. Cheng, X., Yan, X., Lan, Y., Guo, J.: BTM: topic modeling over short texts. IEEE Trans. Knowl. Data Eng. **26**(12), 2928–2941 (2014)
5. Dhillon, I.S., Modha, D.S.: Concept decompositions for large sparse text data using clustering. Mach. Learn. **42**(1–2), 143–175 (2001)
6. Fan, R.-E., Chang, K.-W., Hsieh, C.-J., Wang, X.-R., Lin, C.-J.: Liblinear: a library for large linear classification. J. Mach. Learn. Res. **9**, 1871–1874 (2008)
7. Lai, S., Xu, L., Liu, K., Zhao, J.: Recurrent convolutional neural networks for text classification. In: AAAI, pp. 2267–2273 (2015)
8. Landeiro, V., Culotta, A.: Robust text classification in the presence of confounding bias (2016)
9. Liu, C.-L., Hsiao, W.-H., Lee, C.-H., Chang, T.-H., Kuo, T.-H.: Semi-supervised text classification with universum learning. IEEE Trans. Cybern. **46**(2), 462–473 (2015)
10. Luo, L., Li, L.: Defining and evaluating classification algorithm for high-dimensional data based on latent topics. PloS one **9**(1), e82119 (2014)
11. Luss, R., d'Aspremont, A.: Predicting abnormal returns from news using text classification. Quant. Financ. **15**(6), 999–1012 (2015)
12. Minh, H.Q., Niyogi, P., Yao, Y.: Mercer's theorem, feature maps, and smoothing. In: Lugosi, G., Simon, H.U. (eds.) COLT 2006. LNCS (LNAI), vol. 4005, pp. 154–168. Springer, Heidelberg (2006). doi:10.1007/11776420_14
13. Moura, S., Partalas, I., Amini, M.-R.: Sparsification of linear models for large-scale text classification. In: Conférence sur l'APprentissage automatique (CAp 2015) (2015)
14. Nguyen, V.T., Huy, H.N.K., Tai, P.T., Hung, H.A.: Improving multi-class text classification method combined the svm classifier with oao and ddag strategies. J. Convergence Inf. Technol. **10**(2), 62–70 (2015)
15. Phan, X.-H., Nguyen, L.-M., Horiguchi, S.: Learning to classify short and sparse text & web with hidden topics from large-scale data collections. In: Proceedings of the 17th International Conference on World Wide Web, pp. 91–100. ACM (2008)
16. Seetha, H., Murty, M.N., Saravanan, R.: Effective feature selection technique for text classification. Int. J. Data Min. Model. Manag. **7**(3), 165–184 (2015)
17. Shalev-Shwartz, S., Singer, Y., Srebro, N., Cotter, A.: Pegasos: primal estimated sub-gradient solver for svm. Math. Program. **127**(1), 3–30 (2011)

18. Song J., Zhang P., Qin S., Gong, J.: A method of the feature selection in hierarchical text classification based on the category discrimination and position information. In: 2015 International Conference on Industrial Informatics-Computing Technology, Intelligent Technology, Industrial Information Integration (ICIICII), pp. 132–135. IEEE (2015)
19. Wang, J., Li, L., Tan, F., Zhu, Y., Feng, W.: Detecting hotspot information using multi-attribute based topic model. PloS one **10**(10), e0140539 (2015)
20. Xia, C.-Y., Wang, Z., Sanz, J., Meloni, S., Moreno, Y.: Effects of delayed recovery and nonuniform transmission on the spreading of diseases in complex networks. Phys. A: Stat. Mech. Appl. **392**(7), 1577–1585 (2013)
21. Yan, X., Guo, J., Lan, Y., Cheng, X.: A biterm topic model for short texts. In: Proceedings of the 22nd International Conference on World Wide Web, pp. 1445–1456. International World Wide Web Conferences Steering Committee (2013)
22. Yin, C., Xiang, J., Zhang, H., Wang, J., Yin, Z., Kim, J.-U.: A new svm method for short text classification based on semi-supervised learning. In: 2015 4th International Conference on Advanced Information Technology and Sensor Application (AITS), pp. 100–103. IEEE (2015)
23. Zhang, H., Zhong, G.: Improving short text classification by learning vector representations of both words and hidden topics. Knowl.-Based Syst. **102**, 76–86 (2016)

Human Activity Recognition in a Smart Home Environment with Stacked Denoising Autoencoders

Aiguo Wang[1,2], Guilin Chen[1(✉)], Cuijuan Shang[1], Miaofei Zhang[1], and Li Liu[3]

[1] School of Computer and Information Engineering, Chuzhou University,
Chuzhou 239000, China
{glchen,shangcuijuan,zhangmiaofci}@chzu.edu.cn,
wangaiguo2546@163.com

[2] School of Computer and Information, Hefei University of Technology, Hefei 230009, China
[3] School of Software Engineering, Chongqing University, Chongqing 400044, China
dcsliuli@cqu.edu.cn

Abstract. Activity recognition is an important step towards automatically measuring the functional health of individuals in smart home settings. Since the inherent nature of human activities is characterized by a high degree of complexity and uncertainty, it poses a great challenge to build a robust activity recognition model. This study aims to exploit deep learning techniques to learn high-level features from the binary sensor data under the assumption that there exist discriminant latent patterns inherent in the low-level features. Specifically, we first adopt a stacked autoencoder to extract high-level features, and then integrate feature extraction and classifier training into a unified framework to obtain a jointly optimized activity recognizer. We use three benchmark datasets to evaluate our method, and investigate two different original sensor data representations. Experimental results show that the proposed method achieves better recognition rate and generalizes better across different original feature representations compared with other four competing methods.

Keywords: Activity recognition · Smart homes · Deep learning · Autoencoder · Shallow structure model

1 Introduction

The rapid development of machine learning and mobile computing technologies makes it possible for researchers to customize and provide pervasive and context-aware services to individuals living in smart homes [1]. On the other hand, due to the ever increasing aging population all over the world and the high expenditure of healthcare cost, the elderly healthcare raises us a serious social and fiscal problem. With the growing desire of subjects to remain independent in their own homes, ambient assisted living (AAL) systems that can perceive the states of an individual and corresponding context and act on physical surroundings using different types of sensors and automatically recognize human activities of daily living (ADLs) are in great needs [2, 3]. In such systems, accurately recognizing human activities such as cooking, eating, drinking, grooming and sleeping is an important step towards independent living, which can be achieved by

© Springer International Publishing AG 2016
S. Song and Y. Tong (Eds.): WAIM 2016 Workshops, LNCS 9998, pp. 29–40, 2016.
DOI: 10.1007/978-3-319-47121-1_3

monitoring the function ability of the residents using various sensor technologies. Also, activity recognition can potentially facilitate a number of applications in a home setting such as fall detection, activity reminder, and welling evaluation [4, 5].

Activity recognition (AR) is a challenging and active research area [6], and different types of sensing technologies have been explored by researchers to improve the recognition rate and adapt to different application scenarios. Generally, they can be mainly grouped into three categories: vision-based (e.g. camera, video), wearable/carriable sensor-based (e.g. accelerometer, gyroscope), and environment interactive sensor-based methods (e.g. motion detector, pressure sensor, contact sensor) [7, 8]. Due to the inherent non-intrusiveness, flexibility, low cost, and easy deployment, environment sensor-based approaches are considered a promising way to assess individual physical and cognitive health when privacy and user acceptance issues are considered [1]. Approaches belonging to this category infer the ADLs performed by an individual by capturing the interactions between an individual and a specific object. For example, we can use a contact sensor to record whenever the medicine container is open or closed for the application of adherence to medication. In sensor-based activity recognition, the output of an AR system is a stream of sensor activations [7, 9]. We can then treat activity recognition as a time series analysis problem, and the aim is to identify a continuous portion of sensor data stream associated with one of the preselected known activities. The widely used approach to AR is to apply the supervised learning with an explicit training phase, which mainly consists of three stages [10, 11]. First, a stream of sensor data is divided into segments, in which a sliding window technique is often used. Specifically, a window with a fixed time length or fixed number of sensor events is shifted along the stream with (non-) overlapping between adjacent segments. The next step is to extract features from the segments and transform the raw signal data into feature vectors, followed by the classifier construction with these features. The last task, called recognition phase, is to use the trained classifier to associate a stream of sensor data with a predefined activity. From the view of pattern recognition and machine learning, appropriate feature representation of sensor data, suitable choice of classifier and its parameter settings are crucial factors that determine the performance of AR [12]. Although researchers have proposed a number of models to recognize ADLs, however, most of existing AR approaches usually rely on hand-crafted features such as mean, variance, correlation coefficients and entropy, and this may result in loss of information. Also, most classifiers used have been shown to have shallow structures, hence it is difficult for them to discover the latent non-linear relations inherent in features [13]. Furthermore, in most studies, feature extraction and classifier training are treated as two separate steps, so they are not jointly optimized. Consequently, without the guidance of classification performance, the best way to design and choose feature descriptors is not clear, and we may fail to obtain satisfactory accuracy without the exploration of feature extraction.

In recent years, deep learning techniques have gained great popularity and been successfully applied in various fields such as speech recognition and face recognition due to its representational power. These techniques enable the automatic extraction of features from the original low-level features without any specific domain knowledge but with a general-purpose learning procedure. In this study, to improve the activity recognition performance, we propose to exploit deep learning techniques to discover the latent useful information inherent in the original features, and integrate feature learning and classifier training into an architecture to jointly optimize them. Specifically, we use a denoising autoencoder to learn the underlying feature representation from unlabeled data, and the obtained features are then used as the inputs of a top classifier. This enables us to unify feature learning and classifier training in a single pipeline and further to fine-tune the model parameters using labeled data in order to obtain a robust model.

The rest of this paper is structured as follows. Section 2 briefly reviews related work in activity recognition. We then illustrate the autoencoder model, the pre-training and fine-tuning scheme, and the proposed activity model in Sect. 3. In Sect. 4, experimental setup and results are presented. The last section concludes this study with a short summary and discussion.

2 Related Work

To improve the performance of activity recognition and enable its wide applications in real world scenarios, researchers have conducted considerable work in exploring various sensing technologies and designing a number of methods to model and recognize human activities [7]. It has been shown that different types of sensor modalities are effective for recognizing different activities. Vision-based approaches can provide a better recognition rate, but the use of camera or video is not practical in many indoor environments particularly when the privacy issue is considered [14]. Moreover, vision-based approaches face technical challenges arising from light, distance from cameras, occlusion and low object recognition rate, which largely hinder their wide use. In the past few years, due to the rapid development of information technology, a variety of sensors are designed and used for human activity recognition due to their flexibility, low cost, and less intrusiveness [15]. These sensors can be categorized into wearable sensors and environment interactive sensors. In the former case, commonly used sensors that can be worn or carriable include accelerometer, gyroscope, GPS, and RFID-readers (used together with RFID tags). For example, Bao and Intille used five small biaxial accelerometers that were worn simultaneously on different parts of the body to recognize twenty activities. By collecting experimental data from twenty volunteers and extracting time-domain and frequency-domain features, they compared the recognition rate of three different classifiers and showed that the decision tree algorithm achieved the best performance with an accuracy of 84.0 % [16]. With the increasing processing and communication power of mobiles devices, most smartphones that are embedded with built-in GPS, accelerometers and gyroscopes are used for activity recognition due to the fact that they are less intrusive to subjects and that no additional equipment is required for data collection and procession [17, 18]. For example, Dernbach et al. demonstrated the possibility of using the inertial sensor data collected from android-based smart phones to recognize simple activities such as biking, climbing, driving, lying, sitting, walking, running and standing, as well as complex activities such as cleaning, cooking, medication, sweeping, washing and watering [19]. Besides these, RFID technology provides a solution to activity recognition as well, because they can capture the interaction between an individual and the objects. For example, Kim et al. built an indoor healthcare monitoring system to locate and track the elderly in real time by capturing the interaction between subjects (an individual wearing a RFID reader) and the tagged objects with RFID technology [20]. Philipose et al. applied RFID technology, data mining and a probabilistic engine for fine-grained activity recognition based on the interaction between objects and subjects [9].

Although wearable sensor based approaches can obtain satisfactory performance, it is difficult for them to be widely applied in residences because this kind of method requires users to wear or carry corresponding sensors all the time. Therefore, they are actually intrusive and may bring inconvenience to individuals when performing ADLs. In contrast, environment interactive sensors with inherent non-intrusive characteristics have proven applicable to the home setting when privacy and user acceptance are concerned [1]. For example, Tapia et al. built an activity recognition system installed with a set of simple state-change sensors, and then deployed their system in two houses equipped with seventy-seven and eighty-four sensors, respectively, and collected data for fourteen days to show its feasibility in AR [1]. van Kasteren et al. carried out a research to recognize seven different activities in a home setting via fourteen binary sensors and obtained an accuracy of 79.4 % [21]. In different studies, several models have been used to recognize activity such as Naïve Bayes [1], hidden markov model [2], support vector machine [22], Bayesian networks [23], and sparse coding [24]. One common feature of these models is that they all have shallow structures and may not capture the complex non-linear relations among features [13]. Also, to analyze the complex human activities, it is expected to extract over-complete and discriminant features from sensor data, and traditional methods rely on domain knowledge to extract features and few consider to learn features from data [12]. Moreover, feature extraction and classifier training are taken as two separate steps and not jointly optimized in most of these methods. All of these issues motivate us to explore new ways to improve the performance of activity recognition.

3   Proposed Method for Activity Recognition

3.1   Autoencoder

The autoencoder is a type of artificial neural networks that consist of three layers: input layer, hidden layer and output layer (see Fig. 1), with the constraint that the target values of the output layer are equal or approximate to the inputs during training. An autoencoder aims to learn a latent representation $h(x)$ of the input vector x. Suppose N and k denote the number of input units and the number of hidden units, respectively. Given a N-dimensional input vector x, the autoencoder transforms it to a latent representation $h(x) \in \mathbf{R}^k$ through a deterministic mapping (1),

$$h(x) = f\left(W^{(1)}x + b^{(1)}\right), \tag{1}$$

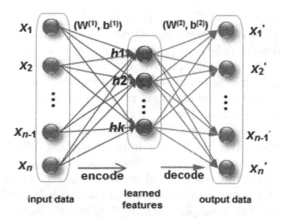

Fig. 1. The autoencoder architecture. The number of units in hidden layer is not necessarily less than that in the input layer.

where $W^{(1)} \in \mathbf{R}^{(k \times N)}$ is a matrix containing the weights from the input units to the hidden units, $b^{(1)}$ represents the bias of the hidden units, and $f(\cdot)$ is the activation function in each units. One of the most commonly used non-linear activation functions is sigmoid function shown mathematically as,

$$f(x) = 1/(1 + \exp(-x)). \tag{2}$$

We then reconstruct the input x from the latent representation $h(x)$ using (3) and try to minimize the difference between x and x'.

$$x' = f(W^{(2)}h((x) + b^{(2)}), \tag{3}$$

where $W^{(2)} \in \mathbf{R}^{(N \times k)}$ contains the weights from the hidden units to the output units, and $b^{(2)}$ represents the bias of the output units. In such way, we can obtain a new feature representation $h(x)$ of x. Of note, the number of units in the hidden layers can be larger or less than the input dimension, enabling a larger exploration of non-linear relations.

With the aim to obtain a robust feature representation, Vincent et al. proposed the denoising autoencoders that try to reconstruct original data from a corrupted input with a local denoising criterion [25]. The corrupted inputs can be generated by adding random noises to the original inputs or randomly choosing a proportion of them and setting them to be zero. In this study, we use the denoising autoencoder as the building block of the proposed activity recognition model.

3.2 Stacked Autoencoder

Recent advances in deep learning show that a deep or hierarchical architecture can contribute to obtaining more complex and non-linear relations underlying in data when compared with these models with shallow structures that contain zero or only one hidden layer [26]. A stacked autoencoder (SAE) is such a hierarchy model, in which an autoencoder is a building block [27, 28]. In SAE, each layer is fully connected to its adjacent layer and there is no connection between units in each layer. In such architecture, the objective function of SAE is to reconstruct the inputs at the output layer. Similar to the autoencoder, each hidden layer of SAE is actually a high-level representation of the input. Interestingly, the number of units in a hidden layer can be equal to, larger or less than the input dimension. This enables us to sufficiently explore different high-level feature representations in a flexible way.

In training a stacked autoencoder, conventional gradient-based optimization methods, such as SGD and L-BFGS, suffer from the gradient diffusion and can easily be trapped into a poor local optimum on a network with randomly initialized weights and biases. To alleviate this problem and improve convergence rate, Hinton et al. proposed a greedy layer-wise learning process to learning a deep belief network and experimentally showed its good performance [27]. In such methods, we train each network separately rather than train them together, and the output of one network is the input of its following network. Specifically, we use the training data as inputs of an autoencoder to learn the first hidden layer, and then use the first hidden layer as input to learn the second hidden layer, and so on. Generally, assume that there is a stacked autoencoder with n layers and the first layer is the original data (training set). For the k-th autoencoder, $W^{(k)}$ are the weights from the input units to the hidden units, and $b^{(k)}$ are the biases of the hidden layer. The greedy layer-wise scheme performs the following two steps iteratively.

$$a^{(m)} = f\left(z^{(m)}\right), \tag{4}$$

$$z^{(m+1)} = W^{(m)}a^{(m)} + b^{(m)}, \tag{5}$$

where $z^{(m)}$ is the input of the m-th layer, $a^{(m)}$ is the activation of the m-th layer, and $a^{(1)} = x$ when $m = 1$. Obviously, $a^{(n)}$ is the inner-most feature representation of interest. The above process is called *pre-training* because it works in an unsupervised way (without using corresponding labels).

3.3   Fine-Tuning the Activity Recognition Model

In order to perform activity recognition, the features learned in the stacked autoencoder are used with a set of labeled data to build a classifier. Accordingly, we can stack another output layer (classifier layer) on top of the SAE to classify an input. In this case, the feature vector encoded in the last hidden layer is the input of a learning algorithm in the classifier layer, and various classifiers are available for use. Figure 2 presents the overall architecture of an activity recognition model when the softmax classifier is used, in which the number of units in the classifier layer equals the number of activity classes.

To improve the performance of activity recognition, we further optimize the activity recognition model in a supervised manner. Specifically, we initialize the weights and biases of the deep network with values obtained in the *pre-training* process, and use the back propagation with gradient descent algorithm to optimize the model parameters. Prior researches show that such a strategy helps escape from the poor local optimum

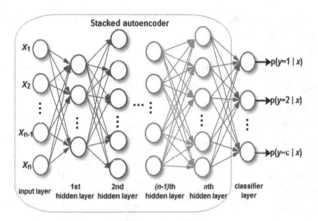

Fig. 2. Illustration to the activity recognition model with a stacked autoencoder and a softmax classifier. The last layer is the classifier layer, and the number of units equals to the number of activities of interest. The probability output determines the label of an input, where c indicates the c-th label. $x_1, x_2, ..., x_{n-1}$ and x_n are a dimension of the original feature representation, each hidden layer is a high-level representation of the original data, and the last hidden layer is retained as the input of the classifier layer.

and improve the time performance [28]. This procedure is called *fine-tuning* and works in a supervised manner (with labeled data involved).

To determine the optimal learning parameters and the network layout (e.g. how many hidden layers and the number of units in each hidden layer), besides the fine-tuning, we employ the grid search strategy and choose the best network structure as the final AR model via cross validation.

4   Experimental Results and Analysis

4.1   Experimental Datasets

To evaluate the performance of the proposed activity recognition model built on deep learning techniques, we conducted experiments on three publicly available datasets collected from three smart homes equipped with various simple sensors, respectively. Each of the smart homes housed one resident performing ADLs in it. For the first smart home (D1), there are three rooms equipped with fourteen sensors in total. Sensor data stream was collected over a period of twenty-five days and ten activities were observed, resulting in 1229 sensor events and 292 activity instances. For the second smart home (D2), thirteen activities were observed during a period of fourteen days in an apartment installed with twenty-three sensors. As a result, there are totally 200 activity instances consisting of 19,075 sensor events. The third smart home (D3) was monitored for nineteen days, and 344 activity instances and 22,700 sensor events were collected. All information regarding the experimental dataset used in this study is briefly summarized in Table 1, and can be found in [29] for other details. Noticeably, all the sensors used are simple state-change sensors, including motion detector sensor, mercury contact, contact switch sensor, pressure mat, and float sensor. So, each dataset consists of binary temporal data that denote the activation of sensors.

Table 1. Experimental dataset description

Dataset	D1	D2	D3
Setting	Apartment	Apartment	House
#resident	1	1	1
Resident age	26	28	57
#rooms	3	2	6
#days monitored	25	14	19
#sensors	14	23	21
#activities	10	13	16
#sensor events	1229	19,075	22,700
#activity instances	292	200	344

4.2 Experimental Setup and Results

The sensor data stream was first divided into segments by shifting a fixed length, non-overlapping sliding window of sixty seconds as suggested by van Kastern et al. [21]. Then a N-dimensional feature vector $x_t = (x_t^1, x_t^2, \ldots, x_t^{N-1}, x_t^N)$ was extracted from each segment, in which N is the total number of sensors installed in a smart home and each dimension of x_t corresponds to a physical sensor. In our experiments, the original features of the sensor data can be represented in two forms: binary representation and numerical representation. The numerical representation method records the number of firings of a sensor during a specific time slice, while the binary representation method records whether a sensor fired at least once during the interval, and the value of a dimension is one if the corresponding sensor fired and zero otherwise. For the evaluation, we performed leave one day out cross validation, in which one full day of sensor data is used to test the classifier performance and sensor data of the remaining days are used for classifier training. We repeat the above process the number of days times and report the average results. Specifically, we evaluate the performance of the proposed model in terms of the following two metrics, the time-slice accuracy and the class accuracy, which can be calculated as follows.

$$\text{timeslice\_accuracy} = \frac{\sum_{n=1}^{M} I(inferred(n) == true(n))}{M} \tag{6}$$

$$\text{class\_accuracy} = \frac{1}{C} \sum_{c=1}^{C} \{ \frac{\sum_{n=1}^{N_c} I(inferred(n) == true(n))}{N_c} \} \tag{7}$$

where $I(a == b)$ is the indicator function returning 1 if a equals b and 0 otherwise, M is the total number of sensor data segments in the test data, N_c denotes the number of segments belonging to class c, $inferred(n)$ is the inferred label of segment n, and $true(n)$ is the true label of segment n. In our study, rather than explore a large number of autoencoders, a two-layer stacked denoising autoencoder (SDAE) is used. Also, we set the number of units in the hidden layer ranging from five to one hundred with a step size of five, and set the percentage of masking noise being 0.5. In addition, we compare the proposed activity recognition model with other four commonly used baselines, including Naïve bayes (NB), hidden markov model (HMM), 1-nearest-neighbor (KNN), and support vector machine with linear kernel (SVM). These four predictors with shallow structures directly use the binary representation and numerical representation rather than the learned high-level features as the inputs. For KNN, we use one nearest neighbor to decide the label of a test sample.

Tables 2, 3 and 4 present the experimental results on the three datasets, respectively. For each, we studied two different original feature representations and reported the average time-slice accuracy and class accuracy of the leave one day out cross validation. From Table 2, we observe that SDAE outperforms other four methods in terms of both time-slice accuracy and class accuracy whichever the original feature representation is adopted. Specifically, SADE obtained a time-slice accuracy of 85.32 % and a class accuracy of 49.91 % compared to the 59.11 % time-slice accuracy and 48.46 % class accuracy of the commonly used probability-based HMM in the case of binary representation. When using the numerical representation, SDAE achieved 85.52 % time-slice accuracy and 53.42 % class accuracy compared to the 59.73 % time-slice accuracy and 43.3 % class accuracy of HMM. For NB classifier, which is built on the basis of conditional independence among features, it performed poorly whichever feature representation was used. This indicates that there exist underlying relations among these features. Also, instance-based learning method KNN also failed to give good results, and consistently obtains the lowest time-slice accuracy and class accuracy. From Table 3, we can observe that deep learning based approaches outperformed their competing methods in time-slice accuracy. Although they failed to achieve the best class accuracy, their difference is quite small. For instance, SDAE obtained a class accuracy of 43.30 %, which was 1.21 % less than the best 44.51 % of SVM. Similar conclusions can be drawn from Table 4.

Table 2. Experimental results on dataset D1.

Method		NB	HMM	1NN	SVM	SDAE
Binary	Time-slice (%)	77.14	59.11	33.10	83.88	85.32
	Class (%)	42.62	45.48	32.43	48.14	49.91
Numerical	Time-slice (%)	77.03	59.73	33.30	83.95	85.52
	Class (%)	38.43	43.35	33.06	48.18	53.42

Table 3. Experimental results on dataset D2.

Method		NB	HMM	1NN	SVM	SDAE
Binary	Time-slice (%)	80.35	63.23	55.73	82.60	84.16
	Class (%)	32.47	44.66	30.47	41.76	39.92
Numerical	Time-slice (%)	80.50	66.79	59.03	81.82	85.61
	Class (%)	24.83	28.79	39.46	44.51	43.30

Table 4. Experimental results on dataset D3.

Method		NB	HMM	1NN	SVM	SDAE
Binary	Time-slice (%)	46.47	26.48	27.73	44.56	50.04
	Class (%)	16.84	17.22	19.81	21.33	21.14
Numerical	Time-slice (%)	41.44	27.37	30.82	42.70	54.82
	Class (%)	11.17	11.21	24.98	25.45	22.08

Overall, we can see that: (1) SDAE outperforms other four competing methods in terms of time-slice accuracy. In class accuracy, deep learning methods achieve better performance than NB, HMM, and 1NN when numerical representation is adopted, and obtain similar performance to others in the case of binary representation. (2) Deep learning based approaches are more robust to the choice of the original feature representation in comparison with other activity recognition models. For example, HMM obtained a class accuracy of 35.8 % in binary representation, decreased by 8.0 % compared to that of the numerical representation. This indicates that deep learning techniques generalize better across different original feature representations and can potentially relieve users of the reliance on domain knowledge to design and select features. Particularly, it should be noted that in this study, we do not fully explore the power of latent feature learning, since we only explore the deep learning architecture with two hidden layers and small number of units in each layer.

5   Conclusions

Wireless sensor network technology has great potential to be widely used in smart homes for human-centric applications due to its non-intrusiveness, low cost, and easy deployment. In activity recognition, researchers have conducted a wealth of work and proposed various models, while few explore how to learn useful features and to jointly optimize feature extraction and classifier construction. In this study, we present a deep learning based activity recognition model that uses an autoencoder to learn useful features from sensor data stream and unifies feature extraction and activity recognition in a single framework. To demonstrate the effectiveness of the proposed approach in activity recognition, we conducted experiments on three publicly available human activity recognition datasets and compared it with other four traditional methods in terms of time-slice accuracy and class accuracy. Experimental results show that our proposed method outperforms the competing methods, indicating its potential in human activity recognition. For the future work, we plan to further optimize the proposed model by varying the number of hidden layers and units in each layer, and study other feature learning methods.

Acknowledgments. This work was supported by the Natural Science Foundation of China (No. 61472057) and China Postdoctoral Science Foundation (No. 2016M592046).

References

1. Tapia, E.M., Intille, S.S., Larson, K.: Activity recognition in the home using simple and ubiquitous sensors. In: Ferscha, A., Mattern, F. (eds.) PERVASIVE 2004. LNCS, vol. 3001, pp. 158–175. Springer, Heidelberg (2004)
2. Cook, D.J.: Learning setting-generalized activity models for smart spaces. IEEE Intell. Syst. **27**, 32–38 (2010)
3. Ordóñez, F., de Toledo, P., Sanchis, A.: Sensor-based Bayesian detection of anomalous living patterns in a home setting. Pers. Ubiquit. Comput. **19**, 259–270 (2015)
4. Suryadevara, N.K., Mukhopadhyay, S.C.: Determining wellness through an ambient assisted living environment. IEEE Intell. Syst. **29**, 30–37 (2014)
5. Liu, L., Peng, Y.X., Wang, S., Huang, Z.G., Liu, M.: Complex activity recognition using time series pattern dictionary learned from ubiquitous sensors. Inf. Sci. **340–341**, 41–57 (2016)
6. Cook, D.J., Krishnan, N.C., Rashidi, P.: Activity discovery and activity recognition: a new partnership. IEEE Trans. Cybern. **43**, 820–828 (2013)
7. Krishnan, N.C., Cook, D.J.: Activity recognition on streaming sensor data. Pervasive Mob. Comput. **10**, 138–154 (2014)
8. Tapia, E., Intille, S., Haskell, W., Larson, K., Wright, J., King, A., Friedman, R.: Real-time recognition of physical activities and their intensities using wireless accelerometers and a heart rate monitor. In: 11th IEEE International Symposium on Wearable Computers, pp. 37–40. IEEE Press, New York (2007)
9. Philipose, M., Fishkin, K.P., Perkowitz, M., Patterson, D.J., Fox, D., Kautz, H., Hähnel, D.: Inferring activities from interactions with objects. IEEE Pervas. Comput. **3**, 50–57 (2004)
10. van Kasteren, T., Englebienne, G., Kröse, B.: An activity monitoring system for elderly care using generative and discriminative models. Pers. Ubiquit. Comput. **14**, 489–498 (2010)
11. Liu, L., Peng, Y.X., Huang, Z.G., Liu, M.: Sensor-based human activity recognition system with a multilayered model using time series shapelets. Knowl. Based Syst. **90**, 138–152 (2015)
12. Plötz, T., Hammerla, N.Y., Olivier, P.: Feature learning for activity recognition in ubiquitous computing. In: Proceedings of the 22nd International Joint Conference on Artificial Intelligence, pp. 1729–1734. AAAI Press, California (2011)
13. Bengio, Y.: Learning deep architectures for AI. Found. Trends Mach. Learn. **2**, 1–127 (2009)
14. Chen, L., Hoey, J., Nugent, C.D., Cook, D.J., Yu, Z.: Sensor-based activity recognition. IEEE Trans. Syst. Man Cybern. Part C **42**, 790–808 (2012)
15. Figo, D., Diniz, P.C., Ferreira, D.R., Cardoso, J.M.: Preprocessing techniques for context recognition from accelerometer data. Pers. Ubiquit. Comput. **14**, 645–662 (2010)
16. Bao, L., Intille, S.S.: Activity recognition from user-annotated acceleration data. In: Ferscha, A., Mattern, F. (eds.) PERVASIVE 2004. LNCS, vol. 3001, pp. 1–17. Springer, Heidelberg (2004)
17. Reiss, A., Hendeby, G., Stricker, D.: A competitive approach for human activity recognition on smartphones. In: European Symposium on Artificial Neural Networks, Computational Intelligence and Machine Learning, ESANN, Belgium, pp. 455–460 (2013)
18. Wang, A.G., Chen, G.L., Yang, J., Zhao, S.H., Chang, C.Y.: A comparative study on human activity recognition using inertial sensors in a smartphone. IEEE Sens. J. **16**, 4566–4578 (2016)
19. Dernbach, S., Das, B., Krishnan, N.C., Thomas, B.L., Cook, D.J.: Simple and complex activity recognition through smart phones. In: 8th International Conference on Intelligent Environments, pp. 214–221. IEEE Press, New York (2012)
20. Kim, S.C., Jeong, Y.S., Park, S.O.: RFID-based indoor location tracking to ensure the safety of the elderly in smart home environments. Pers. Ubiquit. Comput. **17**, 1699–1707 (2013)

21. Van Kasteren, T., Noulas, A., Englebienne, G., Kröse, B.: Accurate activity recognition in a home setting. In Proceedings of the 10th International Conference on Ubiquitous Computing, pp. 1–9, ACM Press, New York (2008)
22. Fleury, A., Vacher, M., Noury, N.: SVM-based multimodal classification of activities of daily living in health smart homes: sensors, algorithms, and first experimental results. IEEE Trans. Inf. Technol. Biol. **14**, 274–283 (2010)
23. Wilson, D.H., Atkeson, C.G.: Simultaneous tracking and activity recognition (star) using many anonymous, binary sensors. In: Gellersen, H.-W., Want, R., Schmidt, A. (eds.) PERVASIVE 2005. LNCS, vol. 3468, pp. 62–79. Springer, Heidelberg (2005)
24. Bhattacharya, S., Nurmi, P., Hammerla, N., Plötz, T.: Using unlabeled data in a sparse coding framework for human activity recognition. Pervasive Mob. Comput. **15**, 242–262 (2014)
25. Vincent, P., Larochelle, H., Bengio, Y., Manzagol, P.A.: Extracting and composing robust features with denoising autoencoders. In: Proceedings of the 25th International Conference on Machine Learning, pp. 1096–1103. ACM Press, New York (2008)
26. Hinton, G.E., Salakhutdinov, R.R.: Reducing the dimensionality of data with neural networks. Science **313**, 504–507 (2006)
27. Hinton, G.E., Osindero, S., Teh, Y.W.: A fast learning algorithm for deep belief nets. Neural Comput. **18**, 1527–1554 (2006)
28. Bengio, Y., Lamblin, P., Popovici, D., Larochelle, H.: Greedy layer-wise training of deep networks. In: Schölkopf, B., Platt, J., Hoffman, T. (eds.) Advances in Neural Information Processing Systems, pp. 153–160. MIT Press, Cambridge (2007)
29. van Kasteren, T., Englebienne, G., Kröse, J.A.B.: Human activity recognition from wireless sensor network data: benchmark and software. In: Chen, L., Nugent, C., Biswas, J., Hoey, J. (eds.) Activity Recognition in Pervasive Intelligent Environments. Atlantis Ambient and Pervasive Intelligence, vol. 4, pp. 165–186. Atlantis, Amsterdam (2011)

Ranking Online Services by Aggregating Ordinal Preferences

Ying Chen[1], Xiao-dong Fu[1,2(✉)], Kun Yue[3], Li Liu[1,2], and Li-jun Liu[1,2]

[1] Kunming University of Science and Technology, Kunming, Yunnan, China
578922532@qq.com, xiaodong_fu@hotmail.com
[2] Yunnan Provincial Key Laboratory of Computer Application, Kunming, Yunnan, China
[3] Yunnan University, Kunming, Yunnan, China
kyue@ync.edu.cn

Abstract. With the increase of the number of online services, it is more and more difficult for customers to make the decision on service selection. Services ranking mechanism is an important service for e-commerce that facilitates consumers' decision-making process. Traditional services ranking methods ignore the fact that customers cannot rate services under the same criteria, which leads to the ratings are actually incomparable. In this paper, we propose to exploit the ordinal preferences rather than cardinal ratings to rank online services. Ordinal preferences are elicited from filled ratings so that the intensity of ratings is ignored. We construct a directed graph to depict the set of pairwise preferences between services. Then, the strongest paths of the directed graph are identified and the evaluation values of services are calculated based on the strongest paths. We prove our method satisfies some important conditions that a reasonable services ranking method should satisfy. Experiments using real data of movie ratings demonstrate that the proposed method is advantageous over previous methods, and so the proposed method can rank services effectively even the ratings are given by customers with inconsistent criteria. In addition, the experiments also verify that it is more difficult to manipulate the ranking result of our method than existing methods.

Keywords: Online services ranking · Ordinal preference · Inconsistent rating criteria · Manipulation

1 Introduction

The services in this paper refer to all kinds of online services, e.g., Web services, e-business services, mobile services and cloud services. Mobile service or electronic service using mobile devices and wireless telecommunications networks has become a hot topic in the information systems and marketing research community. And the great development of the Internet promotes the increase of services and competition dramatically. A consequence is that there are many services sharing the same or similar functions. Moreover, online service provision commonly takes place between parties who have never transacted with each other before, in an environment where the service consumer often has insufficient information about the service provider, and about the goods and services offered [1]. Then how to select

S. Song and Y. Tong (Eds.): WAIM 2016 Workshops, LNCS 9998, pp. 41–53, 2016.
DOI: 10.1007/978-3-319-47121-1_4

an optimal service becomes an important yet difficult problem to consumers. This problem can be mitigated through the use of service ranking. Therefore, reliable ranking measures of services are crucial for business applications.

Some people select services in accordance with the rank of reputation values. But, previous research on services ranking that based on reputation has included counting approach, probabilistic approach, fuzzy approach, flow approach [2, 3] etc. However, there is no consideration on users' preference or evaluation inconsistency in these methods. Some users prefer to give a higher rating to a certain service, while others prefer to give a lower rating because of the difference of consumers' psychology and background, which makes it impossible to compare one service's ratings from different users. For example, services with the same performance may get different ratings, while the previous researches make the services with the same performance get different evaluation results. Therefore, the ranking result of online services based on previous approach is actually incomparable because it lacks the objective representation of services quality. Only when the relationship of ratings assigned to different services by users is considered, will a comparable evaluation method be set up. In addition, the previous research needs only to give a high (or low) rating to one service repeatedly so that it can achieve the purpose of handling the service. Therefore, the ability of the previous research to prevent manipulation is small.

The contributions of the paper can be summarized as follows.

(1) We explore the problem of ranking online services without considering the inconsistence of users' rating criteria and state the idea to address it by using Social Choice Theory. In this way, users' preference can be integrated with the way of aggregating ordinal preference without considering the problem that the ratings are given by users with inconsistent criteria.
(2) We propose a Social Choice based services ranking algorithm, hereinafter to be referred to SCBSR algorithm. The proposed algorithm aggregates the ordinal preferences and returns the ranking results quickly to the user. We further prove that the result of proposed algorithm is hard to manipulate.
(3) We prove some useful theorems so that the Social Choice Theory can be used to rank services with ordinal preferences. These theorems are closely related to users' consumption psychology, so that the efficiency of the algorithm can be improved.
(4) We evaluate the rationality and validity of the proposed algorithms through a comprehensive experimental study and performance analysis.

The rest of paper is organized as follows. In Sect. 2 we review related work. In Sect. 3 we introduce some basic concepts and present a motivating example. In Sect. 4, we introduce the method in details and propose an algorithm to rank services. In Sect. 5, we prove some related theorems. In Sect. 6, we show a systematic empirical study. In Sect. 7, we conclude and discuss the future work.

2 Related Work

With the abundance and fast growth of online services available, how to rank services effectively has been a pressing problem which attracts lots of researchers doing valuable and interesting research [4, 5]. Related work in this area can be classified into two main streams [4]: ranking method based on attribute where the ranking is only related to the features of services, such as sales, price, reputation and ratings; and personalized ranking method.

(1) Attribute-Based Ranking. Kai Hwang [6] combines the services attributes such as sale price, quantity ordered, delivery time, seller trust, and service quality to ranks sellers. S.N. Junaini [5] proposes a framework consisting of the usability factors such as simplicity, attractiveness, effectiveness, service image, site information, service details and so on to rank online services. Services are ranked based on each considered service feature, using text analysis techniques to extract condensed information from massive customer reviews in [7–9]. Some people also select services in accordance with the rank of reputation values [10]. Jiliang Tang obtains the user preference from users' reviews with changes over time, thereby compute the trust values of services and rank the services. *EBay*[1] computes the reputation of the service provide through collecting the feedback information from users after each transaction. Feedback information provided by users includes positive rating, neutral rating and negative rating. The reputation value is the result that the number of total positive ratings minus the number of total negative ratings. *Amazon*[2] also uses the feedback information to compute the reputation value. The only difference is the reputation value is the result of the average of all ratings. However, all of the researches mentioned above, whether based on the ranking of service feature or based on the ranking of reputation, involve the users' feedback information, but ignore the fact that the feedback information are given by customers actually incomparable. Moreover, some users maybe provide dishonest opinions. As a result, the result of ranking is subject to manipulation.

(2) Personalized Ranking. In [11], services are ranked based on the users' own preferences and also on the information in the different search engines about the services. Ghose et al. [12] proposes a 'utility-preserving' ranking strategy from an economic perspective which takes multi-dimensional preference and customer heterogeneity into consideration. In [13], a new personalized service ranking method is proposed based on estimating consumer information search benefits and considering the uncertainty and confidence. All of these works ignore the relationship between different services, although some works take the users' preference into account. Different from these works, the approach proposed in our paper processes the ratings data firstly, rather than use it directly. Additionally, the approach takes the relationship among different services into account.

---

[1] http://www.ebay.com/.
[2] http://www.amazon.cn/.

3   Problem Description

In this section, we first describe the solution of the problem. Then we present a motivating example to highlight the focus of our study.

Definition 1: We let $U = \{u_1, u_2, \ldots, u_m\}$ denote the set of users, $S = \{s_1, s_2, \ldots, s_n\}$ denote the set of services; m is the number of users, n is the number of services.

Definition 2: $R = [r_{ik}]_{m \times n}$ denotes the ratings matrix of user-service, r_{ik} is the rating that user u_i rate the service s_k. When $r_{ik} = 0$, it indicates that user u_i has not yet rated the service s_k.

Definition 3: $O = \{o_k | k = 1, 2, \ldots n\}$ denotes the overall ranking of services. If $o_k > o_l$, it means the service s_k is better than s_l.

Clearly, an effective online services ranking method should satisfy the following criteria: (1) Condorcet: we define a Condorcet winner to be a service which beats every other service in pairwise matchups. (2) Monotonicity: increase one user's rating to a service shouldn't decrease their final ranking. (3) Non-Dictatorship: one user's preference shouldn't determine the outcome of the ranking. (4) Manipulation of Complexity: for any service s_l, the increase of users who only give the service s_l high rating and don't evaluate other services wouldn't change the result. (5) Majority Rule: the service s_k will be defined as the better one if the number of users who prefer service s_k to s_l are more than the number of users who prefer service s_l to s_k.

Example 1. As an example, we assume that there are 4 users (u_1, u_2, u_3, u_4) co-evaluating 3 services (s_1, s_2, s_3). The ratings matrix $R = [r_{ik}]_{4 \times 3}$ is showed in Table 1.

Table 1. Ratings matrix

r_{ik}	s_1	s_2	s_3
u_1	4	3	5
u_2	2	1	3
u_3	1	2	2
u_4	5	4	1

Using the summation method (referred to SUM) and the average method (referred to AVG) which are widely used in eBay, Amazon, Epinions, Taobao etc. well-known sites as comparison in the remainder of this paper. The results are shown in Table 2.

Table 2. The result of evaluation

	o_1	o_2	o_3	Ranking
Sum	0	−1	−1	$s_1 \succ s_2 \sim s_3$
Avg	3	2.5	2.75	$s_1 \succ s_3 \succ s_2$

With Table 1, we can know there are more than half of users preferring service s_3 to every other service, while only one user think service s_1 is better than service s_3. According to the Condorcet, service s_3 is the best and should be ranked first. However, the results of SUM approach and AVG approach are different, believing s_1 is the first. Simultaneously, they violate the non-dictatorship. Moreover, if we add a new user who gives the service s_3 five points, then the service s_3 is ranked first. The non-manipulable is weak. Consequently, users cannot gain an accurate and stable ranking result. Thus, it is desired to develop a ranking approach that meets users preference and difficult to manipulate.

4   Ranking Online Services

Due to the inconsistent rating criteria and users' different subjective preferences, ratings given by different users are actually incomparable. And, users' rating for services is a personal utility. It cannot be measured or aggregated. So, the evaluation information that gathered by different user's rating cannot have cardinality. For all this, without considering the cardinal utility, we need to know the relationship of users' ordinal preference for services. Thus we should change the current social rating systems to enable each user to conduct a preference ranking of services based on evaluation instead of simply giving a rating after a transaction. But it will cost too much, and for users, require users to express the complete preference for all services will necessarily lead to problems of cost, cognitive, communication and privacy etc. Users cannot and will not provide the complete preference information.

Based on this, we firstly transfer the ratings into ordinal preference ranking for services through calculation, and then the number of users preference is obtained through aggregating the users preference. In this way, we avoid the problem that the ratings given by different users are actually incomparable. Lastly, we use the Schulze social choice function generates the online services ranking. Schulze social choice function is a voting algorithm proposed by Schulze Markus. The method can be used to generate a list of winners from users preference. In recent years, the Pirate Party of Sweden, the Debian project, the Wikimedia Foundation, the Gentoo project and other private organizations adopted the function for internal elections and referendum [14]. In this section, we firstly introduce the users preference obtaining. Then we describe our method in details and present the algorithm. Analysis and verification of the properties are provided in the next section.

4.1   Users Preference Obtaining

Since the services that users co-evaluated are few, the user-service ratings matrix is generally an incomplete set of entries. However, our method is based on the model of pair-wise comparisons, so it is necessary to fill the incomplete matrix. Considering the method of collaborative filtering is widely used, and different users have different preferences is an implicit assumption in the method [15], so we use the method to fill the incomplete matrix.

Based on the complete user-service ratings matrix, we calculate preference relations of each user for services, and establish preference matrix for each user. We express it as $LM_i[lm_{kl}]$, which is defined as:

$$lm_{kl} = \begin{cases} 1, & \text{if } r_{ik} > r_{il} \\ 0, & \text{if } r_{ik} = r_{il} \\ -1 & \text{if } r_{ik} < r_{il} \end{cases} \tag{1}$$

lm_{kl} denotes the users preference relationship between service s_k and service s_l. When $lm_{kl} = 1$, it means users prefer s_k to s_l. When $lm_{kl} = -1$, it means users prefer s_l to s_k. When $lm_{kl} = 0$, it means users think s_k and s_l have no difference.

According to the preference matrix of each user, we count the number of $lm_{kl} = 1$ in m users so that we can get the ordinal preferences. And we express it as comparative matrix of service-service $CM[cm_{kl}]_{n\times n}$ $(k,l = 1,2,...,n; k \neq l)$, which is defined as:

$$cm_{kl} = \sum_{i=1}^{m} lm_{kl}. \tag{2}$$

cm_{kl} denotes the amount of users who prefers s_k to s_l.

4.2 Social Choice Based Services Ranking Algorithm

For making the algorithm clear and understandable, we introduce some more definitions as follows:

Definition 4: We define a directed graph $G = (V,E)$ such that $V = S$ is the set of vertices, and E is the set of edges, which is defined as the set $\{(s_k,s_l) \in V \times V\}$. The weight of edge is the value of cm_{kl} and $cm_{kl} \geq cm_{lk}$.

Definition 5: We use a sequence of services set $\{s(1),s(2),...,s(t)\} \subseteq S$ denotes a path from service s_k to service s_l and it must satisfy the following characteristics:

① $s(1) = s_k, s(t) = s_l$
② $0 \leq t \leq n$
③ $\forall t = 1,...,n-1: cm_{s(t),s(t+1)} > cm_{s(t+1),s(t)}$ and $s(t) \neq s(t+1)$

The algorithm starts with the comparative matrix of service-service CM we calculated above. Then, based on the Definitions 4 and 5 we look for the strongest path from service s_k to s_l and express it as the strongest path length matrix $PM[pm_{kl}]$. The values of the strongest path pm_{kl} have three cases, as follows:

① If there is no path from vertex s_k to vertex s_l, then there is also no strongest path from service s_k to service s_l and $pm_{kl} = 0$;
② If there is only one path can be reached from vertex s_k to s_l, then this path is the strongest path from the service s_k to service s_l and the minimum weight in this path is the value of the strongest path, that is $pm_{kl} = min(cm_{s(i),s(i+1)})$, $i = 1,...t-1$;

③ If there are multiple paths to reach from vertex s_k to vertex s_l, then compared the minimum weight of each path, the maximum value of that path is the strongest path from service s_k to service s_l. So $pm_{kl} = max\{min(cm_{s(i),s(l+1)})\}$, $i = 1,...t-1$.

Then we calculate the evaluation values of services based on the strongest path length matrix, that is $o_k = \sum_{l=1}^{n} (I(pm_{kl} > pm_{lk}))$, $l = 1,2,...n$, and $I(*)$ is a indicate function.

$$I(pm_{kl} > pm_{lk}) = \begin{cases} 1, \text{ if } pm_{kl} > pm_{lk} \\ 0, \text{ others} \end{cases} \tag{3}$$

As can be seen from above, the evaluation results of the services depend on the quantity of this service's strongest path value more than others. Finally, we can get the whole ranking of services by sorting the evaluation values of each service. Algorithm 1 is our SCBSR algorithm.

Algorithm 1: SCBSR algorithm

```
cm_{k1}, the value of comparison of two services from item
k to item 1 (Input)
O, the overall ranking of services(Output)
  begin
    for k from 1 to n
      for l from 1 to n
        if(k≠l)then
          if(cm_{k1}≥cm_{1k})then
            pm_{k1}:=cm_{k1};
          else pm_{k1}:=0;
    for t from 1 to n
      for k from 1 to n
        if(t≠k)then
          for l from 1 to n
            if(t≠l and k≠l) then
              pm_{k1}: =max(pm_{k1}, min(pm_{kt}, pm_{t1}));
    for k from 1 to n
      count:=0;
      for l from 1 to n
        if(pm_{k1}≠0 or pm_{1k}≠0)then
          if(pm_{k1}>pm_{1k})then
            count:=count+1;
      o_i:=count;
    sort(o_i);
  end.
```

Applying the above algorithm to Example 1 in Sect. 3, the ranking result is $s_3 > s_1 > s_2$. Table 1 shows that more than half of users think that service s_3 is better than the other services, while only one user think service s_1 is better than service s_3. According to the Condorcet and Non-dictatorship, service s_3 is the best and should be ranked first, consistent with the results. So, Example 1 satisfies Condorcet and Non-dictatorship.

5 Theoretical Analysis of Ranking Model

In this section we prove our method satisfies Condorcet, monotonicity, and non-dictatorship condition. The complexity of manipulation of the rank results also be analyzed.

Theorem 1 (Condorcet). For any service s_k, when more than half of users think that service s_k is better than every other service in pairwise matchups, then s_k is Condorcet candidate.

Proof: When more than half of users think that service s_k is better than other services, it is obvious that $cm_{kl} > cm_{lk}$. So there is no path that treats other services as vertex point to service s_k. So $pm_{kl} > pm_{lk}$ for any $l = 1,2,\ldots n$, namely, service s_k is the best, featuring Condorcet. Condorcet meets most people's preference, embodying the will of the majority of people. So we can also treat the Condorcet candidate as evaluation result to recommend.

Theorem 2 (Monotonicity). When users' preference are invariable and their ranking for any service s_k and s_l is as $s_k \succ s_l$ (denotes service s_k is better than s_l). If a certain user upgrades his/her rating for service s_k, then the ranking of service s_k will be unchanged or move ahead compared to s_l, featuring monotonicity.

Proof: If a certain user upgrades his/her rating for service s_k, then the number of users who think the service s_k is better than s_l increases. So the value of cm_{kl} increases. Then the minimum strength of the path that starts with the service s_k point to service s_l is impossible to reduce. Thus the value of pm_{kl} is impossible to reduce. So $pm_{kl} > pm_{lk}$ is unchanged, namely, the entire ranking is maintaining the ordering as $s_k \succ s_l$. Thus this method satisfies monotonicity. Monotonicity highlights the stability of ranking, and it wouldn't decrease the rankings of service by increasing the number of users who think a service is better than other services.

Theorem 3 (Non-dictatorship). If and only if there is one people believes s_k is better than s_l, while others believe that s_l is better than s_k. When the result is the ranking of s_l no worse than s_k, the non-dictatorship can be satisfied.

Proof: When there is only one people thinking s_k is better than s_l, according to Formulas (1) and (2), $cm_{kl} = 1, cm_{lk} > 1$. So the strength of the path that begin with the service s_l point to service s_k is greater than 1. Thus $pm_{lk} > 1$, $pm_{lk} \geq pm_{kl}$, so service s_l will not inferior to s_k. Thus this method satisfies non-dictatorship. Non-dictatorship guarantees that the users' specific preferences don't affect overall evaluation results.

Theorem 4 (Manipulation of Complexity). For any service s_l, the increase of users who only give the service s_l high rating and don't evaluate other services wouldn't change the result.

Proof: According to the ranking method of this paper, we need to fill the user-service ratings matrix of incomplete. However, according to the method of collaborative filtering, the filling data by calculated is equal to the rating that new users have given to service s_l, so the value of cm_{kl} is unchanged; the value of pm_{kl} is also unchanged. Thus the ranking result maintains unchanged. Therefore, when you tend to manipulate one service, the method of service evaluation in this paper is more complex than the existing methods.

6   Experiment and Analysis

In this section, we through verifying the Condorcet, monotonicity, majority rule, and complex of manipulation of the ranking results to test the effectiveness and rationality of online services rank method. All experiments were conducted on a PC with Intel Core i3 3.50 GHz CPU and 8 GB of RAM. The algorithms are implemented in MyEclipse 2015.

To verify the validity of method, we have used a movie data set with the real ratings: MovieLens[3] which has 1682 movies, 943 users and 100 thousand ratings. In the experiment, to ensure the movie that users have common rated have a certain number, and in order to compare the advantages and disadvantages with other methods on different sparsties, it is necessary to filter on datasets. Five groups of different datasets were obtained after processing the initial data.

6.1   Condorcet

If the first service ranked by this method is equal to the Condorcet candidate, then this method satisfies Condorcet property. Aggregating all users' preferences, we can know that more than half of users think the service 47# is better than any other services, so the service 47# is Condorcet candidate. The result has been shown in Fig. 1, the evaluation value of service 47# is the biggest, ranking first. Consistent with theoretical analysis result, so this method satisfies the Condorcet property.

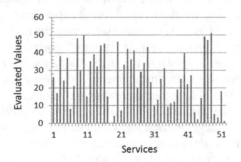

Fig. 1.   Condorcet verification

---

[3]   http://www.grouplens.org/node/73.

6.2 Monotonicity

In order to verify the monotonicity of our method, the paper selects a service randomly and increases its rating, to judge whether the ranking of the service is changed. If the place of this service is ahead or unchanged, then the monotonicity of this method is verified. In the experiment, taking service 1# and service 2# as examples, service 1# is better than service 2#. After we increase the rating rated by users to service 1#, evaluation value of service 1# is increases, while the evaluation value of service 2# and other services remains unchanged. The result has been shown in Fig. 2. So the final ordering of service 1# is still ahead of service 2#, and compared to the rankings of other services, the ranking of service 1# can only be ahead or unchanged but not lower, verifying the monotonicity of this method.

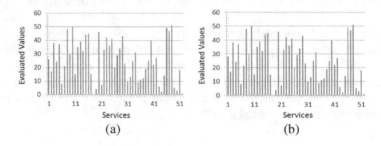

Fig. 2. Monotonicity verification. (a)Before change (b) After change

6.3 The Complexity of Manipulation

In order to verify the complexity of manipulation of the method, the paper chooses a service randomly and increases several users' rating for the service. Compared with other methods, if our method has the same results and the other methods make the ranking values of this service move ahead, then our method satisfies the complexity of manipulation. In this section, we conduct two sets of experiments. A set of experiments assumes that the number of users is constant, increasing the rating. So we assume that 10 users give the service 7# 5 points, 4 points, 3 points, 2 points, 1 point in turn to observe the rankings of the service, as shown in Fig. 3(a). Another set of experiments assumes that the rating is constant, increasing the number of dishonest users. So we assume these users give the service 7# 5 points, then increasing the number of dishonest users. The result of the rankings of the service has shown in Fig. 3(b). According to Fig. 3, we know no matter how we try to change, the ranking of service 7# remains unchanged, while the rankings in other two methods is increase or decrease, achieving the purpose of manipulation. So our method is more difficult to manipulate than the others.

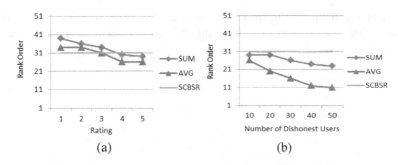

Fig. 3. The complexity of manipulation verification. (a)Increase ratings (b) Increase the number of dishonest users

6.4 Majority Rule

According to the herd mentality of the consumer, we verify the majority rule of economics theory. That is to say, when all of services are for pairwise comparison, the result of ranking is s_l superior to s_k if the number of users who prefer service s_k to service s_l are more than the number of users who prefer service s_l to service s_k. But because of Condorcet's Paradox, the results cannot be reached 100 %. In this section, we count Majority Rule ratio of three methods by testing five groups of different datasets. As shown in Fig. 4, our method achieving on 99 % accuracy, higher than the other two methods.

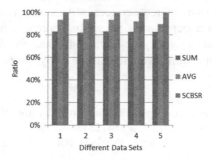

Fig. 4. Majority rule verification

6.5 Response Time

When the number of services is fixed, we increase the number of users in turn, and record the response time of the system for ranking all of services. As shown in Fig. 5, with the increase of the number of users, the response time of the system is approximately in a linear speed increase. As the number of users increases, the amount of calculation for the overall ranking will increase, but our ranking method did not lead to the response time increasing in exponentially, so our method is more efficient. And the Fig. 5 shows that our method has a long system response time compared with the other two methods.

On the other hand, manipulating the ranking results of our method takes longer than the other two methods, so our method is more difficult to manipulate than the others.

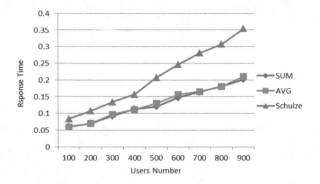

Fig. 5. The response time of system

7 Conclusion

In this paper, we address the problem of users rating criteria inconsistent-based services ranking. We firstly transfer the ratings in current social rating systems into users' preference ranking for services through calculation. Then we propose an algorithm based on social choice theory to get the services ranking. By this way, we get the ordinal preferences without changing the current ratings system. Additionally, the approaches take the relationship between different services into account, rather than some separate services as the evaluation objects. Lastly, we apply theoretical analysis and experiment to verify the Condorcet, monotonicity, un-dictatorship and complex of manipulation of our approach. In the future work, we will further study the ranking services method under the condition of not filling the matrix.

Acknowledgment. This work was partially supported by the National Natural Science Foundation of China (Grand 61462056, 61472345, 61462051, 81560296), the Applied Fundamental Research Project of Yunnan Province (Grand No. 2014FA028, 2014FA023, 2014FB133).

References

1. Jøsang, A., Ismail, R., Boyd, C.: A survey of trust and reputation systems for online service provision. Decis. Support Syst. **43**, 618–644 (2007)
2. Yao, Y., Ruohomaa, S., Xu, F.: Addressing common vulnerabilities of reputation systems for electronic commerce. J. Theor. Appl. Electr. Commer. Res. **7**(1), 1–20 (2012)
3. Malaga, R.A.: Web-based reputation management systems: problems and suggested solutions. Electr. Commer. Res. **1**(4), 403–417 (2001)
4. Abdul-Muhmin, A.G.: Contingent decision behavior: effect of number of alternatives to be selected on consumers' decision processes. J. Consum. Psychol. **8**(1), 91–111 (1999)

5. Zhang, K,, Narayanan, R., Choudhary, A.: Voice of the customers: mining online customer reviews for product feature-based ranking. In: Proceedings of the 3rd Conference on Online Social Networks, p. 11 (2010)
6. Feng, Q., Hwang, K., Dai, Y.: Rainbow product ranking for upgrading e-commerce. IEEE Internet Comput. 13(5), 72–80 (2009)
7. Najmi, E., Hashmi, K., Malik, Z., et al.: CAPRA: a comprehensive approach to product ranking using customer reviews. Computing 97(8), 843–867 (2015)
8. Gangothri, V., et al.: Engender product ranking and recommendation using customer feedback. In: Suresh, L.V., Panigrahi, B.K. (eds.) Proceedings of the International Conference on Soft Computing Systems, pp. 851–859. Springer, India (2016)
9. Najmi, E., Hashmi, K., Malik, Z., et al.: CAPRA: a comprehensive approach to product ranking using customer reviews. Computing 1–25 (2015)
10. Tang, J., Gao, H., Dassarma, A., et al.: Trust evolution: modeling and its applications (2013)
11. Mohanty, B.K., Passi, K.: Web based information for product ranking in e-business: a fuzzy approach. In: Proceedings of the 8th International Conference on Electronic commerce: The New e-Commerce: Innovations for Conquering Current Barriers, Obstacles and Limitations to Conducting Successful Business on the Internet, pp. 558–563. ACM (2006)
12. Ghose, A., Ipeirotis, P.G., Li, B.: Designing ranking systems for hotels on travel search engines by mining user-generated and crowd sourced content. Mark. Sci. 31(3), 493–520 (2012)
13. Zhang, M., Guo, X., Chen, G., et al.: Predicting consumer information search benefits for personalized online product ranking: a confidence-based approach (2014)
14. Schulze, M.: A new monotonic, clone-independent, reversal symmetric, and condorcet consistent single-winner election method. Soc. Choice Welfare 36(2), 267–303 (2011)
15. Jøsang, A., Guo, G., Pini, M.S., Santini, F., Xu, Y.: Combining recommender and reputation systems to produce better online advice. In: Torra, V., Narukawa, Y., Navarro-Arribas, G., Meg\'ıas, D. (eds.) MDAI 2013. LNCS, vol. 8234, pp. 126–138. Springer, Heidelberg (2013)

DroidDelver: An Android Malware Detection System Using Deep Belief Network Based on API Call Blocks

Shifu Hou[1], Aaron Saas[1], Yanfang Ye[1(✉)], and Lifei Chen[2]

[1] Department of Computer Science and Electrical Engineering,
West Virginia University, Morgantown, WV 26506, USA
{shhou,asaas}@mix.wvu.edu, yanfang.ye@mail.wvu.edu
[2] School of Mathematics and Computer Science, Fujian Normal University,
Fuzhou 350117, China
clfei@fjnu.edu.cn

Abstract. Because of the explosive growth of Android malware and due to the severity of its damages, the detection of Android malware has become an increasing important topic in cyber security. Currently, the major defense against Android malware is commercial mobile security products which mainly use signature-based method for detection. However, attackers can easily devise methods, such as obfuscation and repackaging, to evade the detection, which calls for new defensive techniques that are harder to evade. In this paper, resting on the analysis of Application Programming Interface (API) calls extracted from the smali files, we further categorize the API calls which belong to the some method in the smali code into a block. Based on the generated code blocks, we then apply a deep learning framework (i.e., Deep Belief Network) for newly unknown Android malware detection. Using a real sample collection from Comodo Cloud Security Center, a comprehensive experimental study is performed to compare various malware detection approaches. Promising experimental results demonstrate that *DroidDelver* which integrates our proposed method outperform other alternative Android malware detection techniques.

Keywords: Android malware detection · API call block · Deep belief network

1 Introduction

Smart phones have been widely used in people's daily life, such as paying bills, controlling smart homes, and entertainment. Due to their mobility and ever expanding capabilities, smart phones have experienced an explosive growth rate. It is estimated that, in 2019, 77.7% of all devices connected to the Internet will be smart phones, leaving PCs falling behind at 4.8% [14]. Android is an open

---

Y. Ye—This work is partially supported by the U.S. National Science Foundation under grant CNS-1618629.

S. Song and Y. Tong (Eds.): WAIM 2016 Workshops, LNCS 9998, pp. 54–66, 2016.
DOI: 10.1007/978-3-319-47121-1_5

source and customizable operating system for smart phones which is currently dominating the smart phone market at 82.8 % [15]. Due to its large market share and openness of development, Android is an ideal platform for legitimate developers, but also for attackers creating malware (*mal*icious soft *ware*). Many examples of Android malware have already been released in the market (e.g., Geinimi, DriodKungfu and Hongtoutou) [16] which pose serious threats to the smart phone users, including autodialing premium numbers, stealing sensitive information, and sending SMS messages and advertisements without user's concern [2]. In 2015, over 750,000 new Android malware samples were identified, a 32 % increase from 2014 [9], and this trend is only expected to worsen. To protect smart phone users from the attacks of Android malware, currently, the major defense is mobile security products, such as Lookout and Comodo Mobile Security, which mainly use the signature-based method to recognize threats. However, attackers can easily use the techniques such as code obfuscation, repackaging to evade the signature-based detection.

In order to help combat the malware threats, in this paper, resting on the analysis of Application Programing Interface (API) calls extracted from the smali files, we further categorize the API calls which belong to the some method in the smali code into a block. Based on the generated code blocks, we then apply a deep learning framework (i.e., Deep Belief Network) for newly unknown Android malware detection. Using a real sample collection from Comodo Cloud Security Center, a comprehensive experimental study is performed to compare various malware detection approaches. Promising experimental results demonstrate that *DroidDelver* which integrates our proposed method outperform other alternative Android malware detection techniques. The major contributions of our work can be summarized as follows:

- *A novel feature representation method for Android malware detection.* Instead of using API calls directly, we further categorize the API calls which belong to the same method in the smali code into a block. For each Android application (app), it will finally be represented by blocks of API calls. The API call block can well represent the complete function in an app, which creates a higher-level representation than a simple list of API calls and requires more efforts for attackers to evade the detection.
- *An exploration of deep neural network.* Deep neural network has been shown to have better feature learning capability due to its multilayer architecture. Based on the generated API call blocks, a deep belief network is explored to learn the patterns associated with malicious and benign apps in order to detect newly unknown Android Malware.
- *Comprehensive experimental study on a real world sample collection.* We develop an Android malware detection system *DroidDelver* which integrates our proposed method and provide a comprehensive experimental study on the real sample collection from Comodo Cloud Security Center, which consists of 2,500 benign apps and 2,500 Android malware including the families of DriodKungfu, MonkeyTest, WifiPassword, and FakePlayer etc.

The remainder of this paper is organized as follows. Section 2 discusses the related work. Section 3 presents an overview of our system architecture. Section 4 introduces our proposed method in detail. In Sect. 5, based on the real sample collection from Comodo Cloud Security Center, we systematically evaluate the performance of our developed Android malware detection system *DroidDelver* in comparison with other alternative detection methods. Section 6 concludes.

2   Related Work

In recent years, there have been many researches on developing intelligent malware detection systems using machine learning and data mining techniques [8,13,18,19,23]. In these systems, the detection process is generally divided into two steps: *feature extraction* and *classification*. For feature extraction, there are generally two kinds of approaches: *dynamic analysis* and *static analysis*. Dynamic analysis is the process of evaluating an app during execution, which is commonly performed in a virtual environment to prevent a malware infection on the host. DroidDolphin [23] used a dynamic analysis framework including Droid-Box [7] and APE [1] to record 13 activity features from the collected Android apps, and then applied Support Vector Machine (SVM) to build a malware prediction model. Crowdroid [13] also performed dynamic analysis for Android malware detection which extracted API system calls as the feature set for k-means clustering. CopperDroid [17], an automatic Virtual Machine Introspection (VMI) based dynamic analysis system, extracted operating system interactions (e.g., file and process creation), as well as intra- and inter-process communications (e.g., SMS reception) as the features to represent the behaviors of the Android apps. Though dynamic extraction is more resilient to low level obfuscation, it is computationally expensive to perform and requires simulation of user interactions. On the contrast, static analysis focuses on analyzing the internal components of an app without executing it. This makes it much cheaper to perform than dynamic analysis. DroidMat [8] performed static analysis on Android apps to extract API calls, permissions and intent messages as the input features for k-means clustering and finally k-NN classification. DroidMiner [4] also extracted API calls, but then transformed them into modalities for associative classifier. Peiravian and Zhu [20] analyzed Android apps creating a feature set consisting of API calls and permission requests that they then fed to SVM, Decision Tree, and ensemble classifiers. Due to its high efficiency in feature extraction, in this paper, we choose to use static analysis for feature representation of Android apps. We first extract API calls from the smali files. Different from the existing works, we then further categorize the API calls which belong to the same method in the smali code into a block, since it provides a higher-level representation than a simple list of API calls and requires more efforts for attackers to evade the detection.

Although shallow learning methods such as Support Vector Machine (SVM), Artificial Neural Network (ANN), Naïve Bayes (NB), and Decision Tree (DT) have successfully applied in Android malware detection [8,18–20,23], they have limitations that leave a large room for improvement. Deep learning, a new frontier in machine learning and data mining, is starting to be leveraged in industrial and academic research for different applications (e.g., Computer Vision) [26,27]. In this paper, based on our collected sample set and extracted features, we intend to explore a deep learning framework for newly unknown Android malware detection.

3   System Architecture

In this paper, based on the collected Android apps, we extract the API call blocks as the features and apply a deep learning framework (i.e., Deep Belief Network) for newly unknown Android malware detection. Figure 1 shows the system architecture of our developed Android malware detection system *DroidDelver*, which consists of the following five major components.

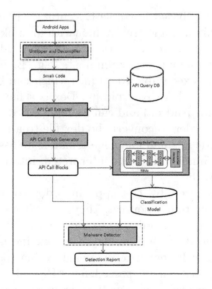

Fig. 1. System architecture of *DroidDelver*

- *Unzipper and Decompiler:* The APKTool [3] is used to unzip the APKs and decompile the dex files to smali codes. (See Sect. 4.1 for details.)
- *API Call Extractor:* It automatically extracts the API calls from the smali codes. Through the API query database, the API calls extracted from the smali codes will be converted to a group of global integer IDs which represents the static execution sequence of the corresponding API calls. (See Sect. 4.1 for details.)

- *API Call Block Generator:* Based on the extracted API calls from the smali code, the APIs which belong to the some method (between each pair of "*.methods*" and "*.end method*") in the smali code will be further categorized into a block. For each Android app, it will finally be represented by blocks of API calls. (See Sect. 4.2 for details.)
- *Deep Belief Network Classifier:* Resting on the generated API call blocks, a Deep Belief Network (DBN) is used for model construction and thus for newly unknown Android malware detection. (See Sect. 4.3 for details.)
- *Malware Detector:* For each new collected unknown Android app, it will be first parsed through the unzipper and decompiler to get the smali codes, then its API calls will be extracted from the smali codes, and finally the API call blocks will be generated as the feature vector. By using the constructed classification model, this app will be labeled either benign or malicious.

4 Proposed Method

4.1 Android API Call Extraction

Unlike traditional desktop based Portable Executable (PE) file, Android app is compiled and packaged in a single archive file (with an .apk suffix) that contains the dex file (app code), resources, assets, and manifest file. Android apps are developed with Java. Development environments (e.g., Eclipse) convert the Java source codes into Dalvik executable (dex) files which can be run on the Dalvik Virtual Machine (DalvikVM) in Android [6]. Dex is a file format that contains compiled code written for Android and can be interpreted by the DalvikVM [5], but it is unreadable. In order to convert the dex file to a readable format, smali provides us readable code in smali language. Smali is an assembler/disassembler for the dex format [5]. Smali code is the interpreted, intermediate code between Java and the DalvikVM [6].

Since API calls are used by the apps in order to access operating system functionality and system resources, they can be used as representations of the behaviors of an app. In order to extract the API calls, the app is first unzipped to provide the dex file, and then the dex file is further decompiled into smali code using a well-known reverse engineering tool APKTool [3]. The converted smali code can then be parsed for API call extraction. Listing 1.1 shows a segment of the converted smali code from "winterbird.apk" (MD5: *53cec6444101d1976af1b253ff5b2226*) which is a theme wallpaper app embedded with malicious code that can steal user's credential. In this Smali file, the API calls of "*Lorg/apache/http/HttpRequest; →containsHeader*" and "*Lorg/apache/http/HttpRequest; →addHeader*" will be extracted.

As not all of the extracted API calls are contributing to malware detection [24,28], in this paper, we further apply *Max-Relevance* algorithm [12] to select a set of API calls with the highest relevance to the target classes (either benign or malicious).

Listing 1.1. Example of smali code

```
# direct methods
.method process()V
  .locals 2
  .parameter ''request''
  .parameter ''context''
  const-string v0, ''Accept-Encoding''
  invoke-interface{p1,v0},Lorg/apache/http/HttpRequest;
  ->containsHeader(Ljava/lang/String;)Z
  const-string v0, ''Accept-Encoding''
  const-string v1, ''gzip''
  invoke-interface{p1,v0,v1},Lorg/apache/http/HttpRequest;
  ->addHeader(Ljava/lang/String;Ljava/lang/String;)V
  return-void
.end method
```

4.2   API Call Block Generation

API calls can be used to represent the behaviors of an Android app. However, if using API calls only, sometimes it's not enough to detect a malicious app. For example, Fig. 2 shows the screen shots of "Locker.apk" (MD5: *f836f5c6267f13bf9f6109a6b8d79175*), which is a ransomware [21] that will lock the smart phone's screen (as shown in Fig. 2(c)) after installation (Fig. 2(a) and (b) display its installation and running interfaces). If the smart phone is infected by this malware, the victim is demanded to pay a ransom to the attackers to unlock his/her smart phone. After further analysis, we find that the API calls of "*Ljava/io/FileOutputStream-write*", "*Ljava/io/IOException-printStackTrace*", and "*Ljava/lang/System-load*" together in the method of "loadLibs" in the converted smali code indicate this ransomware intends to write malicious code into the system kernel. Though it may be common to use

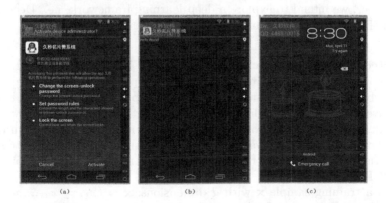

Fig. 2. Screen shots of the ransomware "Locker.apk"

them individually in benign apps, they three together rarely appear in the same method in benign apps. Thus, this API call block containing those three sensitive API calls can be used as a feature to detect this ransomware. In this paper, after API call extraction, we further categorize the API calls which belong to the same method in the smali code into a block. For each Android app, it will finally be represented by blocks of API calls. The API call block can well represent the complete function in an app, which creates a higher-level representation than a simple list of API calls and requires more efforts for attackers to evade the detection since it may result in execution collapse if the attackers add several non-associated API calls in the block.

To generate the API call blocks from the extracted API calls, we define a *block* as the code between a pair of ".*method*" and ".*endmethod*" in the smali file. Listing 1.1 shows an example of a block, which includes two API calls "*Lorg/apache/http/HttpRequest;* →*containsHeader*" and "*Lorg/a-pache/http/HttpRequest;* →*addHeader*". By querying the API database, these two API calls will be mapped to global integer IDs of 520 and 560. Thus, this API call block can be represented as $< 520, 560 >$. In this paper, we don't consider the order and frequency of the API calls. For the example of "winterbird.apk", the API call blocks are recorded as following: $\{< 520, 560 >, < 520, 580, 582 >, < 520, 528, 580 >, < 520, 580, 582 >, ...\}$, where 580, 582, and 528 represent the API calls of "*Landroid/os/Handler;*→*post*", "*Ljava/lang/Exception;* →*toString*", and "*Landroid/util/Log;* →*w*" respectively. After transformation, this app can finally represented by a call block vector $< B1, B2, B3, B2, ... >$.

4.3   Deep Belief Network for Android Malware Detection

Although classification methods based on shallow learning architectures, such as Support Vector Machine (SVM), Artificial Neural Network (ANN), Naïve Bayes (NB), and Decision Tree (DT), can be used to solve the Android malware detection problem [8,18–20,23], deep learning has shown great performance in classification tasks such as image recognition and natural language processing [11,22]. In this paper, we intend to explore a deep learning framework, i.e., Deep Belief Network (DBN), for Android malware detection.

Restricted Boltzmann Machine. A Restricted Boltzmann Machine (RBM) is a type of neural network with an energy-based probabilistic model [11]. An RBM is a bipartite graph with a visible layer (an input layer, denoted as v) and a hidden layer (denoted as h), where the only interactions are formed between the layers, but not between units of the same layer. If all nodes are random binary variable node (i.e., only take value 0 or 1), and assume probability distribution $p(v, h)$ satisfies Boltzmann distribution, we call this model Restricted Boltzmann Machine (RBM).

Given a training sample $\mathbf{x}$ with d dimensions: $\mathbf{x} = (x_1, x_2, x_3, ..., x_d)$, as a special form of Boltzmann Machines, the energy function of RBM is bilinear, which can be denoted as [11]

$$E_\theta(\mathbf{x}, \mathbf{h}) = -\sum_{i=1}^{d}\sum_{j=1}^{|H|} w_{ij}x_i h_j - \sum_{i=1}^{d} b'_i x_i - \sum_{j=1}^{|H|} b_j h_j, \tag{1}$$

where the hidden layer $\mathbf{h}$ includes $|H|$ hidden units: $\mathbf{h} = (h_1, h_2, h_3, ..., h_{|H|})$, and the mapping parameter set θ is composed of weight matrix $\mathbf{w}$, the offset vector $\mathbf{b}$ and $\mathbf{b'}$ (i.e., $\theta = (\mathbf{w}, \mathbf{b}, \mathbf{b'})$).

When $\mathbf{x}$ is fed to the visible layer, the RBM will turn on or off each hidden unit based on its conditional probability, and visa versa. In our application, we use sigmoid function to compute $P(h_j|\mathbf{x})$ and $P(x_i|\mathbf{h})$ [11]:

$$P(h_j|\mathbf{x}) = s\left(\sum_{i=1}^{d} w_{ij}x_i + \mathbf{b}\right), \ \ P(x_i|\mathbf{h}) = s\left(\sum_{j=1}^{|H|} w_{ij}h_j + \mathbf{b'}\right). \tag{2}$$

The process of training RBM, in fact, is to find the best probability distribution to reproduce each input vector.

Deep Belief Network. Deep Belief Network (DBN) is built on a stack of RBMs, where the trained activations of one RBM are used as the inputs of the next RBM [11]. It first uses a layer by layer approach to greedily pre-train the network (i.e., initialize the parameters of the network through unsupervised training), and then further fine-tune all the parameters with respect to the labeled information through associative memory. Figure 3 shows the architecture of the DBN in our application.

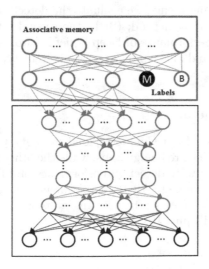

Fig. 3. Deep Belief Network (DBN) for android malware detection

The training process of DBN in our application for Android malware detection is summarized as follows:

Pre-training Process. The first layer RBM is at first fully trained. Then the states of the hidden units in the first hidden layer is used as the input for the second RBM. After fully training the second RBM, it will be stacked on top of the first RBM. The above steps are repeated to build a multi-layer RBM neural network. In the associative memory, at the top layer of the RBMs, two nodes are added to represent the class labels (either benign or malicious) for fine-tuning process.

Fine-Tuning Process. After unsupervised pre-training, DBN further uses a contrastive version of the wake-sleep algorithm [10] to fine-tune the model. Except the top-layer RBM, each other layer of the RBMs includes upward cognitive weights and downward generation weights. For the *wake* phase (also called cognitive process), it uses input features and upward cognitive weights of each layer to produce a compressed representation (node statuses), and applies gradient descent method to modify the downward generation weights between layers; For the *sleep* phase (also called generation process), it uses the top layer representation and downward weights to generate underlying states, while modifies the upward cognitive weights between layers.

5   Experimental Results and Analysis

In this section, two sets of experimental studies are conducted using a real sample collection obtained from Comodo Cloud Security Center to fully evaluate the performance of our developed Android malware detection system *DroidDelver*: (1) In the first set of experiments, we evaluate the detection performance of our proposed feature extraction method; (2) In the second set of experiments, we evaluate the detection performance of Deep Belief Networks (DBNs) with different sets of parameters and also compare with different typical shallow learning methods.

5.1   Experimental Setup

The sample set obtained from Comodo Cloud Security Center includes 5,000 android apps, half of which are benign apps, while the other half are malware. We evaluate the Android malware detection performance of different methods using the measures shown in Table 1. All the experiments are performed under the environment of 64 Bit Windows 8.1 operating system with Intel(R) Core(TM) i7-4790 CPU @ 3.60 GHZ plus 16 GB of RAM.

Table 1. Performance indices of android malware detection

Indices	Specification
True Positive (TP)	Number of apps correctly classified as malicious
True Negative (TN)	Number of apps correctly classified as benign
False Positive (FP)	Number of apps mistakenly classified as malicious
False Negative (FN)	Number of apps mistakenly classified as benign
TP Rate (TPR)	$\frac{TP}{TP+FN}$
FP Rate (FPR)	$\frac{FP}{FP+TN}$
Accuracy (ACY)	$\frac{TP+TN}{TP+TN+FP+FN}$

Table 2. Comparisons between Android API calls and API call blocks

Method	Accuracy	TP	FN	TN	FP
API Calls+SVM (A_SVM)	91.1 %	2280	220	2275	225
API Calls+ANN (A_ANN)	90.22 %	2239	261	2272	228
API Calls+NB (A_NB)	79.44 %	1976	524	1996	504
API Calls+DT (A_DT)	90.74 %	2262	238	2275	225
API Call Blocks+SVM (B_SVM)	94.24 %	2348	152	2364	136
API Call Blocks+ANN (B_ANN)	93.68 %	2322	178	2362	138
API Call Blocks+NB (B_NB)	80.02 %	2009	491	1992	508
API Call Blocks+DT (B_DT)	94.08 %	2349	151	2355	145

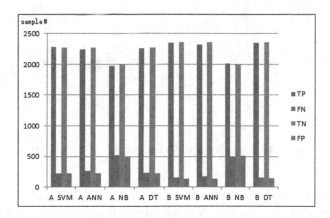

Fig. 4. Comparisons between Android API calls and API call blocks

5.2 Comparisons of Detection Performance Between Android API Calls and API Call Blocks

In this set of experiments, the performance of Android malware detection is compared between the features of Android API calls and API call blocks, using

Table 3. Comparisons of different Deep Belief Network (DBN) constructions

Layer #	Neuron #	Accuracy	TP	FN	TN	FP
2	[300,300]	95.58%	2376	124	2403	97
2	[200,200]	95.72%	2396	104	2390	110
2	[100,100]	95.96%	2400	100	2398	102
3	[300,300,300]	96.2%	2397	103	2413	87
3	**[200,200,200]**	**96.66%**	**2419**	**81**	**2414**	**86**
3	[100,100,100]	95.84%	2393	107	2399	101
4	[200,200,200,200]	95.44%	2392	108	2380	120
5	[200,200,200,200,200]	95.26%	2373	127	2390	110

Table 4. Comparisons between DBN and shallow learning models

Method	Accuracy	TP	FN	TN	FP
API Call Blocks+SVM (B_SVM)	94.24%	2348	152	2364	136
API Call Blocks+ANN (B_ANN)	93.68%	2322	178	2362	138
API Call Blocks+NB (B_NB)	80.02%	2009	491	1992	508
API Call Blocks+DT (B_DT)	94.08%	2349	151	2355	145
API Call Blocks+DBN (B_DBN)	**96.66%**	**2419**	**81**	**2414**	**86**

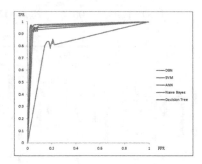

Fig. 5. ROC curves of different classification models

four typical shallow learning methods (i.e., SVM, ANN, NB and DT). Based on
the sample set described in Sect. 5.1, we conduct 10-fold cross validations for
the evaluation. The experiment results shown in Table 2 and Fig. 4 demonstrate
that using API call blocks as the features outperform using API calls directly for
Android malware detection. The API call block features will therefore be used
for the remaining experiments.

5.3   Evaluation of Deep Belief Network in Android Malware Detection

In this set of experiments, based on the sample set described in Sect. 5.1, using the features of API call blocks, we further evaluate the detection performance of Deep Belief Network (DBN). While building the deep learning model, there are two key parameters: the number of hidden layers and the number of neurons in each hidden layer. Table 4 shows the detection performance changes with different DBN constructions. Comparisons are also conducted between DBN and other four typical shallow learning classification models as shown in Table 4. Figure 5 shows the ROC curves of different classification methods. From Table 4 and Fig. 5, it can be seen that, compared with the typical shallow learning methods, the Android malware detection performance is greatly improved by using DBN.

6   Conclusion

In this paper, an automatic Android malware detection system, *DroidDelver*, is developed using a deep learning framework (i.e., Deep Belief Network) based on the API call block features extracted from the smali code. The proposed feature representation statically extracts API call blocks from the smali codes maintaining the inherent relationship existing within the API calls. A comprehensive experimental study is performed on a real sample collection from Comodo Cloud Security Center. Promising experimental results demonstrate that *DroidDelver* which integrates our proposed method outperform other alternative Android malware detection techniques. In our future works, we will further improve *DroidDelver* by enhancing the semantic depth of the feature extraction, and explore how other deep learning frameworks perform in Android malware detection (Table 3).

References

1. APE: a smart automatic testing environment for android malware. https://books.google.com/books?id=hUYDrgEACAAJ
2. Felt, A.P., Finifter, M., Chin, E., Hanna, S., Wagner, D.: A survey of mobile malware in the wild. In: SPSM (2011)
3. APKTool. http://ibotpeaches.github.io/Apktool/
4. Yang, C., Xu, Z., Gu, G., Yegneswaran, V., Porras, P.: DroidMiner: automated mining and characterization of fine-grained malicious behaviors in android applications. In: Kutyłowski, M., Vaidya, J. (eds.) ICAIS 2014, Part I. LNCS, vol. 8712, pp. 163–182. Springer, Heidelberg (2014)
5. Dalvik Opcodes. http://pallergabor.uw.hu/androidblog/dalvik_opcodes.html
6. Dex. http://www.openthefile.net/extension/dex
7. DroidBox. https://github.com/pjlantz/droidbox
8. Wu, D., Mao, C., Wei, T., Lee, H., DroidMat, K.: Android malware detection through manifest and API calls tracing. In: ASIA JCIS (2012)

9. G DATA. Mobile malware report for the fourth quarter of 2015. https://www.gdata-software.com
10. Hinton, G.E., Dayan, P., Frey, B.J., Neal, R.M.: The wake-sleep algorithm for unsupervised neural networks. Science **268**, 1158–1161 (1995)
11. Hinton, G.E., Osindero, S., Teh, Y.: A fast learning algorithm for deep belief nets. Neural Comput. **18**, 1527–1554 (2006)
12. Peng, H., Long, F., Ding, C.: Feature selection based on mutualinformation: criteria of max-dependency, max-relevance, and min-redundancy. In: TPAMI (2005)
13. Burguera, I., Zurutuza, U., Nadjm-Tehrani, S.: Crowdroid: behavior-based malware detection system for Android. In: SPSM (2011)
14. IDC. http://www.idc.com/getdoc.jsp?containerId=prUS25500515
15. IDC. http://www.idc.com/prodserv/smartphone-os-market-share.jsp
16. Xu, J., Yu, Y., Chen, Z., Cao, B., Dong, W., Guo, Y., Cao, J.: MobSafe: cloud computing based forensic analysis for massive mobile applications using data mining. Tsinghua Sci. Technol. **18**, 418–427 (2013)
17. Tam, K., Khan, S.J., Fattori, A., Cavallaro, L.: CopperDroid: automatic reconstruction of android malware behaviors. In: NDSS (2015)
18. Dimjasevic, M., Atzeni, S., Ugrina, I., Rakamaric, Z.: Evaluation of android malware detection based on system calls. In: IWSPA (2016)
19. Dimjasevic, M., Atzeni, S., Ugrina, I., Rakamaric Z.: Android malware detection based on system calls. In: UUCS (2015)
20. Peiravian, N., Zhu, X.: Machine learning for android malware detection using permission and API calls. In: ICDM (2013)
21. Ransomware. https://containment.comodo.com/why-comodo/ransomware.php
22. Collobert, R., Weston, J.: A unified architecture for natural language processing: deep neural networks with multitask learning. In: ICML (2008)
23. Wu, W., Hung, S.: DroidDolphin: a dynamic Android malware detection framework using big data and machine learning. In: RACS (2014)
24. Xu, J., Sung, A., Chavez, P., Mukkamala, S.: Polymorphic malicious executable scanner by API sequence analysis. In: HIS (2004)
25. Bengio, Y.: Learning deep architectures for AI. Found. Trends Mach. Learn. **2**(1), 1–127 (2009)
26. Bengio, Y., Lamblin, P., Popovici, D., Larochelle, H.: Greedy layer-wise training of deep networks. In: NIPS (2007)
27. Lv, Y., Duan, Y., Kang, W., Li, Z., Wang, F.: Traffic flow prediction with big data: a deep learning approach. Intell. Transp. Syst. **16**(2), 1–9 (2014)
28. Ye, Y., Wang, D., Li, T., Ye, D.: IMDS: intelligent malware detection system. In: SIGKD (2007)

A Novel Feature Extraction Method on Activity Recognition Using Smartphone

Dachuan Wang[1], Li Liu[2], Xianlong Wang[1], and Yonggang Lu[1(✉)]

[1] School of Information Science and Engineering, Lanzhou University,
Gansu 730000, China
ylu@lzu.edu.cn
[2] School of Software Engineering, Chongqing University, Chongqing 400044, China

Abstract. In the last few years, research on human activity recognition using the built-in sensors of smartphones instead of the body-worn sensors has received much attention. Accelerometer is the most commonly used sensor of smartphone for the application. An important step in activity recognition is feature extraction from the raw acceleration data. In this work, a novel feature extraction method which considers both the distribution and the rate of change of the raw acceleration data is proposed. The raw time series liner acceleration data was collected by a smartphone application developed by ourselves. The proposed feature extraction method is compared with a previously proposed statistics-based feature extraction method using two evaluation methods: (a) distance matrix before clustering, (b) ARI and FM-index after clustering using MCODE. Both results show that the newly proposed feature extraction method is more effective for daily activity recognition than the previously proposed method.

Keywords: Activity recognition · Feature extraction · Smartphone · Unsupervised classification

1 Introduction

At present, smartphones are so popular that people carry with them almost anywhere and anytime. Most of the smartphones are equipped with a rich set of embedded sensors, such as accelerometer, GPS sensor, gyroscope, etc. In the last few years, some works using the built-in sensors of smartphones for human activity recognition (HAR) have been proposed. Human activity recognition is one of the important and challenging research areas in ubiquitous computing since it has a wide range of applications including security, healthcare, lifestyle analysis, smart environments, surveillance, etc. Camera-based computer vision systems have been widely used for human activity tracking, but they mostly require infrastructure support, for example, complete camera coverage in the monitoring areas [1]. Alternatively, inertial sensor-based systems for human activity recognition has become an active field of research in the domain of pervasive and mobile computing. Within various sensors, accelerometer is the most commonly used

© Springer International Publishing AG 2016
S. Song and Y. Tong (Eds.): WAIM 2016 Workshops, LNCS 9998, pp. 67–76, 2016.
DOI: 10.1007/978-3-319-47121-1_6

sensor for recoding body motion signals, because daily activities such as walking, standing, sitting and jogging can be clearly defined by the motion of the body parts. In the work of Bao et al. [2], five biaxial accelerometers are placed in five locations on the user's body to monitor 20 types of activities, trained with 20 users using well-known machine learning classifiers. However, these special sensors are usually not available for the common users. In [3], the acceleration data collected from the embedded triaxial accelerometers of smartphones have been investigated for HAR. As in those different approaches to activity recognition, standard classification algorithms cannot be applied directly on raw time series data. Usually, feature extraction methods have to been used to produce a new data representation (called features) before the classification. Popular features computed from the acceleration data are mean, variance or standard deviation, energy, entropy, correlation between axes or discrete FFT coefficients [4]. In the work by Ravi et al. [5], four features (mean, standard deviation, energy and correlation) are extracted from each axis of a single triaxial accelerometer to recognize eight activities.

In this work, we have proposed a novel feature extraction method which creates a feature vector from the raw time series acceleration data aiming to enhance activity recognition rate especially for the activities ascending stairs and descending stairs. Both the distribution and the rate of the change of the acceleration data are considered in the feature extraction. Molecular Complex Detection (MCODE) clustering method is used to classify these human daily activities including walking, jogging, ascending stairs, descending stairs, sitting and standing. MCODE is an unsupervised clustering method, which has been successfully applied to distinguish race walking from normal walking and running in our previous work [6].

2   Related Work

Many human-activity recognition systems have been proposed such as Camera-based computer vision systems and inertial sensor-based systems. In computer vision-based activity recognition, the common approach is to extract image features from the images or video and to issue a corresponding activity class label [1]. In general, the computer vision-based techniques for human activity tracking often work well in a laboratory or well-controlled environment. However, they require cameras to be placed beforehand at the predetermined points of interest, and they can be influenced by the lighting conditions. Hence, these techniques are not appropriate for highly varied activities that take place in the natural environments. Alternatively, many human activity recognition systems based on the inertial sensor have been developed in recent years. Some of the earliest works based on the inertial sensor focus on the use of multiple accelerometers and possibly other sensors. Bao et al. [2] have used five biaxial accelerometers worn on the user's right hip, wrist, upper arm, ankle and thigh to collect acceleration data from 20 subjects while performing 20 activities. Using several classifiers, they create models to recognize twenty daily activities. It is shown that the decision tree based classifier shows the best performance for recognizing daily activities, which can produce an overall accuracy rate of 84 %. Furthermore, their results show that placing accelerometers on the subjects thigh is the most effective way for distinguishing the set of 20 activities. Ravi et al. [5] have used a single triaxial accelerometer worn near the pelvic region to distinguish 8 activities: standing, walking, running, ascending stairs, descending stairs, vacuuming, brushing, and situps. They have run several supervised classifiers on data sets in four different settings. It is shown that ascending stairs and descending stairs are hard to be distinguished. In the work of Kwapisz et al. [3], smartphone accelerometers are used to perform activity recognition with data collected from 29 subjects, each carrying a smartphone in their pocket as they perform six activities which are walking, jogging, ascending stairs, descending stairs, sitting and standing, using three supervised classification methods (Decision Trees, Regression, Neural Network). The results show that the activities except ascending stairs and descending stairs can be recognized correctly with a accuracy over 90 %, while the two activities ascending stairs and descending stairs are difficult to be distinguished. All the above studies are using supervised classification methods to perform activity recognition, while in our previous work [6] an unsupervised method named MCODE is used for recognizing race walking using smartphone sensors. The experimental results show that MCODE is effective to distinguish race walking from normal walking and running using smartphone accelerometers.

As mentioned above, in both papers [3,5], it is found that two activities ascending stairs and descending stairs are difficult to be classified. So we have proposed the feature extraction method for improving activity recognition accuracy, especially for the activities ascending stairs and descending stairs.

3 Method

3.1 Data Collection

In order to collect data for our experiment, we have developed a simple Android application that runs on the smartphone. The application permits us to start or stop the data collection and label the activity through a simple graphical user interface. When the data collection is stopped, the data will be saved in a textfile. By setting the application to record the data of linear acceleration sensor type, we can collect linear acceleration data along the x-axis, y-axis and z-axis of the device, not including gravity. Six healthy subjects with ages from 22 to 28 volunteer for the study. Each of them is instructed to carry a Samsung Galaxy SIII smartphone in their front pants leg pocket while performing a specific set of activities for a certain time. These activities are walking, jogging, ascending stairs, descending stairs, sitting and standing. The data collection process is under supervision by one of our team members to ensure the quality of the data.

3.2 Feature Extraction

In the data collection, we have obtained triaxial linear acceleration time series along x-axis, y-axis and z-axis. Figure 1 demonstrates these axes relative to a

Fig. 1. Axes of motion relative to user

user. In this section, we present a novel method to transform the time series data into a feature vector. At first, the sliding window approach is employed to divide time series data into smaller time segments using a window size of 300 with 150 samples overlapping between consecutive windows, where each segment represents an instance of certain activity. At a sampling frequency of 50 Hz using the smartphone, each instance contains 6-s data readings. We choose the window size of 6 s because it can sufficiently capture cycles in these activities for each segment.

In our feature extraction method (the proposed method), for each segment, a 32-dimensional feature vector is created, denoted by $\boldsymbol{F}$ with $\boldsymbol{F}_i$ indicating the ith dimension value of $\boldsymbol{F}$. Let $\boldsymbol{Y}$ denote the 300 samples along y-axis of each segment, with $\boldsymbol{Y}_j$ indicating the jth dimension value of $\boldsymbol{Y}$ in the unit of m/s^2. Let $\boldsymbol{Z}$ denote the 300 samples along z-axis of each segment, with $\boldsymbol{Z}_j$ indicating the jth dimension value of $\boldsymbol{Z}$ in the unit of m/s^2. The $\boldsymbol{F}$ is calculated as follow:

$$\boldsymbol{F}_i = \sqrt{\sum_{j=1}^{300} \exp(-|\boldsymbol{Y}_j - i + 15|)} \quad i \in [1, 2, ..., 30]$$

$$\boldsymbol{F}_{31} = \sqrt{\sum_{j=2}^{300} |\boldsymbol{Y}_j - \boldsymbol{Y}_{j-1}|}$$

$$\boldsymbol{F}_{32} = \sqrt{\sum_{j=2}^{300} |\boldsymbol{Z}_j - \boldsymbol{Z}_{j-1}|}$$

The first formula gives 30 features which represent the distribution of the acceleration data around 30 distinct acceleration values from -15 m/s^2 to 14 m/s^2 along y-axis, the second and the third formulas give the 31th and 32th features which represent the rate of change of the acceleration data along y-axis and z-axis respectively.

3.3   MCODE

After obtaining the features, the MCODE-based method was utilized for activity recognition. MCODE is an efficient graph clustering method and is first applied to large protein-protein networks [7]. The input to MCODE is an undirected graph and the output are clusters of instances. In order to utilize the MCODE clustering algorithm for activity recognition, all the feature vectors of each subjects data need to be mapped into an undirected graph, denoted by G. The graph G is a complete graph and each vertex in the graph represents a feature vector, namely, an instance of certain activity. The set of all vertices and all edges in G are denoted as V and E, respectively. The length of each edge in E refers to the Euclidean distance between the two corresponding vertices and the edge weight is the reciprocal of the length of the edge. For using the MCODE clustering algorithm effectively, some edges in the complete graph G need to be deleted, which is accomplished by defining a threshold for the edge weight. The value of the threshold is the averaged weight of E multiplied by a parameter named EWP. If the weight of an edge is less than the threshold, the edge will be deleted from the complete graph.

MCODE algorithm operates in three stages, vertex weighting, cluster finding and optionally post-processing. Post-processing is used to assign vertices to multiple clusters and delete the clusters with a small number of vertices [7]. In our study, every vertex belongs to one and only one cluster, so the stage of post-processing is not required. The first stage of MCODE, vertex weighting, weights all vertices based on their local network density using the highest k-core of the vertex neighborhood. Density of a graph, $G = (V, E)$, with number of vertices $|V|$ and number of edges $|E|$, is defined as $|E|$ divided by the number of edges of a complete graph composed of V. Thus, density of G, $DG = |E|/(|V|(|V|-1)/2)$. A k-core is a graph of minimal degree k (graph G, for all v in G, $deg(v) \geq k$). For any vertex $v \in V$, the highest k-core called M is found from the local network N_v that is composed of v and its neighbors. Finally, the weight of v is the product of Ms k-core and Ms density. The second stage, cluster finding, takes as input the vertex weighted graph, seeds a cluster with the highest weighted vertex and set the vertex with the highest weight as seed vertex. Then for each neighbor v of the seed vertex, if the weight of v is greater than a given threshold, which is the weight of the seed vertex multiplied by a parameter named VWP, the v will be classified into the cluster which the seed belongs to and set as new seed. Then the neighbours of the new seed will be recursively checked in the same manner. A vertex is not checked more than once and this process stops once no more vertices can be added to the cluster based on the given threshold and is repeated for the next highest unseen weighted vertex in the graph.

4   Results and Discussion

In our previously proposed method [6], statistic-based features are extracted, which includes mean, standard deviation, variance, skewness, kurtosis, signal magnitude area and correlation between axis pair. In this work, we compared the previous method with the newly proposed feature extraction method. Adjusted Rand Index (ARI) [8] and Fowlkes-Mallows Index (FM-index) [9] were used to evaluate the experimental results. The ARI and FM-index are both used to measure the similarity between the clustering result and the benchmark. The maximum value of the ARI and FM-index are both 1, which means that the clustering result are exactly the same as the benchmark, while the minimum value of them are both 0, which means the clustering result is a random result. For both the ARI and FM-index, a larger value means a better similarity between the clustering result and the benchmark.

4.1   Parameter Selection Using ARI

As mentioned above, two parameters, EWP and VWP, need to be provided for applying the MCODE method for activity recognition. In order to find the best parameters for both methods, we have tested various parameter combinations of EWP and VWP on the dataset of 6 subjects. Obviously, the largest sum of ARI values of 6 subjects indicates the best parameter setting. Thus, with each combination of EWP and VWP, for each subjects data, we have computed the ARI value of clustering result by applying the proposed method and the previous method, respectively. Figure 1(a) shows the sum of ARI values of different combinations of EWP and VWP produced by the proposed method, while Fig. 1(b) shows the sum of ARI values produced by the previous method. As shown in Fig. 2, when EWP = 1.6 and VWP = 0.1, the sum of ARI values reaches the largest for the proposed method, whereas the parameter setting of EWP = 1.8 and VWP = 0.1 leads to the largest sum of ARI values for the previous method. In the experiments, it is found that the best parameter settings given by FM-index are exactly the same as the results listed above. Therefore, we have selected EWP = 1.6 and VWP = 0.1 for the proposed method, while EWP = 1.8 and VWP = 0.1 are used for the previous method.

4.2   Evaluation

Evaluation Using Distance Matrix. A good feature extraction method for activity recognition ought to make sure that the exacted features of the same activity are similar while the features of different activities are dissimilar. In other words, the more similar the exacted features of the same activity, and the more dissimilar the features of the different activities, the better the feature extraction method is. So, the pairwise Euclidean distance matrix of the feature vectors can be used to measure the similarities between features and thus can be used to evaluate the feature extraction method. Here the Euclidean distance matrixes are used to compare the proposed feature extraction method with previous method. For each subjects data, we have constructed two Euclidean distance matrixes produced using the proposed method and previous method respectively. Each matrix is normalized by dividing the mean value of the matrix. All the matrixes are shown as color map in Fig. 3 through Fig. 8. In all the figures, a pixel represents the distance between the two corresponding instances of activities so that

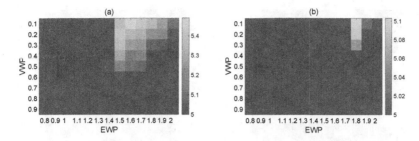

Fig. 2. Sum of ARI values for different combinations of EWP and VWP. (a) Proposed Method; (b) Previous Method

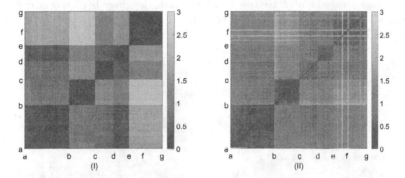

Fig. 3. Distance matrix of activities (walking: a–b, jogging: b–c, ascending stairs: c–d, descending stairs: d–e, sitting: e–f, standing: f–g) of *Subject 1* with (I) Proposed Method; (II) Previous Method

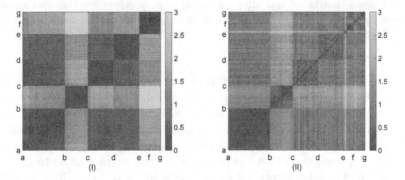

Fig. 4. Distance matrix of activities (walking: a–b, jogging: b–c, ascending stairs: c–d, descending stairs: d–e, sitting: e–f, standing: f–g) of *Subject 2* with (I) Proposed Method; (II) Previous Method

the area of the squares of each activity is determined by the square of the number of instances of the activity.

Obviously, the proposed feature extraction method outperforms previous method as shown in each figure. It is clearly shown that the similarities within

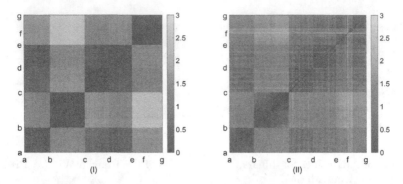

Fig. 5. Distance matrix of activities (walking: a–b, jogging: b–c, ascending stairs: c–d, descending stairs: d–e, sitting: e–f, standing: f–g) of *Subject 3* with (I) Proposed Method; (II) Previous Method

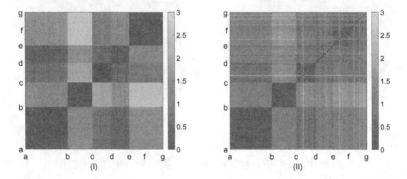

Fig. 6. Distance matrix of activities (walking: a–b, jogging: b–c, ascending stairs: c–d, descending stairs: d–e, sitting: e–f, standing: f–g) of *Subject 4* with (I) Proposed Method; (II) Previous Method

the same activity for the proposed method is apparently higher than the previous method in all the figures. For the proposed method, all the figures show that the activities including walking, jogging, ascending stairs and descending stairs are all distinct from each others, while for the previous method, ascending stairs and descending stairs cannot be easily distinguished in some cases, such as in Figs. 5, 6 and 8. These results show that the proposed feature extraction method is more effective for distinguishing between ascending stairs and descending stairs. It is also found that the sitting and standing are not distinct from each other in the results of both methods. It is well known that both sitting and standing are static, namely, the liner acceleration of them are all near 0, and our approach only uses the single liner acceleration readings of activities, which may explain the failure for distinguishing the two activities, which is also the case for the previous method. For both of the two methods, it can be seen from Fig. 4 that the similarities among the walking, ascending stairs and descending stairs are relatively high, which is also mentioned in the work [10].

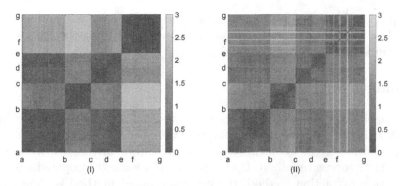

Fig. 7. Distance matrix of activities (walking: a–b, jogging: b–c, ascending stairs: c–d, descending stairs: d–e, sitting: e–f, standing: f–g) of *Subject 5* with (I) Proposed Method; (II) Previous Method

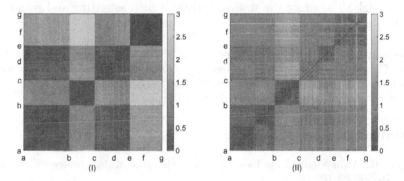

Fig. 8. Distance matrix of activities (walking: a–b, jogging: b–c, ascending stairs: c–d, descending stairs: d–e, sitting: e–f, standing: f–g) of *Subject 6* with (I) Proposed Method; (II) Previous Method

Table 1. Comparison of the proposed feature extraction method with the previous method using ARI and FM-index

		Subject1	Subject2	Subject3	Subject4	Subject5	Subject6
ARI	Proposed Method	0.9119	0.9363	0.9264	0.9487	0.8859	0.8818
	Previous Method	0.8841	0.9135	0.7443	0.8313	0.8700	0.8597
FM-index	Proposed Method	0.9326	0.9490	0.9420	0.9589	0.9120	0.9068
	Previous Method	0.9065	0.9308	0.8044	0.8630	0.8963	0.8874

Evaluation Using ARI and FM-index. The best parameters selected for both the proposed method and the previous method in Sect. 4.1 are used in the evaluation, which are EWP = 1.6 and VWP = 0.1 for the proposed method, and EWP = 1.8 and VWP = 0.1 for the previous method. The ARI and FM-index produced by the two methods for 6 subjects are listed in Table 1. It can be seen that for each subject, both the ARI value and the FM-index value produced by the proposed method are higher than these of the previous method. These results show that the proposed method is more effective for the daily activity recognition than the previous method (Fig. 7).

5   Conclusion

In this paper a novel feature extraction method for activity recognition based on the distribution and the rate of change of the linear acceleration data is proposed. Activities except for sitting and standing can be distinguished from each other by the proposed method. Compared to a previously proposed statistics-based feature extraction method, the newly proposed method is shown to be more effective for daily activity recognition, especially for distinguishing between ascending stairs and descending stairs. In the future work, we will involve more phone-based sensors than linear acceleration senor because the liner acceleration are not enough to separate the statical activities such as sitting and standing of which the liner acceleration are both close to zero.

Acknowledgments. This work is supported by the National Natural Science Foundation of China (Grants No. 61272213) and the Fundamental Research Funds for the Central Universities (Grants No. lzujbky-2016-k07). The authors want to thank the volunteers for their time and effort to help us collecting data.

References

1. Poppe, R.: A survey on vision-based human action recognition. Image Vis. Comput. **28**, 976–990 (2010)
2. Bao, L., Intille, S.S.: Activity recognition from user-annotated acceleration data. In: Ferscha, A., Mattern, F. (eds.) Pervasive 2004. LNCS, vol. 3001, pp. 1–17. Springer, Heidelberg (2004). doi:10.1007/978-3-540-24646-6_1
3. Kwapisz, J., Weiss, G., Moore, S.: Activity recognition using cell phone accelerometers. ACM SigKDD Explor. Newslett. **12**(2), 74–82 (2011)
4. Huynh, T., Schiele, B.: Analyzing features for activity recognition. In: Proceedings Conference on Smart Objects and Ambient Intelligence, pp. 159–163. ACM, New York (2005)
5. Ravi, N., Dandekar, N., Mysore, P., Littman, M.: Activity recognition from accelerometer data. Am. Assoc. Artif. Intell. **05**, 1541–1546 (2005)
6. Wei, Y., Liu, L., Zhong, J., Lu, Y., Sun, L.: Unsupervised race walking recognition using smartphone accelerometers. In: Zhang, S., Wirsing, M., Zhang, Z. (eds.) KSEM 2015. LNCS (LNAI), vol. 9403, pp. 691–702. Springer, Heidelberg (2015). doi:10.1007/978-3-319-25159-2_63
7. Bader, G., Hogue, C.: An automated method for finding molecular complexes in large protein interaction networks. BMC Bioinf. **4**(1), 2 (2003)
8. Lawrence, H., Arabic, P.: Comparing partitions. J. Classif. **2**(1), 193–218 (1985)
9. Fowlkes, E., Mallows, C.: A method for comparing two hierarchical clusterings. J. Am. Stat. Assoc. **78**(383), 553–569 (1983)
10. He, Y., Li, Y.: Physical activity recognition utilizing the built-in kinematic sensors of a smartphone. Int. J. Distrib. Sens. Netw. **2013**, 10 pages (2013). doi:10.1155/2013/481580. Article ID 481580

Fault-Tolerant Adaptive Routing in n-D Mesh

Meirun Chen[1($\boxtimes$)] and Yi Yang[2]

[1] School of Applied Mathematics, Xiamen University of Technology,
Xiamen 361024, Fujian, China
`mrchen@xmut.edu.cn`
[2] School of Information Science and Engineering, Lanzhou University,
Lanzhou 730000, Gansu, China

Abstract. Utilizing mobile cloud computing has had numerous advantages such as the reduction of costs, development of efficiency, unlimited storage capacity, easy access to services at any time and from any location. It should be mentioned that the provision of various mobile cloud computing services is faced with problems and challenges that the fault tolerance can be mentioned as the main restrictions. In this paper, we consider fault-tolerant adaptive routing in n-D mesh network. We propose a new faulty block model. All the faulty nodes and faulty links are surrounded in some block, which is a convex structure, in order to avoid routing livelock. The new model constructs the interior spanning forest for each block in order to keep in touch with the nodes inside of each block. The procedure for block construction is dynamically, it will adjust its scale in accordance with the situation of networks, either the fault emergence or the fault recovery. In our new adaptive routing, the message can be sent to its destination whether the destination is inside or outside a block if and only if the destination nodes keep connecting with the mesh networks.

Keywords: Adaptive routing · Faulty block · Cracky faulty block · Livelock · n-D mesh

1 Introduction

Nowadays, smartphones and tablet PCs are more wide spread and users are relying more on them compared to conventional personal computers. This has led to an unprecedented growth in the mobile devices market ushering us into what is known as the era of mobile computing. However, mobile devices suffer from many limitations related to their limited resources. Limitations include: connectivity issues, limited bandwidth, security vulnerabilities, applications compatibility, and a restricted power source since they are mostly powered by batteries. A lot of work about mobile computing has been done, see [1,4,5]. In this paper, we consider the fault tolerant problem in mobile computing network. The mesh structure is one of the most important interconnection network models. As a topology to interconnect multiprocessor computer systems, it has been proved to possess many attractive properties. Parallel computers using mesh as their

© Springer International Publishing AG 2016
S. Song and Y. Tong (Eds.): WAIM 2016 Workshops, LNCS 9998, pp. 77–87, 2016.
DOI: 10.1007/978-3-319-47121-1_7

underlying architecture have been around for years [6]. A very important aspect of mesh fault tolerance is its ability to route from a source node to a destination. Routing is a process to send messages, which can be either data or instructions, from a source node to a destination node, passing some intermediate nodes. There are basically two types of routing: *deterministic routing* and *adaptive routing*. In deterministic routing, a fixed path is used to send/receive messages for a particular pair of source/destination. The obvious advantages of this routing are its simplicity and ease of implementation. However, it suffers the shortcoming of weak fault tolerability. In adaptive routing, there is no dedicated path for a pair of source and destination. The path is adaptive constructed in the process of routing. As a result, adaptive routing can tolerate more faulty nodes than deterministic routing.

Fault-tolerant routing has been studied extensively. In fault-tolerant routing on mesh, most work uses rectangular fault block model [2,7,9]. In rectangular model, all faulty nodes are grouped in disjointed, rectangular areas, called fault blocks. A fault block is constructed in a way that includes as few non-faulty nodes as possible while maintaining rectangular shape. All nodes in these fault blocks, whether they are faulty or non-faulty, are to be bypassed in any routes. However, as [8]'s experiments show that the distribution of faulty vertices has the tendency to make the whole mesh one "big block". In consequence, the whole mesh becomes useless because this big faulty block occupies the entire mesh region.

In our previous work [10], we proposed a new model, cracky rectangular block, for fault-tolerant adaptive routing in two-dimensional (2-D for abbreviation) mesh. The cracky rectangular block is a rectangular block with spanning forest internal induced by all the connected vertices inside of this block. As a result, all connected vertices that would have been included in original rectangular fault block now can become candidate routing vertices. In our new cracky rectangular block, the message can be sent to its destination whether the destination is inside or outside a block. This is a noticeable overall improvement of fault-tolerability of the system.

The 2-D mesh topology has been adopted by Symult 2010 [3], Intel Touchstone DELTA [6] and Intel paragon; the MIT J-machine adopts three-dimensional (3-D for abbreviation) mesh topology. In this paper, we extend the cracky rectangular model and the corresponding adaptive routing in 2-D mesh to multidimensional mesh. Due to the popularity of low dimensional mesh networks, we will take the 3-D mesh as a special case in our paper.

The rest of this paper is organized as follows. Section 2 describes the basic routing function in multi-dimension mesh. Section 3 proposes the cracky block and the adaptive fault-tolerant routing in multidimensional mesh.

2   Basic Routing Function in Multidimensional Mesh

2.1 n-D Mesh

We would like to represent a n-D mesh ($n \geq 2$) with graph definitions first. Let $G = (V, E)$ be a undirected graph. The set V of vertices of graph G represents nodes of the network. The set E of edges of graph G represents links between the nodes. The distance of a shortest path between nodes X and Y, denoted by $dist(X, Y)$, is the length (number of links) of a shortest path between X and Y.

The n-D mesh $M(d_0, d_1, \cdots, d_{n-1})$, where $d_0, d_1, \cdots, d_{n-1}$ are integers and $d_i \geq 2$ for any $i \in \{0, 1, \cdots, n-1\}$, is defined as follows:

- $V(M(d_0, d_1, \cdots, d_{n-1}))$ is the set of all pairs in $Z_{d_0} \times Z_{d_1} \times \cdots \times Z_{d_{n-1}}$;
- there is a link between two different nodes $X = (x_0, x_1, \cdots, x_{n-1})$ and $Y = (y_0, y_1, \cdots, y_{n-1})$ if and only if $x_i = y_i$ for all i, $0 \leq i \leq n-1$, except one j, where $y_j = x_j \pm 1$, the addition and subtraction are modulo d_j.

For any node $X = (x_0, x_1, \cdots, x_{n-1})$, let $\Gamma(X)$ be the set of nodes adjacent to X in $M(d_0, d_1, \cdots, d_{n-1})$. If $1 \leq x_i \leq d_i - 2$, then X has two neighbors in dimension i, there are $(x_0, \cdots, x_i + 1, \cdots, x_{n-1})$ (denoted by X_+^i for short) and $(x_0, \cdots, x_i - 1, \cdots, x_{n-1})$ (denoted by X_-^i for short). Otherwise, $x_i = 0$ (resp. $x_i = d_i - 1$), then X has one neighbor in dimension i, it is $(x_0, \cdots, x_i + 1, \cdots, x_{n-1})$ (denoted by X_+^i for short) (resp. $(x_0, \cdots, x_i - 1, \cdots, x_{n-1})$ (denoted by X_-^i for short)). Thus, every node has from n to $2n$ neighbors, depending on its location in the mesh.

In the following, we will use the symbol $X_{\phi(i)}^i$ to denote the neighbor of X in dimension i, where $\phi(i) \in \{+, -\}$, $i = 0, 1, \cdots, n-1$. We will also use the symbol $!\phi(i)$, the relationship between $!\phi(i)$ and $\phi(i)$ is

$$!\phi(i) = \begin{cases} + & \text{if } \phi(i) = -, \\ - & \text{if } \phi(i) = +. \end{cases}$$

For each pair of nodes $X = (x_0, x_1, \cdots, x_{n-1})$ and $Y = (y_0, y_1, \cdots, y_{n-1})$, we define the i-distance ($i = 0, 1, \cdots, n-1$) between X and Y by $d_i(X, Y) = |x_i - y_i|$.

2.2   A Basic Routing Function in N-D Mesh

The basic routing function in n-D mesh is a classical greedy routing function as follows. A message m with destination Y received by the router of $X = (x_0, x_1, \cdots, x_{n-1})$, if $X \neq Y$ then let i be the first subscript such that $d_i(X, Y) = \max\{d_j(X, Y) | j = 0, 1, \cdots, n-1\}$, the routing function $f(X) = X_{\phi(i)}^i$, where $d(X_{\phi(i)}^i, Y) = d(X, Y) - 1$; otherwise, the message will be absorbed by the router of X.

It is easy to know that if there is no fault in the mesh, the routing function provided is livelock-free since the function sends m more closer to the destination than before.

2.3   Livelock Situations

With the increasing of the scale of the massively parallel computers, the probability of link and node failure is also increasing. As we will show below, in case of some links fail which don't disconnect the n-D mesh, using the basic routing

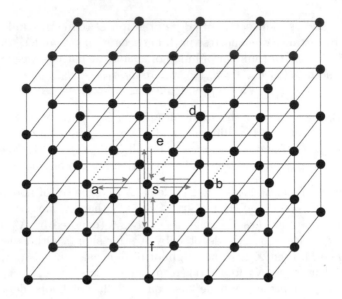

Fig. 1. Livelock situation in 3-D mesh, node s tries to send a message m to node d, but m is in a livelock induced by nodes s, a, b, e, f. The dotted line denotes the faulty link

function does not guarantee that m will reach its destination. It can be blocked in a part of the mesh.

As shown in Fig. 1, the basic routing function can unfortunately lead to blocking situations due to some properties of the structure of the faulty links. Clearly, in this example the message will always in the subgraph induced by nodes s, a, b, e, f and will never reach its destination node d. Actually, this message m is in *livelock* situation, which keeps a message moving indefinitely without reaching the destination.

It is well known that the adaptive routing may cause livelock problems. Therefore, routing without livelock is one of the most important design issues for communication operations in multicomputer systems. The novel cracky faulty block strategy which will be stated in Sect. 3 will avoid the livelock situation.

3   The Fault-Tolerant Adaptive Routing in n-D Meshes

In order to solve the livelock situation and the big block problem, we propose a novel strategy for fault-tolerant routing. We use the cracky block to avoid livelock and traverse block's every connecting internal node if needed. Therefore, we can transmit each message to any node not only outside of a block but also inside of the block, and the message can reach the inside nodes which are forbidden in the original rectangular block.

A block $B((l_0, h_0), (l_1, h_1), \cdots, (l_{n-1}, h_{n-1}))$ is a submesh of the mesh $M(d_0, d_1, \cdots, d_{n-1})$, contains all the nodes in the set $V(B) = \{(x_0, x_1, \cdots, x_{n-1}) | l_i \leq x_i \leq h_i, i = 0, 1, \cdots, n-1\}$ and every link that connects any two nodes in $V(B)$. A node $X = (x_0, x_1, \cdots, x_{n-1})$ is said to be a boundary node of the block $B((l_0, h_0), (l_1, h_1), \cdots, (l_{n-1}, h_{n-1}))$ if $x_i = h_i$ or $x_i = l_i$ for some i.

3.1 The Forming of the Disjoint and Minimal Faulty Blocks in nD Mesh

Each node's activities are based on message-driven mechanism. There are two types of messages routed in mesh. One is entity message (message for short), which is routed between any node pair. The other one is system message, this type of message can only be sent between neighbors, and their contents are mainly about the status of themselves, such as the nodes faulty degree and its detailed situation of faulty links. The first one is the entity for computing or communication, and the later one concentrates on maintaining the usability and robustness of networks; in other words, it is for constructing the cracky block when some faults occur in this mesh in order to avoid the livelock situation as mentioned above.

In the beginning, all the nodes work well; that is, there does not exist any faulty node or faulty link. Any node can both receive the message from any of its neighbors, and vice versa, of course, depending on the basic routing strategy. When some nodes or links are ruined because of some reasons, these failed nodes or nodes incident with failed links will judge their current status immediately, and then they send the system messages as soon as possible including their status to its connected neighbors to tell them what have happened in detail. For neighbors, once they receive the system messages, they judge their current status depending on their latest status and the received system messages at once. Of course, they will notice their connected neighbors about current status if and only if the current status is different from previous status. Finally, the construction of a stable cracky block is implemented by the above system messages exchange.

In this subsection, we will construct the disjoint and minimal blocks to contain all the faults inside the blocks. To construct the disjoint and minimal faulty blocks in mesh, we have to give some definitions first.

The degree $d(X)$ (resp. faulty degree $fd(X)$) of a node X in mesh is the number of links (resp. faulty links) of $M(d_0, d_1, \cdots, d_{n-1})$ incident with X.

Let $S = \{\mathcal{I}\} \cup \{\emptyset\} \cup \{\{0^{\psi(0)}, 1^{\psi(1)}, \cdots, (n-1)^{\psi(n-1)}\}\}$ be the status set, for any i ($i \in \{0, 1, \cdots, n-1\}$), $\psi(i) \in \{+, -, 0\}$. If $\psi(i) = 0$, then delete the corresponding element $i^{\psi(i)}$ from the set $\{0^{\psi(0)}, 1^{\psi(1)}, \cdots, (n-1)^{\psi(n-1)}\}$; if $\psi(i) = 0$ for any $i \in \{0, 1, \cdots, n-1\}$, then view the set $\{0^{\psi(0)}, 1^{\psi(1)}, \cdots, (n-1)^{\psi(n-1)}\}$ as $\emptyset$. That is, there are $3^n + 1$ elements in S. View $\mathcal{I}$ and $\emptyset$ as universal set and empty set, respectively. For any two elements $s_1, s_2 \in S$, $s_1 \cap s_2$ (resp. $s_1 \cup s_2$) is the intersection (resp. union) of s_1 and s_2 as usual. For any $s \in S \setminus \{\mathcal{I}, \emptyset\}$, let $|s|$ be the number of elements contained in s. Clearly, $1 \leq |s| \leq n$.

Let $\Gamma^*(X)$ be the set of all good neighbors of X. That is, if $Y \in \Gamma^*(X)$ then the link $\langle X, Y \rangle$ is good. Moreover, we set an order in $\Gamma^*(X)$ as follows: $Y_0, Y_1, \cdots, Y_{|\Gamma^*(X)|-1}$.

Denote by $s(X)$ the status of node X, where $s(X) \in S$. Let $M(X \to Y)$ be the content of system message sent by node X to $Y \in \Gamma^*(X)$.

Now, we can introduce the procedure to form the minimal and disjoint blocks containing all faulty links and nodes inside.

Step 1. Let $s(X)$ be the status of node X, initialize every node's status by its faulty degree:

$$s(X) = \begin{cases} \mathcal{I} & \text{if } fd(X) = 0, \\ \emptyset & \text{if } fd(X) \geq 2, \\ \{i^{\phi(i)}\} & \text{if } fd(X) = 1 \text{ and the link} \langle X, X^i_{!\phi(i)} \rangle \text{ is faulty.} \end{cases}$$

Step 2. If $s(X) \neq \mathcal{I}$, then notice all or some nodes of $\Gamma^*(X)$ by sending the system message.

$\quad$ *Case 1.* If $s(X) = \emptyset$, send the system message $M(X \to X^j_{\phi(j)}) = \{j^{\phi(j)}\}$ to every node $X^j_{\phi(j)} \in \Gamma^*(X)$.

$\quad$ *Case 2.* If $s(X) = \{i^{\phi(i)}\}$, send the system message $M(X \to X^j_{\phi(j)}) = s(X) \cup \{j^{\phi(j)}\}$ to every node $X^j_{\phi(j)} \in \Gamma^*(X)$, where $j \in \{0, 1, \cdots, n-1\} \backslash \{i\}$, $\phi(j) \in \{+, -\}$.

Step 3. If X receives the system message $M(Y_i \to X)$ and $M(Y_i \to X) \cap s(X) \neq s(X)$, then update the status of X by setting $s(X) = M(Y_i \to X) \cap s(X)$ and notice its neighbors by the following rule: if $s(X) = \emptyset$, then send system message $M(X \to X^j_{\phi(j)}) = \{j^{\phi(j)}\}$ to every node $X^j_{\phi(j)} \in \Gamma^*(X)$; if $s(X) = \{i_1^{\phi(i_1)}, \cdots, i_k^{\phi(i_k)}\}$ and $|s(X)| < n$, then send system message $M(X \to X^j_{\phi(j)}) = \{j^{\phi(j)}\} \cup s(X)$ to $X^j_{\phi(j)}$, $\phi(j) \in \{+, -\}$, where $j \in \{0, 1, \cdots, n-1\} \backslash \{i_1, \cdots, i_k\}$.

$\quad$ We show the above procedure by Algorithm 1.

$\quad$ See Fig. 2 for an example, the dotted line denotes the faulty link, by Algorithm 1, the block in thick line is the faulty block formed by the faults. Take four nodes E, U, W, Z to show how the algorithm performs. In the first phase, node U will initial its status $s(U) = \emptyset$ as a result of $fd(U) = 2$, node E will initial its status $s(E) = \{0^+\}$ since $fd(E) = 1$ and the link $\langle E, E^0_- \rangle$ is faulty, nodes Z and W will initial their status $s(Z) = s(W) = \mathcal{I}$ because of $fd(Z) = fd(W) = 0$. In the second phase, according to the algorithm, node U will send the system message $M(U \to W) = \{2^-\}$ to W, node E will send the system message $M(E \to Z) = \{0^+\} \cup \{2^-\} = \{0^+, 2^-\}$ to Z. Finally, node W refreshes its status by $s(W) = \mathcal{I} \cap M(U \to W) = \{2^-\}$ and sends system message $M(W \to Z) = \{2^-\} \cup \{0^+\} = \{0^+, 2^-\}$ to Z; node Z refreshes its status by $s(Z) = \mathcal{I} \cap M(E \to Z) \cap M(W \to Z) = \{0^+, 2^-\}$.

$\quad$ It is easy to see that nodes inside a faulty block will get the status $\emptyset$ finally, nodes outside any faulty block get status $\mathcal{I}$ while nodes on the border of the faulty blocks get status from the set $S \backslash \{\mathcal{I}, \emptyset\}$.

3.2 The Forming of the Cracky Blocks in nD Mesh

Say a node X with $s(X) = \emptyset$ is *connected*, if there exist a node Z with $s(Z) \neq \emptyset$ and a (X, Z)-path consisting of good links.

Algorithm 1. Let X be any node in the n-D meshes, $fd(X)$ be the faulty degree of X, $s(X)$ be the status, and $M(X \rightarrow Y_i)$ be the content of system message sent by X to Y_i.

```
1:  procedure INITIAL_STATUS(X)
2:      if f(X) = 0 then
3:          s(X) ← I
4:      else if fd(X) = 1 and ⟨X, X^i_{!ϕ(i)}⟩ is faulty then
5:          s(X) ← {i^{ϕ(i)}}
6:      else if fd(X) ≥ 2 then
7:          s(X) ← ∅
8:      end if
9:      Algorithm_notice_status(X)
10: end procedure
11: procedure UPDATE_STATUS(X)
12:     if ∃ Y_i ∈ Γ*(X) such that s(X) ≠ M(Y_i → X)∩ s(X) then
13:         s(X) ← M(Y_i → X) ∩ s(X)
14:         Algorithm_notice_status(X)
15:     end if
16: end procedure
17: procedure NOTICE_STATUS(X)
18:     if s(X) = ∅ then
19:         send system message M(X  ,  X^j_{ϕ(j)}) − {j^{ϕ(j)}} to every node X^j_{ϕ(j)} ∈
                Γ*(X)
20:     else if s(X) = {i_1^{ϕ(i_1)}, · · · , i_k^{ϕ(i_k)}} and |s(X)| < n then
21:         send system message M(X → X^j_{ϕ(j)}) = {j^{ϕ(j)}} ∪ s(X) to X^j_{ϕ(j)}, ϕ(j) ∈
                {+, −},
22:         j ∈ {0, 1, · · · , n − 1}\{i_1, · · · , i_k}
23:     end if
24: end procedure
```

After constructing the minimal and disjoint faulty blocks in the mesh, we would like to form the cracky blocks by hanging all the connected nodes.

Hang the node W if $s(W) \in S\backslash\{\emptyset\}$. We say the node X with $s(X) = \emptyset$ is hung if and only if it chooses exactly one hung neighbor as predecessor. Denote by $pred(X)$ the predecessor of X. A node which is not hung is free. We denote by $l(X)$ these two boolean states $\{h, f\}$, refer to the hung and free status for each node inside of the faulty block.

The node W with $s(W) \in S\backslash\{\emptyset\}$ can be hung easily, we will show how to hang the connected nodes by Algorithm 2.

For any node X with $s(X) \in S\backslash\{I\}$, define $Succ(X) = \{Y_i \in \Gamma^*(X)|s(Y_i) = \emptyset$ and $X = pred(Y_i)\}$. We consider an order over the elements of $Succ(X)$. Denote by $succ_i(X)$ the ith element of $Succ(X)$, with $1 \le i \le k_X = |Succ(X)|$. A node X with $s(X) \in S\backslash\{I\}$ is said to be $final$ if $Succ(X) = \emptyset$. Take Fig. 2 for an example, every connected node is hung, the arrow points to the predecessor.

A cracky block is obtained from a faulty block, the nodes of the cracky block is composed of the boundary nodes and the interior connected nodes of the faulty

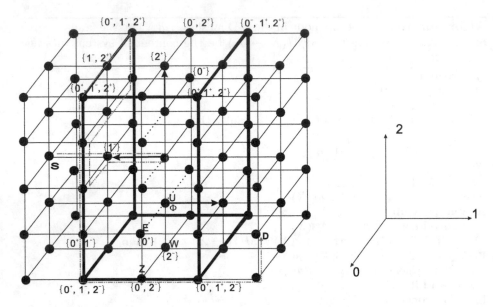

Fig. 2. The faulty block formed by the faults

block while the boundary links of the faulty block and the links connecting the interior connected nodes and their predecessors consists of the links of the cracky block. By Algorithm 2 and the definition of the cracky block, we know that there are trees inside the block with roots on the boundary of the block.

3.3   Adaptive Fault-Tolerant Routing with the Cracky Blocks

In this subsection, we will give the complete fault-tolerant routing strategy for the message sent in the n-D mesh with the cracky blocks.

Let $A = ((a_0, b_0), (a_1, b_1), \cdots, (a_{n-1}, b_{n-1}))$ be any faulty block. Let $I(A)$ be the set of all nodes inside of A, i.e., $\{(x_0, x_1, \cdots, x_{n-1}) : a_i + 1 \le x_i \le b_i - 1, i = 0, \cdots, n - 1\}$. Denote by $S(a_i)$ (resp. $S^*(a_i)$), $0 \le i \le n - 1$, the "a_i"-surface (resp. "a_i"-intersurface) of A, i.e., the induced subgraph by the nodes $(x_0, x_1, \cdots, x_{i-1}, a_i, x_{i+1}, \cdots, x_{n-1})$ with $a_j \le x_j \le b_j$ for any $j \ne i$ (resp. $a_j + 1 \le x_j \le b_j - 1$ for any $j \ne i$). Similarly define $S(b_i)$ and $S^*(b_i)$ the "b_i"-surface and "b_i"-intersurface of A respectively.

In the process of routing, denote the current node by X. Let X_p be the node who sends message to node X, and X_s be the node who will receive the message sent by X.

Case 1. $s(X) = \mathcal{I}$, then route message by the greedy algorithm defined in Sect. 2.2.

Case 2. $s(X) = \emptyset$, then route message by line 4–10 in Algorithm 3.

Case 3. $s(X) \in S \setminus \{\emptyset, \mathcal{I}\}$, then the message locates on the surface of some cracky block, say $\tilde{A}$. Assume $s(X) = \{i_1^{\phi(i_1)}, \cdots, i_k^{\phi(i_k)}\}$, the neighbors of X

Algorithm 2. Let X be a node with $s(X) = \emptyset$, make it hung when it is possible

---

1: **procedure** ALGORITHM_HUNGNODE(X)
2:     **if** X is a node with $s(X) \in S \backslash \{\emptyset\}$ **then**
3:        $l(X) \leftarrow h$
4:        return
5:     **else if** X is a node with $s(X) = \emptyset$ **then**
6:        $l(X) \leftarrow f$
7:     **end if**
8:     $i \leftarrow 0$
9:     $\exists\, Y_i \in \Gamma^*(X)$
10:    **while** $l(Y_i) = f$ **do**
11:       $i \leftarrow (i+1) \mathrm{mod}(|\Gamma^*(X)|)$
12:    **end while**
13:    $l(X) \leftarrow h$
14:    $pred(X) \leftarrow Y_i$
15: **end procedure**

Algorithm 3. The novel routing function based on cracky blocks

---

1: **procedure** ALGORITHM_ROUTING(X)
2:     **if** $s(X) = \mathcal{I}$ **then**
3:        basic routing function with node X
4:     **else if** $s(X) = \emptyset$ **then**
5:        **if** X is final or $X_p = succ_{\kappa_X}(X)$ **then**
6:           $X_s \leftarrow pred(X)$
7:        **else if** $X_p = succ_i(X)$ and $i < \kappa_X$ **then**
8:           $X_s \leftarrow succ_{i+1}(X)$
9:        **else if** $X_p = pred(X)$ **then**
10:       $X_s \leftarrow succ_1(X)$
11:       **end if**
12:    **else if** $s(X) \in S \backslash \{\mathcal{I}, \emptyset\}$ **then**
13:       route m according to Algorithm 4.
14:    **end if**
15: **end procedure**

---

not in $\widetilde{A}$ are $X^{i_l}_{\phi(i_l)}$, where $l = 1, \cdots, k$. If m going from X to $X^{i_l}_{\phi(i_l)}$ can decrease the number of the maximum distance, then m goes from X to $X^{i_l}_{\phi(i_l)}$ and leaves $\widetilde{A}$. At the same time, set the signal $h(m)$ null. Else if X is not final and $s(X_p) = \emptyset$, set $X_s \leftarrow X^{S(m)-1}_{!\varphi(S(m)-1)}$. That is, travel on the surface of $\widetilde{A}$. Else if X is final and $\phi(i_1) = +$, m travels $\widetilde{A}$ along the order $T(b_{i_1}), T(b_{i_1+1}), \cdots, T(b_{n-1})$, $T(a_0), \cdots, T(a_{n-1})$, $T(b_0), \cdots, T(b_{i_1})$; while if X is final and $\phi(i_1) = -$, m travels $\widetilde{A}$ along the order $T(a_{i_1}), T(a_{i_1+1}), \cdots, T(a_{n-1})$, $T(b_0), \cdots, T(b_{n-1})$, $T(a_0), \cdots, T(a_{i_1})$. Else if X is not final and $s(X_p) \neq \emptyset$, then m will traverse the inter-tree with root X by depth first search (DFS).

The parameter $S(m)$ in the signal $h(m)$ denotes the surface m is being routed. The message m is routed on the surface by the following routing, we assume the

Algorithm 4. Node X route a message m with signal $h(m)$

```
1: procedure ALGORITHM_ROUTING AT N-MESH(m, h(m))
2:     if m can be routed out of the block then
3:         out of block and set h(m) null
4:     else if |h(m)| = 2 then
5:         if S(m)^{!\varphi(S(m))} \in s(X) then
6:             h(m) \leftarrow (S(m), \phi(0), ..., \phi(n-1))
7:             N-MESH(m, h(m))
8:         else
9:             X_s \leftarrow X_{!\varphi(S(m))}^{S(m)}
10:        end if
11:    else
12:        i \leftarrow (S(m) - 1) \bmod n
13:        while i^{!\varphi(i)} \in s(X) do
14:            \varphi(i) \leftarrow \phi(i)
15:            i \leftarrow (i - 1) \bmod n
16:        end while
17:        if i = S(m) then
18:            S(m) \leftarrow (S(m) + 1) \bmod n
19:            if (S(m) = 0 \ \& \ \varphi(0) \neq \varphi(n-1)) or (S(m) > 0 \ \& \ \varphi(S(m)) = \varphi(S(m) -
               1)) then
20:                h(m) \leftarrow (S(m), \phi(0), ..., \phi(n-1))
21:                N-MESH(m, h(m))
22:            else if (S(m) = 0 \ \& \ \varphi(0) = \varphi(n-1)) or (S(m) > 0 \ \& \ \varphi(S(m)) \neq
               \varphi(S(m) - 1)) then
23:                h(m) \leftarrow (S(m), \varphi(S(m)))
24:                N-MESH(m, h(m))
25:            end if
26:        else
27:            if X is final then
28:                X_s \leftarrow X_{!\varphi(i)}^i
29:            else if X_p = succ_1(X) then
30:                X_s \leftarrow X_{!\varphi(i)}^i
31:            else
32:                X_s \leftarrow succ_1(X)
33:            end if
34:        end if
35:    end if
36: end procedure
```

addition and subtraction are modulo n. Judge along the order $S(m) - 1, S(m) - 2, \cdots, 0, n - 1, \cdots, S(m) + 1, S(m)$ whether there exists $t \in \{S(m) - 1, S(m) - 2, \cdots, 0, n - 1, \cdots, S(m) + 1, S(m)\}$ and $t^{!\varphi(t)} \in s(X)$. Let l be the first one along the order $S(m) - 1, S(m) - 2, \cdots, 0, n - 1, \cdots, S(m) + 1, S(m)$ which doesn't satisfy the condition. By the assumption, we know that $S(m)^{\varphi(S(m))} = S(m)^{\phi(S(m))} \in s(X)$, so l always exists.

We now give the complete local routing function which runs in each node of the mesh, as shown in Algorithm 3. This algorithm is based on the basic routing function we have defined in Sect. 2.2. See Fig. 2 for an example, the node S wants to send a message to the node D, the routing path is depicted in the figure by our algorithm.

It can be seen in our algorithm that the message goes a Hamiltonian path on the surface except the first touched surface before ending traveling on it. We can prove that the message can be sent to the destination similar to Sect. 4 in our previous work [10]. We omit it here.

Acknowledgments. The authors would like to thank the Natural Science Foundation of China (11101345) and the Natural Science Foundation of Fujian Province (2016J01041).

References

1. Benkhelifa, E., Welsh, T., Tawalbeh, L., et al.: User profiling for energy optimisation in mobile cloud computing. Procedia Comput. Sci. **52**, 1159–1165 (2015)
2. Boppana, R.V., Chalasani, S.: Fault-tolerant wormhole routing algorithms for mesh networks. IEEE Trans. Comput. **44**(7), 848–864 (1995)
3. Intel Corporation, A Touchstone DELTA System Description (1991)
4. Gai, K., Qiu, M., Zhao, H., Tao, L., Zong, Z.: Dynamic energy-aware cloudlet-based mobile cloud computing model for green computing. J. Netw. Comput. Appl. **59**, 46–54 (2016)
5. Liu, W., Nishio, T., Shinkuma, R., Takahashi, T.: Adaptive resource discovery in mobile cloud computing. Comput. Commun. **50**(1), 119–129 (2014)
6. Seitz, C.L.: The architecture and programming of the Amete Series 2010 multicomputer. In: Proceedings of Third Conference on Hypercube Concurrent Computer and Applications, pp. 33–36 (1988)
7. Su, C.C., Shin, K.G.: Adaptive fault-tolerant deadlock-free routing in meshes and hypercubes. IEEE Trans. Comput. **45**(6), 666–683 (1996)
8. Wang, D.: A rectilinear-monotone polygonal fault block model for fault-tolerant minimal routing in mesh. IEEE Trans. Comput. **52**, 310–320 (2003)
9. Wu, J.: Fault-tolerant adaptive and minimal routing in mesh-connected multicomputers using extended safety levels. IEEE Trans. Parallel Distrib. Syst. **11**(2), 149–159 (2000)
10. Yang, Y., Chen, M., Li, H., Li, L.: Adaptive fault-tolerant routing in 2D mesh with cracky rectangular model. J. Appl. Math. **10**, 1–592638 (2014)

An Improved Slope One Algorithm Combining KNN Method Weighted by User Similarity

Songrui Tian and Ling Ou[✉]

School of Computer and Information Science, Southwest University,
Chongqing 400715, China
tsongrui@sina.com, ouling@swu.edu.cn

Abstract. Data sparsity is a main factor affecting the prediction accuracy of collaborative filtering. Based on the simple linear regression model, Slope One algorithm aims to enhance the performance significantly by reducing the response time and maintenance, and overcoming the cold start issue. It uses rating data to do calculation without considering the similarity. In this paper, we proposed an improved algorithm by combining the dynamic k-nearest-neighborhood method and the user similarity generated by the weighted information entropy with Slope One algorithm. Especially, the similarity between users is calculated and added on the fly. Experiments on the MovieLens data set show that the proposed algorithm can achieve better recommendation quality and prediction accuracy. Besides, the stability of the algorithm is also relatively satisfying.

Keywords: Weighted information entropy · K-nearest-neighborhood · Slope One algorithm · Personalized recommendation · Data mining

1 Introduction

With the expanding amount of information and rapid development of economy on the Internet, users are submerged in the vast amounts of information. Personalized recommendation system is an effective tool to solve the problem of information overload. Collaborative filtering (CF) is widely-used in recommendation systems, especially in the electronic commerce [1]. It recommends goods or information to a specific user by using the preferences of group which has common experience. CF can be divided into user-based CF [2] and item-based CF [3,4] according to different types of models. Slope One is an item-based CF algorithm proposed by Lemire and Maclachlan [5]. It has roughly the same recommendation quality as other complex recommendation algorithms while using less time and space. Slope One algorithm calculates the deviation between the scores of items with all the users ratings. However, some of the users have a very low similarity with the current active user. Adding those scores into the forecast without screening will reduce the accuracy. Therefore, we consider the similarity of users and find k nearest neighbors based on the threshold for the

© Springer International Publishing AG 2016
S. Song and Y. Tong (Eds.): WAIM 2016 Workshops, LNCS 9998, pp. 88–98, 2016.
DOI: 10.1007/978-3-319-47121-1_8

current active user. The common method for similarity measurement is Cosine similarity, modified Cosine similarity, Pearson correlation coefficient [6], Spearman similarity...etc. Considering the poor prediction quality existing in traditional similarity calculation with sparse data, we use a method based on the information entropy between differences of users. The information entropy is a concept for the measurement of information chaos degree in information theory. Generally speaking, the more confusing the information is, the bigger the information entropy is. We weighted the entropy by the common and different evaluation and then normalized it to measure the similarity between users. This paper proposed a new algorithm combining KNN method and user similarity produced by weighted information entropy based on Slope One Algorithm. The rest of the paper is organized as follows. In Sect. 2, we describe the related works about Slope One algorithm. In Sect. 3, we introduce several typical Slope One algorithms. Section 4 presents the theoretical foundations followed by the detailed discussion of our approach. Performance evaluation and conclusions are presented in Sects. 5 and 6, respectively.

2   Related Works

The traditional Slope One algorithm only calculates the average difference between users ratings, without taking the intrinsic link between items into account. Aim to solve this problem, Du [7] proposed a Slope One algorithm based on neighbor items, choosing some of the nearest items to the target one to join the calculation. It weighted prediction score by item similarity then. Lin [8] mentioned a developed algorithm using SVD (Single Value Decomposition, SVD) [9] to simplify the rating matrix from the angle of mathematical operation. The above two methods are combined by Chai [10], mending problems that the poor recommendation of the algorithm based on neighbor items when the rating matrix is extremely sparse and the data loss of the algorithm based on SVD. All of those improved algorithms mentioned above retained the original Slope One algorithms characteristic which is purely item-based, excluding the users role. Sun [11] proposed a Slope One algorithm combing neighbor users, which took a certain amount of neighbor users whose similarity meets the conditions into consideration. It removed some of the weak correlation user data to a certain extent, but its calculation of the rating deviation is identical with the original Slope One algorithm; user similarity didn't apply to the actual score prediction. In this paper, we proposed an improved Slope One algorithm considering the effects of both items and users, which can be regarded as a new hybrid collaborative filtering recommendation algorithm.

3   Slope One Algorithm Theory

3.1   Basic Slope One Algorithm

If there are m users and n items in the recommendation system. Constructing two sets $U = \{u_1, u_2, \ldots, u_m\}$ and $I = \{i_1, i_2, \ldots, i_n\}$, U represents the collection

	...	$Item_c$	...	$Item_d$	...
...	...	...	...	...	...
$User_A$	...	$r_{A,c}$	...	$r_{A,d}$	...
...	...	...	...	...	...
$User_B$	...	$r_{B,c}$	...	?	...
...	...	...	...	...	...

Fig. 1. Users' scores to items

of users, I represents the collection of items. Recommendation algorithm usually use user-item matrix $R_{m \times n}$ to express the scores that different users give to each item. The line vector R_m represents each user's score while the column vector R_n represents the score of each item. In order to facilitate the calculation, we used $r_{i,j}$ to represent the score that user i gives to item j ($1 \leqslant i \leqslant m, 1 \leqslant j \leqslant n$). Slope One algorithm adopts the predictor form as $f(x) = x + b$. The parameter b is the mean deviation of ratings that users gave to two items. The predict score user B to item d is $r_{B,d} = r_{B,c} + (r_{A,d} - r_{A,c})$ as shown in Fig. 1.

Slope One algorithm adopts the predictor form as $f(x) = x + b$. It calculates the average deviation of ratings dev_{jk} between target item i_j and other item i_k and then makes the rating prediction $Prediction_{u,j}$ that the current active user u may give to the target item j. We defined S_{jk} as the collection of users who have rating to both item i_j and i_k, R_j as the collection of items which are scored with item i_j at the same time, $count(X)$ as the number of elements in the collection X. The mean deviation dev_{jk} and the rating prediction $Prediction_{u,j}$ are as follows:

$$dev_{jk} = \sum_{u_i \in S_{jk}} \frac{r_{i,j} - r_{i,k}}{count(S_{jk})} \tag{1}$$

$$Prediction_{u,j} = \frac{\sum_{k \in R_j} (dev_{jk} + r_{u,k})}{count(R_j)} \tag{2}$$

3.2 Weighted Slope One Algorithm

In order to balance the effect of each item on the target one, the number of users who rated to both i_j and i_k at the same time is added as the weight value of two items' deviation, which expressed by s_{jk}, $s_{jk} = count(S_{jk})$.

$$Prediction_{u,j}^w = \frac{\sum_{k \in R_j} (dev_{jk} + r_{u,k}) s_{jk}}{\sum_{k \in R_j} s_{jk}} \tag{3}$$

3.3 Bi-Polar Slope One Algorithm

Taking the user preferences for goods into account, the collection of items I_{jk} is divided into two groups according to user's ratings. One is I_{jk}^{like} that user has rated and the score is higher than the user's average score, the other is $I_{jk}^{dislike}$. The liked-deviation dev_{jk}^{like} and the disliked-deviation $dev_{jk}^{dislike}$ are calculated respectively.

$$dev_{jk}^{like} = \sum_{u_i \in S_{jk}^{like}} \frac{r_{i,j} - r_{i,k}}{count(S_{jk}^{like})} \tag{4}$$

Using the same method, we can calculate $dev_{jk}^{dislike}$. Finally, the result of prediction is as follows:

$$Prediction_{u,j}^{BI} = \frac{\sum_{k \in I_{jk}^{like}} \left(dev_{jk}^{like} + r_{u,k}\right) s_{jk}^{like} + \sum_{k \in I_{jk}^{dislike}} \left(dev_{jk}^{dislike} + r_{u,k}\right) s_{jk}^{dislike}}{\sum_{k \in I_{jk}^{like}} s_{jk}^{like} + \sum_{k \in I_{jk}^{dislike}} s_{jk}^{dislike}} \tag{5}$$

4 Rating Prediction

4.1 User Similarity Based on the Weighted Information Entropy

The information entropy increases with the increase of the degree of information chaos. In this paper, we used the information entropy to calculate the similarity between users. For a given sample set X, we defined N as the classification number of X; $p(x_i)$ as the probability of element i in X, the entropy formula is as follows:

$$H(X) = -\sum_{i=1}^{N} p(x_i) \log_2 p(x_i) \tag{6}$$

The bigger the score difference that two users have in the common items, the more confused the difference of the users. In order to calculate the user similarity, we firstly constructing the common item collection I_{ab} for two users, u_a and u_b, $I_{ab} = \{i_1, i_2, \ldots, i_n\}$; the set of ratings to each item in I_{ab} from u_a and u_b are respectively expressed by R_a and R_b, $R_a = \{r_{a,1}, r_{a,2}, \ldots, r_{a,n}\}$, $R_b = \{r_{b,1}, r_{b,2}, \ldots, r_{b,n}\}$; the score difference between u_a and u_b is $Dif(R_a, R_b) = \{def_1, def_2, \ldots, def_n\} = \{r_{a,1} - r_{b,1}, r_{a,2} - r_{b,2}, \ldots, r_{a,n} - r_{b,n}\}$. Then the calculation of difference information entropy is as follows:

$$H(Def(u_a, u_b)) = -\sum_{i=1}^{N} p(def_i) \log_2 p(def_i) \tag{7}$$

In formula (7), N is the classification number of def_i. The similarity shows negative growth with rating difference $|def_i|$. Besides, the number of evaluations that the two users give to the common items, expressed by Num also has influence on the similarity that the latter increases with the former [12]. Adding $|def_i|$ and Num as weight, the new formula is as follows:

$$WDE(u_a, u_b) = -\frac{1}{Num} \sum_{i=1}^{Num} \left[\frac{p(def_i)}{n_i} \times \log_2 p(def_i) \times |def_i| \right] \tag{8}$$

In formula (8), def_i is the score difference, n_i is the number of def_i in the collection of difference D. $WDE(u_a, u_b) \in (0, +\infty)$. Next, $WDE(u_a, u_b)$ is normalized as follows:

$$NWDE_u(i) = \frac{Max(WDE_u) - WDE_u(i)}{Max(WDE_u) - Min(WDE_u)} \tag{9}$$

In formula (9), $NWDE_u$ is the normalized similarity which is between 0 and 1; $Max(WDE_u)$ is the maximum value in collection $NWDE_u$; $Min(WDE_u)$ is the minimum value in collection $NWDE_u$. The bigger $NWDE_u$, the higher similarity between users [12].

4.2 Dynamic K-nearest-neighborhood

Slope One algorithm uses all of the users' rating data in recommendation system to make prediction which means some users who have different even opposite preferences are involved. The data of the users whose interest is not close to the current active user will interfere the accuracy of the judgment. To clean out the noise data, KNN [12] method is used to determine the range of users involved in the calculation of deviation. We sort the user similarity based on the weighted information entropy to find k nearest neighbors for the current active user. A similarity threshold λ is set as well for further optimization.

4.3 NWDE-KNN Slope One Algorithm Design

As introduced in Sect. 3, when there exists a huge difference in users' interests, Slope One algorithm may have a lower accuracy rate, which may lead to deviation in results and have influence on final recommendation. This can be improved by clustering [13]. In this paper, we adopt KNN method to solve this problem. Typical KNN method usually calculate similarities and sort to find k of the largest figures which are greater than the threshold. The k-most-similar users form a group of neighborhood expressed by $KNN(u)$. It improves the accuracy of prediction to a certain extent but the similarity is merely used to select neighbors instead of being added into deviation calculation. However, the rating deviations of neighbors with different similarities generally have different influence. Contrast with the typical KNN and other user-based algorithms, we proposed a new algorithm combining user similarity to deviation calculation. We firstly producing the weighted information entropy according to formula (8), and then normalize it via formula (9). Next the produced user similarity between the current user and neighbors, expressed by $SIM(u, u_i)$, is added to the deviation calculation formula as follows:

$$dev_{jk} = \sum_{u_i \in KNN(u)} \frac{SIM(u, u_i)(r_{i,j} - r_{i,k})}{count(KNN(u)_{jk})} \tag{10}$$

In formula (10), the collection of neighbors $KNN(u) = \{u_i | sim(u, u_i) \geqslant \lambda\}$, $count(KNN(u)_{jk})$ is the number of neighbors scored item i_j and i_k at the same time. The main algorithm is as Algorithm 1.

Algorithm 1. Calculate Prediction Score

Require: User-item rating matrix $R_{m \times n}$, current active user u, target item i_j
Ensure: Predictive score $Prediction_{u,j}$
1: $//U$ represents the collections of all users expect u
2: **for** each $u_i \in U$ **do**
3:     calculate $sim(u, u_i)$
4: **end for**
5: Get user similarity matrix $User - SIM_{m \times m}$
6: sort $sim(u, u_i)$
7: **if** $sim(u, u_i) \geqslant \lambda$ **then**
8:     $//KNN(u)$, the collection of k-nearest neighbors
9:     add u_i to $KNN(u)$
10:     $//I_u$, the collection of items u has scored
11:     **for** each $i_k(k \neq j) \in I_u$ **do**
12:         $numerator = 0$; $denominator = 0$
13:         **for** each $v \in KNN(u)$ **do**
14:             **if** $r_{v,k} \neq 0$ and $r_{v,j} \neq 0$ **then**
15:                 $dev_{jk} += \frac{sim(u,v)(r_{v,j} - r_{v,k})}{count(KNN(u)_{jk})}$
16:             **end if**
17:         **end for**
18:         $numerator +=$
19:         $(dev_{jk} + r_{u,k})count(KNN(u)_{jk})$
20:         $denominator += count(KNN(u)_{jk})$
21:     **end for**
22: **end if**
23: **if** $denominator \neq 0$ **then**
24:     $Prediction_{u,j} = \frac{numerator}{denominator}$
25: **end if**
26: **if** $denominator = 0$ and I_u is $NULL$ **then**
27:     $//avg_j$, the average score of by user j
28:     $Prediction_{u,j} = avg_j$
29: **else**
30:     $//avg_u$, the average score of all scores rated by user u
31:     $Prediction_{u,j} = avg_u$
32: **end if**

---

5   Experimental Result

5.1   Preprocessing

We define SP as the sparsity of the data set, SN as the number of values that have been scored in data set, UN as the number of users, IN as the number of items. The sparsity of the data set is usually calculated using the following formula [14]:

$$SP = \frac{SN}{UN \times IN} \times 100\% \tag{11}$$

The data set for experiment is from the website of MovieLens (http://movielens.umn.edu), including 943 users, 1682 items (movies) and 100000 rating record. The score range from 1 to 5. The higher score means the higher user evaluation. The sparsity of the data is:

$$\frac{100000}{943 \times 1682} \times 100\% = 6.30\%$$

The date from MovieLens is sparse according to the above calculation. During the experiment, 80% of total data is randomly selected according to certain rules to form the train set and the rest 20% constitutes the test set. The train set includes 943 users' rating data while the number of users in the test set is half. The train set is equivalent to be the historical records in the recommendation system. The test set is used to examine the accuracy of the prediction results. Another group of experiments testing algorithm stability was done later. We randomly select from the total data according to a certain rule to make a test set, making the ratios between the number of the total users and the user number of the test set are 2:1, 1.5:1 and 1:1. The number of users in train set maintained at 943, which is the total user number. As above, for each experiment, 80% of total data formed the train set and the rest 20% constituted the test set. When the proportion of the number of users in the test set increase, it means using data of the same amount of users to predict more users' ratings. To facilitate the experiment, we set the KNN threshold λ as 0.2, the number of neighbors k from 100 to 900, each increase of 100.

5.2   Recommendation Quality Metrics

The quality metrics of recommendation system mainly include the following two kinds of measurement method: the statistical precision and the decision support accuracy [15]. In statistical precision measurement method, MAE (Mean Absolute Error, MAE) and RMSE (Root Mean Square Error, RMSE) are more widely used. MAE can directly reflect the actual score deviation in train set while RMSE is sensitive to the very large or small measurement deviation. We used MAE and $RMSE$ as metrics in this paper because of their comprehensibility and intuitive measurement. The quality of recommendation is higher while MAE and $RMSE$ are smaller. We defined two groups P and Q to respectively express the collections of the prediction ratings and the true ratings,

$P = \{p_1, p_2, \ldots, p_N\}$, $Q = \{q_1, q_2, \ldots, q_N\}$. The definitions of MAE and $RMSE$ are as follows:

$$MAE = \frac{1}{N} \sum_{i=1}^{N} |p_i - q_i| \qquad (12)$$

$$RMSE = \sqrt{\frac{1}{N} \sum_{i=1}^{N} (p_i - q_i)^2} \qquad (13)$$

5.3 Experimental Results

In order to gradually explore the respective effect that the user similarity has, we make a test that only adds user similarity to deviation calculation formula without using the KNN method at first. We can call this algorithm "NWDE

Table 1. MAE and RMSE

	MAE	RMSE
Slope One	0.780	0.983
Weighed Slope One	0.775	0.981
WDE Slope One	**0.759**	**0.961**

Table 2. MAE and RMSE of NWDE-KNN Slope One

k	100	200	300	400	500	600	700	800	900
MAE	1.364	0.842	0.774	0.759	**0.756**	0.757	0.758	0.759	0.759
RMSE	1.885	1.129	1.002	0.969	0.960	**0.959**	0.961	0.961	0.960

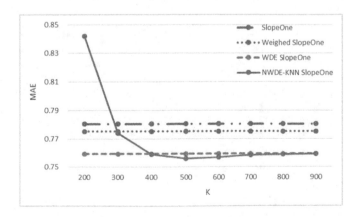

Fig. 2. MAE of different Slope One algorithms

Slope One". After that, we do another test joining the KNN method based on the above one, this new algorithm is called "NWDE-KNN Slope One" in this paper. Next, we compare these two new algorithms with classical Slope One algorithm and weighted Slope One algorithm to test the prediction accuracy is improved. MAE and $RMSE$ for the three algorithms are listed in Table 1.

Influenced by the number of neighbors k, MAE and $RMSE$ for NWDE-KNN Slope One algorithm are listed in Table 2. The more intuitive differences of MAE and $RMSE$ are shown in Figs. 2 and 3 respectively.

The experimental results show that using user similarity based on weighted information entropy can obviously enhance the accuracy of the algorithm. MAE and $RMSE$ of WDE Slope One algorithm are lower than the two typical Slope One algorithms. WDE-KNN Slope One algorithm, the new algorithm proposed in this paper combining both user similarity weight value and KNN method, also has better accuracy when moderate number of neighbors are selected. MAE

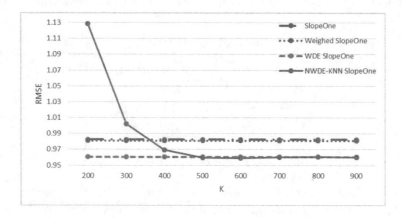

Fig. 3. RMSE of different Slope One algorithms

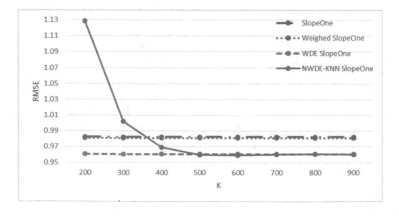

Fig. 4. MAE of different ratios of users in the test set

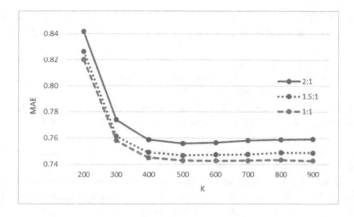

Fig. 5. RMSE of different ratios of users in the test set

of NWDE–KNN Slope One algorithm falls with the increase of k at the first, then rises again and tends to be stable in a certain range. When the number of neighbors is 500, we get the best MAE, which is even smaller than WDE Slope One algorithm's. $RMSE$ of the proposed algorithm also falls with the increase of k and then tends to be stable, which is smaller than typical Slope One algorithms'. With different ratios, the change trends of MAE and $RMSE$ are broadly consistent, which shows the stability of the NWDE-KNN Slope One algorithm is nice. The more intuitive trend of change is showing in Figs. 4 and 5.

6 Conclusion

In this paper, a series of experiments were conducted to compare the rating prediction accuracy of the different Slope One algorithms. Adding user similarity produced by weighted information entropy to weigh the rating deviation calculation can enhance accuracy. Based on the simple and efficient collaborative filtering algorithm Slope One, a new improved algorithm NWDE-KNN Slope One is proposed. It is demonstrated by experiments that NWDE-KNN Slope One algorithm has higher prediction accuracy and lower data sparsity under the condition of choosing the appropriate number of nearest neighbors. Meanwhile, the new algorithm has good stability.

Acknowledgments. This work was supported by National Natural Science Foundations of China (No. 61170192).

References

1. Xu, H.L., Wu, X., Li, X.D., Yan, B.P.: Comparison study of Internet recommendation system. J. Software **20**(2), 350–362 (2009)

2. Herlocker, J., Konstan, J.A., Riedl, J.: An empirical analysis of design choices in neighborhood-based collaborative filtering algorithms. Inf. Retrieval **5**(4), 287–310 (2002)
3. Sarwar, B., Karypis, G., Konstan, J., Riedl, J.: Item-based collaborative filtering recommendation algorithms. In: Proceedings of the 10th International Conference on World Wide Web, pp. 285–295. ACM (2001)
4. Huang, C.G., Yin, J., Wang, J., Liu, Y.B., Wang, J.H.: Uncertain neighbors' collaborative filtering recommendation algorithm. Jisuanji Xuebao (Chin. J. Comput.) **33**(8), 1369–1377 (2010)
5. Lemire, D., Maclachlan, A.: Slope One predictors for online rating-based collaborative filtering. SDM **5**, 1–5 (2005)
6. Pang, H., Zhou, L., Liu, H.: Personalization portal system based on collaborative filtering algorithm. In: Computer, Mechatronics, Control and Electronic Engineering (CMCE), vol. 1, pp. 383–386. IEEE (2010)
7. Du, M., Liu, M., Li, S., Pu, Q.: Slope One collaborative filtering algorithm based on neighboring items. J. Chongqing Univ. Posts: Telecommun. Nat. Sci. Ed. **26**(3), 421–426 (2014)
8. Lin, D.J., Meng, X.W.: Slope One algorithm based on single value decomposition. New Type Industrialization **2**(11), 12–17 (2012)
9. Luo, L., Xie, Y., Zhang, Z., Li, W.J.: Support matrix machines. In: Proceedings of the 32nd International Conference on Machine Learning (ICML-15), pp. 938–947 (2015)
10. Hua, C., Liu, J.: An improved Slope One recommendation algorithm. Netinfo Secur. **2**, 77–81 (2015)
11. Li, J., Sun, L., Wang, J.: A Slope One collaborative filtering recommendation algorithm using uncertain neighbors optimizing. In: Wang, L., Jiang, J., Lu, J., Hong, L., Liu, B. (eds.) WAIM 2011. LNCS, vol. 7142, pp. 160–166. Springer, Heidelberg (2012). doi:10.1007/978-3-642-28635-3_15
12. Liu, W.L., Zhang, G.Y., Chen, Z., Zhu, Q.Q.: Collaborative filtering algorithm based on weighted information entropy similarity. J. Zhengzhou Univ. (Eng. Sci.) **33**(5), 118–120 (2012)
13. Schickel-Zuber, V., Faltings, B.: Using hierarchical clustering for learning theontologies used in recommendation systems. In: Proceedings of the 13th ACM SIGKDD International Conference on Knowledge Discovery and Data Mining, pp. 599–608 (2007)
14. Li, D., Xin, C., Wang, K.: Evaluation of collaborative filtering algorithm based on different data sets. J. Tsinghua Univ. JCR Sci. Ed. **49**(4), 590–594 (2009)
15. Wang, J., De Vries, A.P., Reinders, M.J.: Unifying user-based and item-based collaborative filtering approaches by similarity fusion. In: Proceedings of the 29th Annual International ACM SIGIR Conference on Research and Development in Information Retrieval, pp. 501–508 (2006)

Urban Anomalous Events Analysis Based on Bayes Probabilistic Model from Mobile Phone Records

Rong Xie[✉] and Ming Huang

International School of Software, Wuhan University, 39 Luoyu Road, Wuhan 430079, China
xierong@whu.edu.cn

Abstract. We present an approach to detecting and analyzing urban anomalous events by Bayes Probabilistic Model. Using actual mobile phone data, we compute individual probability and get individual anomalous index under comparing occurrence probability and ordinary probability in a certain region and period. Expanding individual analysis to group analysis, we make statistics on anomalous activities of group and get their regularity so that we can measure the degree of deviation among activities of group during certain period and the regularity and finally judge whether urban anomalous events take place. Taking two areas in Kuming city, China as case study, we demonstrate effectiveness of our approach.

Keywords: Mobile phone data · Bayes probabilistic model · Anomalous event analysis · Individual anomalous analysis · Group anomalous analysis

1 Introduction

Under rapid urbanization and growing urban population, we are now rising to great challenges of urban management. It is becoming more and more important to detect and analyze abnormal events in cities effectively. Mobile phone data, from telecommunication provider, contains a lot of spatial information and temporal information which is close to mobile phone users' daily life, which might be an effective means to study urban activities and regulations.

As the first step for the research of cellphone data mining, Reality Mining project [1] provided a new thinking way. Eagle and Pentland [2] introduced a system for sensing complex social systems with data collected from 100 cellphones over 9 months and demonstrated the ability to use standard Bluetooth-enabled mobile telephones to recognize social patterns in daily user activity and infer relationships, and identify socially significant locations and modeled organizational rhythms. Following their work, there have been some preliminary research results related to analysis of events in cities. Calabrese et al. [3] presented the analysis of crowd mobility during special events by nearly 1 million cell-phone traces and associate their destinations with social events. They showed that the origins of people attending an event were strongly correlated to the type of event, with implications in city management. Using extensive phone records resolved in both time and space, Candia et al. [4] studied the mean collective behavior at large scales and focused on the occurrence of anomalous events. They investigated

© Springer International Publishing AG 2016
S. Song and Y. Tong (Eds.): WAIM 2016 Workshops, LNCS 9998, pp. 99–108, 2016.
DOI: 10.1007/978-3-319-47121-1_9

patterns of calling activity at the individual level and showed that the inter-event time of consecutive calls was heavy-tailed. Bagrow et al. [5] identified real-time changes in communication and mobility patterns in the vicinity of eight emergencies, such as bomb attacks and earthquakes, comparing these with eight non-emergencies, like concerts and sporting events. They found that communication spikes accompanying emergencies were both spatially and temporally localized, but information about emergencies spreads globally, resulting in communication avalanches that engage in a significant manner the social network of eyewitnesses. Ratti [6] analyzed coverage map that users used cellphone at different time period in one day at Milan, Italy and other metropolitan areas, showing spatio-temporal density maps of urban activities and their evolution.

However, there is relatively less research about abnormal events analysis in advance in the current research. In the paper, using mobile phone data, we present an approach to detecting and analyzing urban anomalies by Bayesian uncertainty model.

The rest of the paper is organized as follows. We present our research framework in Sect. 2. Section 3 presents approach to anomalous index calculation, including individual anomalous analysis, as well as group anomalous analysis. Section 4 give case study of abnormal event analysis in Kunming, China. Conclusions and future work are finally presented in Sect. 5.

2   Research Framework

In this paper, we present an approach to detection and analysis of urban abnormal events. Abnormal event, also called hot event, is defined as aggregation of a large number of unconventional groups because some unexpected events happen, such as concerts, sporting events and demonstration etc.

Our research framework is shown in Fig. 1. (1) Data preprocessing on user's mobile phone records. Convert the data into latitude coordinate and longitude coordinate, while filter out the defect records. And then conduct precision transformation and coordinate translation to get data format which is suitable for the subsequent modeling. (2) Classical Bayesian Probability Model is then simplified for our research. (3) Using Bayesian uncertainty model, occurrence probability of individual is calculated in a specific area during a specific time period to obtain a certain amount of data of individual history probability. After comparing current occurrence probability of individual with historical occurrence probability, probability of individual, involved in abnormal events, is calculated. (4) On the basis of analysis of individual anomalies, number of abnormal participants is calculated in a specific area during a specific time period to get historical data set. Comparing the average of current number of anomalies and historical anomalies, we can calculate anomaly index. Threshold is determined for occurrence of abnormal events. At the same time, combined with the original data set and individual abnormal occurrence probability, we can make analysis on information of participants, nonparticipants and others in abnormal events.

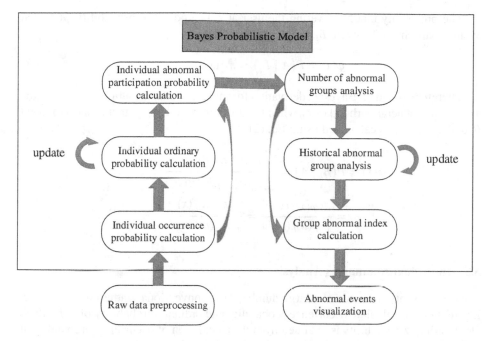

Fig. 1. Research framework.

3 Anomalous Index Calculation

On the basis of Bayesian probability model [7], we make simplification on the model, which can be suitable for our detecting social abnormal events, following the three steps as below. (1) Under the known user's location, calculate probability of user's records by specific base stations. (2) Make transformation using Bayes theorem to obtain probability distribution around base station when the known user is recorded by the base station. (3) By Lebesgue measure, the distribution is converted to integral form.

And then we make calculation of probability an individual occurs in a particular area during a specific time period. Make statistics on individual results in group and analysis on group abnormal situation. Finally obtain group anomaly index, which can be used to determine occurrence of abnormal events.

3.1 Bayesian Probability Model

Suppose the following conditions,

(1) Event A is defined that a specific user is covered by base station X_0.
(2) $\{B_1, B_2, ..., B_n\}$ is a sample space partition divided by neighborhood of X_0, where, event B_i is represented that the user is located at position x, its prior probability is $f(x)$.

The probability that X_0 covering the user at x, i.e., posterior probability $\varphi(x)$, that event A happens under event B_i, is calculated by Eq. (1).

$$\varphi(x) = P(A \mid B_i) = \pi_{X_0}(x). \tag{1}$$

Therefore, based on Bayesian derivation formula, we get distribution probability the user located at neighborhood of X_0 when the user is covered by X_0, that is, $\varphi(x)$ of event B_i under event A, is calculated using Eq. (2).

$$\varphi(x) = P(B_i \mid A) = \frac{P(A \mid B_i)P(B_i)}{\sum_{j=1}^n P(A \mid B_j)P(B_j)}$$

$$= \frac{\varphi(x)f(x)}{\int \varphi(u)f(u)du} = \frac{\pi_{X_0}(x)f(x)}{\int \pi_{X_0}(u)f(u)du} \tag{2}$$

3.2   Individual Anomalous Analysis

Abnormal events are kinds of a large number of unconventional gathering. Therefore, it is required to calculate occurrence probability and ordinary probability of individual. Here, ordinary probability is obtained from the calculation of occurrence probability of historical data. And then, combining with both of them, calculate probability of abnormal event which an individual participates in.

Occurrence Probability. In this paper, abnormal events are detected occurring in a certain area A during a certain day between start time t_s and end time t_e. X_A is defined as a collection of base stations covering A. T is a time period of $[t_s, t_e]$. T_u is the same time period in a day, where, u represents a serial number of day, $u = 1, 2, 3, \dots, W$. From database, select all activity logs during T_u generated by X_A. Suppose a user produces a number of activity records at time $t_1, t_2, \dots, t_c \in T_u$ inside base station $i_1, i_2, \dots, i_c \in X_A$. Here, t and X is not a one to one relationship, that is, user may have some activity records in the same base station at different time. Under the above assumptions, we can compute occurrence probability Pr_p that a user occurs within A during T_u.

Combined with Bayesian probability model, under the above assumptions, if a user produces an activity log c during T_u, then the probability ψ that the user appears in A is calculated using Eq. (3).

$$\psi_{A,c}(t_c) = \int_A \varphi_{ic}(x)dx. \tag{3}$$

Where, t_c is the time when the user's activity record is produced, that is, starting time. i_c is base station. φ is calculated by Bayes formula. As user moves at any time, at a specific time t after starting recording user's activity, the user may move to the other area. When calculating the probability that user appears during T_u, we need to consider

$t(t \neq t_c)$. When $t > t_c$, suppose user leaves the location of its activity records at a constant rate γ, and no longer turn back. For t, Eq. (1) can be rewritten in the form of Eq. (4).

$$\psi_{A,c}(t) = e^{-\gamma|t-t_c|} \int_A \varphi_{i_c}(x)dx. \tag{4}$$

Here, γ allows time difference to be more than 15 min before and after user's activity records, that is, $|t-t_c| > 15$, and probability is reduced of 1 % at t_c. Considering user's all activity records during T_u, normalize them by Eq. (5).

$$Pr_p(A, T_u) = \frac{1}{t_e - t_s} \frac{\max_c(\int_{t_s}^{t_e} \psi_{A,c}(t)dt)}{\max_{i \in X_A} \varphi_i(A)}. \tag{5}$$

Calculate integral value within $[t_s, t_e]$ of user's all activities records and obtain the maximum value. These obtained values may exceed 1, so it is needed to normalize them to the range of $(0, 1)$.

Ordinary Probability. For a given day W, the average of Pr_p is denoted as ordinary probability Pr_o, that the user appears in A during all time period T before W, i.e. ordinary probability, which reflects the user's activities under normal situation, having no relationship with possible abnormal event. Pr_o is calculated as shown in Eq. (6).

$$Pr_o(A, T_W) = \frac{1}{W-1} \sum_{\substack{u=1 \\ u \neq w}}^{W} Pr_p(A, T_u). \tag{6}$$

Where, data for calculation is provided by history collection of occurrence probability. Thus, we can obtain occurrence probability and ordinary probability. Then, combining with these two results, we can calculate probability that user participates in unusual events.

Abnormal Participation Probability. Occurrence probability reflects possibility that user appears in A during a period of T in a day, representing a special situation. Ordinary probability reflects average likelihood that user appears in A during multiple same time periods T before the day, representing a general situation. Differences between these two represents a gap of general situation and special situation. By calculating the difference, we can obtain abnormal participation probability Pr_a which user participates in abnormal events by Eq. (7).

$$Pr_a(A, T_W) = Pr_p(A, T_W)(1 - Pr_o(A, T_W)). \tag{7}$$

As we can see from Eq. (5), when occurrence probability is constant, if the smaller the conventional probability is, the greater the abnormal probability is and vice versa. Therefore, when occurrence probability is low, probability that user participates in abnormal event is also not very high; while, when probability is high, regular probability is low, which indicates that probability of which user appears in A in the past, is low. If the larger the difference is, the greater the multiplied value of occurrence probability is, then the higher the abnormal probability is. Similarly, if the higher the conventional probability is, it indicates probability that user appears in A is higher and the smaller the difference is, then multiplied value of occurrence probability is smaller, abnormal probability is lower.

After individual probability is calculated, we can expand it to group abnormal analysis. According to abnormal participation by groups, abnormal events can be determined to detect urban anomalies.

3.3 Group Anomalous Analysis

A group is generally made up by individuals. Therefore, individual analysis is the basis of group analysis. Based on the results of individual calculation, expand them to group analysis. Similar to individual analysis, we propose a method of abnormal group analysis. Calculate number of abnormal population of group, obtain its historical data. Make statistics on these historical data, calculate number of abnormal results, and compare them to get anomaly index.

Number of Group Abnormal Calculation. In Sect. 3.2, we present calculation method for individual abnormal probability. If the abnormal probability is higher, indicating that the possibility that individuals involve in the abnormal event is also greater. How high the probability is, which can prove that an individual involves in abnormal events, is essential to analyze the abnormal group. Therefore, it is required to provide with a threshold of probability P_γ. If $Pr_a(A, T_W) > P_\gamma$, we can decide that individuals involved in the abnormal event. We propose the following method for determining the threshold. Exception event is generally defined as a rare group event, thus, it is a greater possibility for a small number of individuals to involve abnormal events (even if there is no unusual event). When determining the threshold, it should be able to ensure that only a small part of individuals in group are classified as participants of abnormal event in the normal state; otherwise, if a large number of individuals are classified as participants of abnormal event, such event will become a normal event. Therefore, in the paper, we determine the threshold that only 1 % of individuals in group is determined to be abnormal event participants.

Group Abnormal Index Calculation. According to Sect. 3.2, we can get number of population in group who involve in exception, indicated as N_W, where, W represents a specific day. According to historical exception set of group, we can get the average value μ and standard deviation σ of abnormal population through statistical analysis. And then obtain a calculation method for anomaly index Z, shown in Eq. (8).

$$Z = \frac{N_W - \mu}{\sigma}. \tag{8}$$

Where, Z represents a kind of deviation between number of abnormal population and historical average of number of abnormal population in W. Z is greater, indicating that difference, between the day's activities and historical regularity, is greater, and abnormal events occur more likely. Even if we do not consider deviation situation, N_W itself is also established on the condition that only 1 % of individuals in group meets the exception conditions. It shows that anomaly index may represent possibility of an unusual events.

4   Case Study

4.1   Raw Data

In order to make evaluation on our detection results, two regions in Kunming, China, that is, Tuodong Sports Center and Mianshan Forest Park, are selected as case study.

The raw data for our work is collected by telecommunication service provider. Each record follows a consistent data format, shown in Table 1. When user makes calling, receiving and sending SMS etc., location update will produce a record.

Table 1.   Data format of raw data.

Field	Description
Time	Starting time for communication records, with data format hh:mm:ss, i.e. hour, minute, second, hh taking the 24 h
Type	Type of activity corresponding to communication records, such as calling, SMS or location update
Number	Middle of 5 digits of user's phone number, to ensure that these numbers can identify user uniquely
LAC	Region ID which is covered by base station
CI	Cell tower ID

4.2   Event Detection

Calculated by the Bayes probabilistic model, abnormal indexes of these two regions are given in Figs. 2 and 3, respectively.

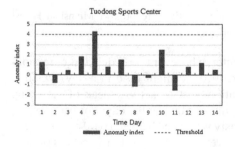

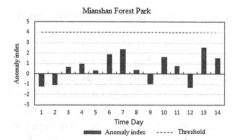

Fig. 2. Anomaly index of Tuodong Sports Center.

Fig. 3. Anomaly index of Mianshan Forest Park.

In the Figs. 2 and 3, the abscissa represents time, starting from Monday, continuing for two weeks. Ordinate represents anomaly index. For Tuodong Sports Center, as shown in Fig. 2, there is no regularity in abnormal indexes. But we can see a clear peak appearing in the first week of Friday, showing that there are abnormal events happened in this day. Through our survey, there is a sporting event held on that day. representing that our results of abnormal event detection are accurate. For Mianshan Forest Park, shown in Fig. 3, abnormal exponential trend is more apparent than regularity in Fig. 2. That is, anomaly index is lower at weekly working days, while, relatively high at weekends. During these days, we can see, there was no unusual index broke the threshold. This reflects that group activities are more frequently on weekends. Therefore, abnormal index will be relatively higher. This is related with pattern of life of group, because groups have more time to choose travel at weekends.

4.3 Activity Analysis

In order to make analysis on abnormal events more detailed, we get participants and non-participants during the time period (set as T) when unusual event happened, and make statistics on activity records at different times in a day, to analyze influence of abnormal events caused by activities, and then take the mean of activity records according to different times in these days. The results are shown in Figs. 4 and 5.

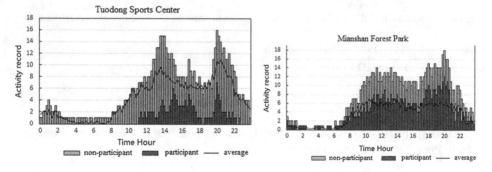

Fig. 4. Activity records of Tuodong Sports Center.

Fig. 5. Activity records of Mianshan Forest Park.

As shown in Fig. 4, for Tuodong Sports Center, whenever for participants or for non-participants, activity records are obviously reduced at night time in a day, which is consistent with the regular pattern of population, that is, activity peak of the group concentrated around 8:00 pm or so and 2:00 pm or so during the day. For participants, It can be seen that activity records were obviously weakened between 6:00 pm–8:00 pm. This can be speculated that it is related to the event of march. Activity records of non-participants during 6:00 pm–8:00 pm are also weakened. Reducing degree is inferior to participants. Comparing with the average, we can see, except the periods during the event, activity records of non-participants are greater than the average. Even after the race, activity records of non-participants are still greater than the historical average, which means that some individuals have been wrongly classified as non-participants. For Mianshan Forest Park, as shown in Fig. 5, all user's activity records are concentrated at the time 8:00 pm–10:00 pm. Here, activity records around 8:00 pm is the peak in a day. 10:00 am–4:00 pm is the second peak that is more concentrated. This may be the reasons that activity records are relative reduced, resulting in two peaks. For participants or non-participants, the trend is similar. On the other hand, In Fig. 5, activity records of participants and non-participants have considerable number. Even the former is slightly larger than the latter, which is different from the results in Fig. 4. This is because the location of the resort is remote, so that usual activities of tourists do not substantially cover here, that is why anomaly index are higher for most tourists. If number of participants are more, active records become more.

Through detecting abnormal events in these two regions, it is seen that our model can capture abnormal event well. Also, from analysis of activity records, our model can reflect the division of participants and non-participants more accurately.

5   Conclusion and Future Work

Through the probabilistic model, we analyze individual occurrence probability and normal occurrence probability in a specific area during a specific time period in order to analyze individual abnormal activity. Expanding from individual to group, we also analyze activities of groups to enable detect and analyze urban abnormal events.

Our contribution is summarized as follows. (1) Using the model, occurrence probability of individual is analyzed within a specific area during a specific time period. We make distinction on the two concepts between occurrence probability and conventional occurrence probability. Through calculating the difference between these two, we obtain probability of individual participants. (2) When handling group analysis, we set threshold to divide participants and non-participants. Through statistical analysis, we get number of group participants, which can analyze historical situation of number of abnormal participants. Comparing number of group abnormal population with historical average, we can obtain anomaly index. (3) We select two areas in Kunming city as an application example to detect abnormal events within these two areas. From the results, we propose a method to better achieve the capture of abnormal events, providing information about participants and non-participants, which can be used to further analyze impact of abnormal events on activity patterns.

Further work shall include as follows. (1) Combined with POI information, we shall introduce offset of location as a priori probability, so that accuracy of the model can be improved. (2) Visualization of heat map on abnormal index shall be handled, making possible to reflect regional anomaly events directly.

Acknowledgments. This work is supported by National Nature Science Foundation of China under grant no. 41231171. The authors would like to thank Xiaoqing Zou at Kunming University of Science and Technology, Kunming, China for providing us with mobile phone data.

References

1. Rojas, F., Calabrese, F., Fiore, F.D., Krishnan, S., Ratti, C.: Real Time Rome. MIT Senseable City Laboratory, Cambridge (2006)
2. Eagle, N., Pentland, A.S.: Reality mining: sensing complex social systems. Pers. Ubiquit. Comput. **10**(4), 255–268 (2006)
3. Calabrese, F., Pereira, F.C., Di Lorenzo, G., Liu, L., Ratti, C.: The geography of taste: analyzing cell-phone mobility and social events. In: Floréen, P., Krüger, A., Spasojevic, M. (eds.) Pervasive 2010. LNCS, vol. 6030, pp. 22–37. Springer, Heidelberg (2010)
4. Candia, J., González, M.C., Wang, P., Schoenharl, T., Madey, G., Barabasi, A.L.: Uncovering individual and collective human dynamics from mobile phone records. J. Phys. A: Math. Theor. **41**(22), 1–16 (2008)
5. Bagrow, J.P., Wang, D., Barabási, A.L.: Collective response of human populations to large-scale emergencies. PLoS One **6**(3), e17680 (2011)
6. Ratti, C., Pulselli, R.M., Williams, S., Frenchman, D.: Mobile landscapes: using location data from cell-phones for urban analysis. Environ. Plan. B: Plan. Des. **33**, 727–748 (2006)
7. Zang, H., Baccelli, F., Bolot, J.: Bayesian inference for localization in cellular networks. In: Proceedings of IEEE INFOCOM, San Diego, California, USA, pp. 1–9. IEEE Press (2010)

A Combined Model Based on Neural Networks, LSSVM and Weight Coefficients Optimization for Short-Term Electric Load Forecasting

Caihong Li[✉], Zhaoshuang He, and Yachen Wang

School of Information Science and Engineering, Lanzhou University,
Lanzhou 730000, People's Republic of China
licaihong@lzu.edu.cn

Abstract. As an essential energy in the daily life, electricity which is difficult to store has become a hot issue in power system. Short-term electric load forecasting (STLF) which is regarded as a vital tool helps electric power companies make good decisions. It can not only guarantee adequate energy supply but also avoid unnecessary wastes. Although there exists quantity of forecasting methods, most of them are not able to make accurate predictions. Therefore, a forecasting method with high accuracy is particularly important. In this paper, a combined model based on neural networks and least squares support vector machine (LSSVM) is proposed to improve the forecasting accuracy. At first, three forecasting methods named generalized regression neural network (GRNN), Elman, LSSVM are utilized to forecast respectively. Among them, simulate anneal (SA) arithmetic is used to optimize GRNN. Then, SA is employed to determine the weight coefficients of each individual method. At last, multiplying all the three forecasting results with the corresponding weights, the final result of the combined model can be attained. Using the electric load data of Queensland of Australia as experimental simulation, case studies show that the proposed combined model works well for STLF and the results prove more accurate.

Keywords: Generalized regression neural network · Elman · Least squares support vector machine simulated annealing algorithm · Short-term electric load forecasting

1 Introduction

Electricity is of vital importance to every region as an essential energy resource in people's daily life. Especially in the underdeveloped areas where electricity is deficient, thus makes it more precious. Short-term electric load forecasting (STLF) is an fundamental tool for transmission dispatching, unit commitment and other public utilities [1]. During the past decades, various models based on mathematics or artificial intelligence [2–7] have been introduced to electric load forecasting. All these methods improve the forecasting performance to a certain degree. However, the individual methods are too simple to comprehensively deal with the forecasting problems. As Moghram and Rahman [8] stated, all the individual methods failed to yield desirable forecasting

© Springer International Publishing AG 2016
S. Song and Y. Tong (Eds.): WAIM 2016 Workshops, LNCS 9998, pp. 109–121, 2016.
DOI: 10.1007/978-3-319-47121-1_10

performance. The forecasting errors will have a tremendous impact on the economic benefits. In the other words, small increasement of forecasting accuracy may save millions of dollors cost which are huge to the electricity market. In order to enhance the forecasting performance, emphases have laid on the hybrid models or combined models. By integrating multiple methods, the forecasting method can process different data with different characteristics and overcome flaws existing in the individual methods. Amina et al. [9] proposed a novel fuzzy wavelet neural network model of Greek Island of Crete. By using the subtractive clustering optimized with the Expectation-Maximization algorithm, the results indicated that it provided significantly better forecasts. Xiao et al. [10] introduced a hybrid forecasting model combined with BP and GRNN. Ahead of forecasting, data preprocessing was used to improve the forecasting accuracy. A hybrid method with empirical mode decomposition (EMD), extended Kalman fitter, extreme learning machine (ELM) and particle swarm optimization (PSO) was utilized by Liu et al. [11] to forecast short-term load. [12–14] proposed the examinations of the combined methods. All these hybrid and combined methods enhance the forecasting performance to a large extend. That makes them the preferred methods when considering the forecasting accuracy.

In this paper, a combined model based on generalized regression neural network (GRNN), Elman, least squares support vector machine (LSSVM) and simulated annealing (SA) is introduced. GRNN is a kind of radial basis function (RBF) network which has a strong nonlinear mapping capacity and high degree of fault tolerance. Its network structure is simple and the calculation results can achieve global convergence. Xia et al. [15] applied GRNN for short-term load forecasting and virtual instrument design. Chelgani et al. [16] used GRNN to predict microwave irradiation pretreatment and peroxyacetic acid desulfurzation of coal. The results turned that GRNN was good at predicting. Li et al. [17] employed a hybrid model based on GRNN which was optimized by fruit fly optimization algorithm. Elman is a dynamic neural network which has a short-term memory. That makes it well accommodate to the time-variant characteristic. Under the influence of temperature, a wavelet Elman neural network for short-term load prediction was proposed by Kelo et al. [18]. Song [19] introduced the Elman networks on the weight convergence. Li et al. [20] utilized Chaotifying linear Elman networks. LSSVM is one of the machine learning methods which has strong nonlinear data processing ability. LSSVM is able to reduce the complexity of the calculation and improve the speed of solving. Shayeghi et al. [21] proposed a hybrid model to forecast day-ahead electricity price. Chaotic gravitational search algorithm was developed to find the optimal parameters of LSSVM. Zhang et al. [22] utilized an unbiased LSSVM with polynomial kernel. Xie et al. [23] applied clustering-LSSVM to forecast electricity price. SA is a stochastic optimization algorithm which based on Monte-Carlo iterative solution strategy. It can effectively avoid falling into local minimum and finally it approaches the global optimal. In this paper, SA not only optimizes the parameters of GRNN, but also determines the weight coefficients of the three individual methods. Hong [24] introduced seasonal SVR which is optimized by simulated annealing algorithm to forecast traffic flow. Yuan et al. [25] proposed the Clound theory-based simulated annealing algorithm and application. Pai et al. [26] utilized support vector machine optimized by simulated annealing algorithm in electricity load forecasting. Firstly, three individual forecasting methods are applied to forecast respectively. Then, SA is employed to determine the weight coefficients of each individual method. At last, multiplying all the three forecasting results with the corresponding weights, the final result of the combined model can be attained.

The rest of this paper is organized as follows. Section 2 introduces the theory of GRNN, Elman, LSSVM and SA which combined the proposed model. The implementation process of the proposed combined model is described in Sect. 3. In Sect. 4, a simulation of electric load forecasting of Australia electric market is shown. The contrasted results demonstrate the preponderance of the proposed method. Finally, Sect. 5 concludes the paper.

2 Methodologies

All the individual methods are introduced in this section, including generalized regression neural network, elman, least squares support vector machine and simulated annealing algorithm.

2.1 Elman Neural Network

The Elman neural network(ElmanNN) first proposed by Elman in 1990 [27], which structure includes the input layer, a particular context nodes input layer, the hidden layer(middle layer) and the output layer, is a feed-forward network with local feedback. The linear or nonlinear function is applied for the transfer function of Elman. The connection of each layer is similar to the feed-forward network and the context layer that can be seen as a one-step delay operator is used to record before moment output values of hidden layer. The framework of ElmanNN is shown in the Fig. 1 and its state space is expressed as:

$$y(k) = g\left(\omega^3 x(k)\right) \tag{1}$$

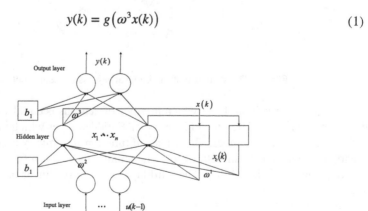

Fig. 1. The framework of a feedback ElmanNN with three layer structure.

$$x(k) = f\left(\omega' x_c(k) + \omega^2(u(k-1))\right) \tag{2}$$

$$x_c(k) = x(k-1) \tag{3}$$

In the Fig. 1, y is m-demension output vector; x is n-demension hidden layer unit vector; u express r-demension input vector; x_c is n-demension feedback state vector; ω^3 is the weight from hidden layer to output layer; ω^2 is the weight from input layer to hidden layer; ω^1 is the weight from context layer to hidden layer; b_1 and b_2 are the threshold value of hidden layer and output layer, respectively; $g(*)$ and $f(*)$ are the transfer function of the output neurons and of hidden neurons, respectively.

2.2   General Regression Neural Network

The general regression neural network (GRNN) proposed by Donald F. Specht in 1991 [28] is a variation of radial basis neural networks which is designed for function approximation and regression. A GRNN consisted of four layers: the input layer, the pattern layer, the summation layer, and the output layer is a novel effective feed-forward neural network model with the standardization of the dot product weight function. Radial Basis Function (RBF) is used as transformation from input layer to hidden layer, while, a special linear transformation is used from hidden layer to output layer. In addition, the network can avoid the impact on the prediction result caused by human subjective assumptions as much as possible, because of that in GRNN there is only one artificially adjust parameters named smooth factor and network learning are all depended on the data sample. Figure 2 shows the structure of GRNN.

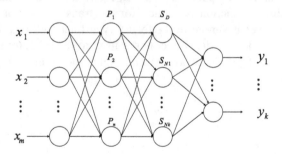

Fig. 2. The structure of the general regression neural network

The first layer is input layer which number of neurons is equal with the number of the input parameters. The second layer is RBF hidden layer which number of neurons is the training sample number. The Gaussian function is normally used to be its transfer function, which contain smooth factor. The smaller the smooth factor, the stronger the function approximation ability. The third layer was simple linear output layer.

2.3 Least Squares Support Vector Machine

Least squares support vector machine (LSSVM) which is developed based on the statistic theory is a new kind of machine learning techniques. Suyken et al. [29] proposed LSSVM by changing the constraint condition and risk function of SVM. The quadratic programming problem can be directly converted into linear equations. That means LSSVM has the ability to accelerate the solving speed and reduce the use of computing resources which can improve the convergence speed.

For a given sample $\{x_i, y_i\}$, where $x_i \in R^n$ is the output vector, $i = 1, 2, \cdots, l, y_i \in R$ is the output. By using the structural risk minimization principle, the optimization problem is expressed as the following constrained optimization problem:

$$\min:J(\omega, e) = \frac{1}{2}\|\omega\|^2 + \frac{1}{2}C\sum_{i=1}^{N} e_i^2 \tag{4}$$

$$s.t.:y_i = \omega^T \varphi(x_i) + b + e_i\, i = 1, 2, \cdots, N \tag{5}$$

Where e_i denotes as the errors, C is the penalty coefficient which is used to control the degree of punishment beyond the errors of the sample.

The optimization problem can be solved by introducing the Lagrange function:

$$L(\omega, b, e, \alpha) = J(\omega, e) - \sum_{i=1}^{N} \alpha_i[\omega^T \phi(x) + b + e_i - y_i] \tag{6}$$

Where $\alpha = (\alpha_1, \alpha_2, \cdots, \alpha_N)^T$ is the Lagrangian multiplier.

After solving the linear equation, the final expression is as follows:

$$f(x) = \sum_{i=1}^{N} \alpha_i k(x_i, x_j) + b \tag{7}$$

$k(x_i, x_j)$ is the inner product kernel function. In this paper, the radial basis function (RBF) kernel function is chose because it is the most effective one to deal with the nonlinear regression problems.

2.4 Simulated Annealing Algorithm

Simulated annealing (SA) is a kind of widely used stochastic optimization algorithm which is based on Monte Carlo iterative solution strategy. By simulating the annealing process of solid material in physics, problems similar to the NP complex can be solved.

Starting from the random feasible solution, for a given initial control parameters, SA keeps carrying out the iterative process of "generate data processing—judgment—accept/abandonment". That is to say when the control parameter of temperature T stays the same, the relatively optimal solution of the combinatorial optimization can be obtained by repeating the Metropolis algorithm. And then reduce the control value of T, repeat the

Metropolis algorithm under different T, when T approaches zero, the final overall optimal solution of the combinatorial optimization problem can be acquired. Four steps of SA are as follows:

1. Initialization: choose the initial temperature and the length of Markov chain which stands for the number of the iteration of Metropolis algorithm.
2. Generate new state: reduce the control temperature according to its attenuation function. Every time the temperature reduces, there exists a random disturbance which generate a new state.
3. Generate new solution: determine whether to accept the new generated state as the new solution according to the accept function.
4. Obtain the optimal solution: the optimal solution will be procured until meeting the stop criterion.

3 The Proposed Method

This paper proposed a new combined predict model based on several methods of power load forecasting.

The details of the combined model are shown as follows:

Step 1. Import the training data. Then build the input set and the output set of the training set and the input set and the output set of the testing set.

Step 2. Build Elman model according to the input set and the output set. Then set the scope of the number of the intermediate layer nodes and to find the optimum value by circulation. Finally, use the training set to train ElmanNN model and get the forecast result set $\hat{Y}_1$.

Step 3. Build the GRNN model according to the input set and the output set. Then choose the smooth factor parameter by using the simulated annealing algorithm. When the simulated annealing algorithm was used, set the parameter of the simulated annealing algorithm. Finally, use the training set to train GRNN model and get the forecast result set $\hat{Y}_2$.

Step 4. Build LSSVM model according to the input set and the output set. Then use the training set to train GRNN model and get the forecast result set $\hat{Y}_3$.

Step 5. Build combined forecast model. Get the linear weighted combination of the above three models.

$$\hat{Y} = \alpha_1 \hat{Y}_1 + \alpha_2 \hat{Y}_2 + \alpha_3 \hat{Y}_3 \tag{8}$$

Where, α_1, α_2 and α_3 are weighted coefficients.

Then, use the simulated annealing algorithm to optimize the weighted coefficients. When the simulated annealing algorithm was used, set the parameter of the simulated annealing algorithm.

Step 6. Testing the trained combined model. Apply the combined model got by step 5 to forecasting.

Step 7. Output the predicted results and calculate the accuracy.

4   Simulation

The main purpose of this section was put forward the simulation of the proposed combined model.

4.1   The Performance Indexes

In this simulation, three generally adopted error indexes, the mean square predict error (MSE), the mean absolute error (MAE) and the mean absolute percent error (MAPE), was used given as follows:

$$MSE = \frac{1}{n} \sum_{i=1}^{n} \left(x_i - \hat{x}_i \right)^2 \tag{9}$$

$$MAE = \frac{1}{n} \sum_{i=1}^{n} \left| x_i - \hat{x}_i \right| \tag{10}$$

$$MAPE = \frac{1}{n} \sum_{i=1}^{n} \left| \frac{x_i - \hat{x}_i}{x_i} \right| \times 100\% \tag{11}$$

Where, x_i is the real data of samples; $\hat{x}_i$ is the predictive values of samples, and n is the number of samples.

4.2   The Process of Simulation

Original data used in this paper come from Queensland(QLD) state of Australia's electricity market, which gathered once every half hour from 0:30 on October 16, 2014 to 0:00 on October 1, 2015, a total of 16800 data items. 16464 data items gathered from 0:30 on October 16, 2014 to 0:00 on September 24, 2015 as the training set, and the remaining 336 data as the test set. The training data of the model was divided into seven groups, and every group were established data combined model. Combined Model is that using the data (48 items) form the previous week Monday to forecast the following Monday's data (48 items), Tuesday forecast Tuesday and soon.

Firstly, Seven ElmanNN models consisting of an inputting layer, an outputting layer, a hidden layer and a recurrent layer was adopted to forecast the whole week electrical load data. Set the number of input layer nodes and output layer nodes of each ElmanNN model were all 48. To achieve better prediction results, the node number of the hidden layer was set from 10 to 20. In this simulation, depending on the MAPE value, the number of hidden layer nodes with the minimum value of MAPE were selected. The optimal value of intermediate layer nodes of each ElmanNN model are shown in the Table 1. We can easily observe that the optimal value of middle layer nodes were 15 for the Sunday and Monday model, while that were 16 for Saturday and Tuesday model. For Wednesday, Thursday and Friday model that were 11, 17 and 12, respectively.

Table 1. The optimal value of intermediate layer nodes of each ElmanNN model.

Time	Thursday	Friday	Saturday	Sunday	Monday	Tuesday	Wednesday
BestHidnum	17	12	16	15	15	16	11
MinMAPE(%)	2.3326	3.1158	1.6596	1.7408	2.1731	1.274	1.8265

Secondly, seven GRNN models consisting of an inputting layer, a pattern layer, a summation layer and an outputting layer was adopted to forecast the whole week electrical load data. Set the input layer nodes was 48 and the number of output layer nodes was 48. The simulated annealing algorithm with the default parameter and 30 annealing chain length was used to find the optimal smoothing parameter.

Thirdly, seven LSSVM models consisting of 48 inputs, 48 outputs and RBF kernel function was adopted to forecast the whole week electrical load data.

Finally, the above three models were combined by linear weighted combination. Combination forecasting model that obtain the combination forecast model in the form of the appropriate weighted average can improve the accuracy of the prediction and reliability. The most concern of combined forecast model is how to calculate the weighted average coefficient to make the combination forecast model more effectively and to improve the prediction precision. Therefore, the simulated annealing algorithm was adopted to optimize the weights. First of all, the initial value of three combination weighted parameters were set to 0.33, 0.33 and 0.33. Then the simulated annealing algorithm with the default parameter and 30 annealing chain length was adopted to optimize the combination weights. The optimized combination weights by the simulated annealing algorithm are given in Table 2, in which α_1, α_2 and α_3 are the combination weights of ElmanNN model, GRNN model and LSSVM model, respectively. It can be clearly seen that on Saturday and Sunday the second parameter α_2 are 0.9956 and 0.9958, respectively, that point out that the results of those two days' data set largely depend on the GRNN model. While, the weights of Friday and Monday shows that the LSSVM model with the weights α_3 of 0.9956 and 0.9858 is extremely important for the combined model, and, the weights α_1 of Thursday shows that the ElmanNN model is a very significantly role. In addition, for the Tuesday and Wednesday, the combined model is affected by the joint ElmanNN model and GRNN model.

Table 2. The optimized combination weights by the simulated annealing algorithm

Time	Thursday	Friday	Saturday	Sunday	Monday	Tuesday	Wednesday
α_1	0.9026	0.0022	0.0432	0.0703	0.0123	0.2950	0.3812
α_2	0.0414	0.0022	0.8208	0.8688	0.0018	0.6612	0.5877
α_3	0.0560	0.9956	0.1359	0.0609	0.9858	0.0438	0.0311

4.3   The Results and Analysis of Simulation

Figure 3 shows the actual values and forecasting values in the whole week of the four methods (Elman, GRNN, LSSVM and the combined model). As seen from Fig. 3, the curve of the combined model is more consistent with the actual data. Although at time of 12:00 to 18:00, the forecasting errors are bigger, it can still reveal that the combined model is better than the other three individual models.

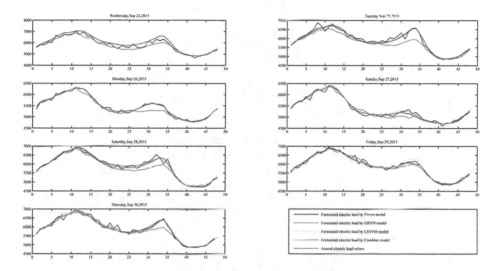

Fig. 3. Final predicted values for each day by the four methods.

Table 3 presents the three indicators of the four forecasting methods in the form of numbers. It can be seen clearly that the average values of the combined model of the whole week are the lowest. The value of MAPE of the combined model is as low as 1.72 %. MAE, MSE and MAPE of the combined model has the lowest value of all the four methods. That means the proposed combined model has the best forecasting performance. When comparing with the other three individual methods, GRNN has the largest forecasting values on Monday and Thursday, LSSVM has the largest values on Tuesday. While, Elman owns the largest forecasting data on Wednesday and Friday. The results display visually that different forecasting methods have different forecasting values which are sometimes meet the demands or sometimes unsatisfactory. Table 3 proved the high forecasting performance of the proposed combined model in another visual angle.

Table 3. Three statistics measures of the four forecasting methods.

Date	MSE (MW) of				MAE(MW) of				MAPE(%) of			
	Elman	GRNN	LSSVM	Combined	Elman	GRNN	LSSVM	Combined	Elman	GRNN	LSSVM	Combined
Thursday	33260.02	85072.50	76671.26	34457.04	134.57	236.91	222.62	137.50	2.33 %	4.04 %	3.81 %	2.38 %
Friday	63008.35	54911.33	49380.66	49394.17	182.83	169.15	161.86	161.90	3.12 %	2.90 %	2.78 %	2.78 %
Saturday	16255.90	12735.84	14882.34	13019.24	89.40	75.63	77.12	75.64	1.66 %	1.41 %	1.43 %	1.41 %
Sunday	14606.58	5421.87	11633.84	5934.85	91.24	53.67	76.32	56.613	1.74 %	1.03 %	1.46 %	1.09 %
Monday	36706.55	31984.93	21479.82	21567.15	125.56	138.75	107.17	107.34	2.17 %	2.37 %	1.84 %	1.85 %
Tuesday	9893.81	7118.90	18808.03	6548.34	75.815	65.29	100.14	62.18	1.27 %	1.09 %	1.67 %	1.04 %
Wednesday	22928.92	20501.05	23259.04	18394.86	108.29	94.71	102.01	88.59	1.83 %	1.59 %	1.71 %	1.50 %
Whole week	28094.3	31106.63	30873.57	21330.81	115.39	119.17	121.03	98.54	2.02 %	2.06 %	2.10 %	1.72 %

5 Conclusions

Interfered by various of factors, time series data has plenty of complex characteristics. Considering that an individual method cannot deal with all kinds of data, a novel combined model for STLF is presented in this paper. The proposed model combined with generalized regression neural network (GRNN), elman and least squares support vector machine (LSSVM). Optimizing by simulated annealing (SA), each individual method is assigned a weight coefficient. Multiplying all the three forecasting results by corresponding weights coefficients, the final forecasting results can be attained. In order to verify the performance of the combined model, electric load data from Queensland of Australia is utilized. The result shows that the average MAPE of the combined model is 1.72 % which is lower than the existing hybrid model named MFES proposed by Zhao et al. [30]. It had reduced MAPE of MFES by 29.04 %. The results of comparison demonstrate the excellent performance of the combined model. The reasons why the combined model has higher forecasting accuracy are listed below. Firstly, the proposed model combines two kinds of artificial neural networks which have strong forecasting performance. Between them, GRNN is optimized by SA, thus makes it more accurate. Secondly, instead of traditional average allocation method, SA is employed to determine weight coefficients of each individual model. Taking advantage of each model, the higher forecasting accuracy can be obtained. In a nutshell, the combined model outperforms other individual models. With higher accuracy, it is a promising tool in the future.

Acknowledgments. The authors would like to thank the Natural Science Foundation of PR of China (61073193,61300230), the Key Science and Technology Foundation of Gansu Province (1102FKDA010), the Natural Science Foundation of Gansu Province (1107RJZA188), and the Science and Technology Support Program of Gansu Province (1104GKCA037) for supporting this research.

References

1. Masa-Bote, D., Castillo-Cagigal, M., et al.: Improving photvoltaics grid integration through short time forecasting and self-consumption. Appl. Energy **125**, 103–113 (2014)
2. Hong, W.C.: Electric load forecasting by seasonal recurrent SVR (support vector regression) with chaotic artificial bee colony algorithm. Energy **36**, 5568–5578 (2011)
3. Li, S., Wang, P., Goel, L.: Short-term load forecasting by wavelet transform and evolutionary extreme learning machine. Electr. Power Syst Res. **122**, 96–103 (2015)
4. Deihimi, A., Showkati, H.: Application of echo state networks in short-term electric load forecasting. Energy **39**, 327–340 (2012)
5. Zhang, R., Dong, Z.Y., Xu, Y., Meng, K., Wong, K.P.: Short-term load forecasting of Australian national electricity market by an ensemble model of extreme learning machine. Gener. Transm. Distrib. IET **7**, 391–397 (2013)
6. Kandil, N., Wamkeue, R., Saad, M., Georges, S.: An efficient approach for short term load forecasting using artificial neural networks. Int. J. Electr. Power Energy Syst. **28**(8), 525–530 (2006)
7. Jin, M., Zhou, X., Zhang, Z.M., Tentzeris, M.M.: Short-term power load forecasting using grey correlation contest modeling. Expert Syst. Appl. **39**, 773–779 (2012)

8. Moghram, I., Rahman, S.: Analysis and evaluation of five short-term load forecasting techniques. IEEE Trans. Power Syst. **4**(4), 1484–1494 (1989)
9. Amina, M., Kodogiannis, V.S., Petrounias, I., Tomtsis, D.: A hybrid intelligent approach for the prediction of electricity consumption. Int. J. Electr. Power Energy Syst. **43**(1), 99–108 (2012)
10. Xiao, L., Wang, J., Yang, X., Xiao, L.: A hybrid model based on data preprocessing for electrical power forecasting. Electr. Power Energy Syst. **64**, 311–327 (2015)
11. Liu, N., Tang, Q., Zhang, J.H., Fan, W., Liu, J.: A hybrid forecasting model with parameter optimization for short-term load forecasting of micro-grid. Appl. Energy **129**, 336–345 (2014)
12. Wang, J.Z., Zhu, S.L., Zhang, W.Y., Lu, H.Y.: Combined modeling for electric load forecasting with adaptive partical swarm optimization. Energy **35**(4), 1671–1678 (2010)
13. Xiao, Y., Liu, J.J., et al.: A neuro-fuzzy combination model based on singular spectrum analysis for air transport demand forecasting. J. Air Transp. Manage. **39**, 1–11 (2014)
14. Souptick, C., Sanjay, G., Dilip, K.P.: A combined neural network and genetic algorithm based approach for optimally designed femoral implant having improved primary stability. Appl. Soft Comput. **38**, 296–307 (2016)
15. Xia, C.H., Lei, B.J., Wang, H.P., Li, J.N.: GRNN short-term load forecasting model and virtual instrument design. Energy Procedia **13**, 9150–9158 (2011)
16. Chelgani, S.C., Jorjani, E.: Microwave irradiation pretreatment and peroxyacetic acid desulfurzation of coal and application of GRNN simultaneous predictor. Fuel **90**(11), 3156–3163 (2011)
17. Li, H.Z., Guo, S., Li, C.J., Sun, J.Q.: A hybrid annual power load forecasting model based on generalized neural network with fruit fly optimization algorithm. Knowl. Based Syst. **37**, 378–387 (2013)
18. Kelo, S., Dudul, S.: A wavelet Elman neural network for short-term load prediction under the influence of temperature. Electr. Power Energy Syst. **43**, 1063–1071 (2012)
19. Song, Q.: On the weight convergence of Elman networks. IEEE Trans. Neural Netw. **21**(3), 96–101 (2010)
20. Li, X., Chen, G., Chen, Z., Yuan, Z.: Chaotifying linear Elman networks. IEEE Trans. Neural Netw. **13**(5), 1193–1199 (2002)
21. Shayeghi, H., Ghasemi, A.: Day-ahead electricity prices forecasting by a modified CGSA technique and hybrid WT in LSSVM based scheme. Energy Convers. Manage. **74**, 482–491 (2013)
22. Zhang, M., Fu, L.: Unbiased least squares support vector machine with polynomial kernel. In: 8th IEEE International Conference on Signal Processing (ICSP-2006), vol. 3, Guilin, China, pp. 16–20 (2006)
23. Xie, L., Zheng, H., Zhang, L.Z.: Electricity price forecasting by clustering-LSSVM. In: Proceedings of the International Power, Engineering Conference, pp. 697–702 (2007)
24. Hong, W.C.: Traffic flow forecasting by seasonal SVR with chaotic simulated annealing algorithm. Neurocomputing **74**, 2096–2107 (2011)
25. Lv, P., Yuan, L., Zhang, J.: Clound theory-based simulated annealing algorithm and application. Eng. Appl. Artif. Intell. **22**, 742–749 (2009)
26. Pai, P.F., Hong, W.C.: Support vector machines with simulated annealing algorithms in electricity load forecasting. Energy Convers. Manage. **46**(17), 2669–2688 (2005)
27. Chandra, R.: Competition and collaboration in cooperative coevolution of Elman recurrent neural networks for time-series prediction. IEEE Trans. Neural Netw. Learn. Syst. **26**, 3123–3136 (2015)
28. Specht, D.F.: A general regression neural network. IEEE Trans. Neural Netw. **2**(6), 568–576 (1991)

29. Suykens, J.A.K., Van Gestel, T., De Brabanter, J., et al.: Least Square Support Vector Machines. World Scientific, Singapore (2002)
30. An, N., Zhao, W., Wang, J., et al.: Using multi-output feedforward nerual network with empirical mode decomposition based signal filtering for electricity demand forecasting. Energy **49**, 279–288 (2013)

SDMMW 2016

Efficient Context-Aware Nested Complex Event Processing over RFID Streams

Shanglian Peng[1,2(✉)] and Jia He[2]

[1] School of Information Science and Technology, Southwest Jiaotong University,
Chengdu 610031, China
psl@cuit.edu.cn
[2] College of Computer Science, Chengdu University of Information Technology,
Chengdu 610225, China
hejia@cuit.edu.cn

Abstract. With large scale of utilization of monitoring devices such as RFID, sensors and mobile phones, events are generated in a high-speed fashion. Decisions should be made in real time during business processes. Complex Event Processing (CEP) has become increasingly important for tracking and monitoring anomalies and trends in event streams. Nested event detection of RFID event stream is one of the most import class of queries. Current optimization of nested RFID event detection mainly considers caching intermediate results to reduce re-computation of similar results for nested subexpression. In this paper, we use context information of an RFID scenario to optimize nested event detection. We formalize context of an RFID scenario as spatial and temporal constraints and transform these constraints into rules over a nested NFA. Further, we present rewriting context rules to optimize nested event query plan. Experimental results show that with context information introduced, response time had been reduced greatly compared with counterpart methods.

Keywords: Context aware · Complex event processing · Nested pattern · NFA · Data stream · RFID

1 Introduction

Radio Frequency IDentification (RFID) has been extensively used in monitoring scenarios including logistics, health care monitoring, supply chain management and asset tracking, etc. These systems depend heavily on real time analysis of event streams to make decisions. Complex Event Processing (CEP) [1] has become one of the most critical technologies in an RFID enabled system. Application systems utilize CEP to work through many layers. Patterns are typically specified as regular expressions over event attributes, then predicates and correlations are defined over the patterns. So pattern queries can be complex (for example, pattern length can be large, pattern form can be sequential or nested, etc.), incurring great computational complexity to the CEP engine that are running the event queries.

© Springer International Publishing AG 2016
S. Song and Y. Tong (Eds.): WAIM 2016 Workshops, LNCS 9998, pp. 125–136, 2016.
DOI: 10.1007/978-3-319-47121-1_11

The state-of-the-art CEP models such as SASE [2–5] and ZStream [6] do not support definition of nested queries. Though the Cayuga system [7,8] proposes complex nested queries, they process negation filter only over single primitive event type within the SEQ query. CEDR [9] allows applying negation operator over complex event types, but the authors do not present details of the execution model for such nested queries. NEEL [10–14] is a nested CEP language that supports nested operators of SEQ, AND, OR and Negation. The authors also present an iterative nested execution strategy for processing nested event queries expressed in NEEL. Also, the authors proposed caching to optimize execution of the nested CEP. However, as the query window (sliding window) slides continuously over the RFID event streams, instances valid for a certain sliding window are possibly valid in the next window. Although the authors propose caching and query sharing methods to reduce complex event processing overhead, we propose to introduce context during nested CEP query evaluation which is not considered in many current CEP engines.

In this paper, we aim to exploit context aware nested complex event processing over RFID event streams. We make the following contributions: (1) We introduce context model into definition of nested event queries. (2) To evaluate context aware nested CEP queries, we transform a context aware nested event query into corresponding Non-determined finite automata (NFA), context information is transformed into context constraints. (3) To reduce partial instances (partial matches), we propose efficient context aware query plan rewritten rules to optimize query execution. Experimental comparisons with methods proposed in NEEL over different data sets verify effectiveness of our method.

Organization of this paper is as follows: related complex event processing works are introduced in Sect. 2; the event model and context model used in this paper are presented in Sect. 3; the context-aware nested CEP evaluation model is described in Sect. 4; experimental studies of the proposed method compared with NEEL are shown in Sect. 5; finally, the work is concluded in Sect. 6.

2   Related Works

CEP has been extensively studied in active database [15,16]. There are many event processing engines with different evaluation models. For instance, SASE [2,3] utilizes NFA to evaluate an event query. In SASE, event queries are parsed into different NFAs, and for a coming event, it may trigger many runs of different instances. ZStream [6] evaluates event queries over a tree model. Concerning RFID CEP optimization algorithms, Hirzel et al. [17] utilizes partition constructs to parallelize event detection. Wu [18] partitions events into round-robin manner so that each operator has an access to a shared state. Schneider et al. [19] evaluate event processing queries across a cluster of machines based on Cayuga. NEEL [10–14] proposes to define and evaluate nested CEP queries systematically. But their optimization based on subexpression sharing and caching still do not work well in a sophisticated RFID scenario especially when the context is of critical in system monitoring.

Recently some work about context-aware event processing has been proposed. Opher et al. [20,21] describe an event processing framework with context support in a common event processing system and they also define many kinds of context in CEP which we would utilize in nested CEP definition. Kulkarni [22] proposes context aware CEP framework and methods which utilizes ontology to model context. Teymourian et al. [23] also introduce ontology and declarative rules into event processing engines to detect complex event more intelligently. Cao et al. [24] focus on context-aware distributed complex event processing in applications of Internet of Things, but their model is a probabilistic model. Most of the works on context-aware RFID CEP lack a detail model and evaluation of context in the CEP engines.

3   Event Model and Context Model

3.1   Event Model

In an RFID application, an event is defined as occurrence of a reading of an RFID over a RFID tag. RFID event is usually in the form of <RID, TID, timestamp, otherattributs>, where RID is the reader identifier, TID is the tag identifier, timestamp is the reading time of the tag, and otherattributs are other attributes of the event. If an event cannot be divided into smaller events, it is called a basic/simple event. For example, an RFID reading e1 = (Shelf1, Tag101, 2016-01-01 20:35:21) is a basic event that indicates a tag, namely, Tag101 is read by reader at Shelf1 at the time 2016-01-01 20:35:21. This event cannot be divided into smaller events. Event type is one of the attributes of a simple event that indicates a specific type of an event. Take the above RFID event. The RFID identifier Shelf1 is the event type which means where the event occurs. In this paper, we simply use uppercase letters such as A, B, or C to denote event type.

Complex event operators are used to connect basic/complex events in order to form a new complex event. Generally, complex event operators used in CEP include: logic AND, logic OR, NOT, SEQ (sequence). The NOT operator (always uses "!" in event definition) constructor is a unary operator, AND and OR operators are binary operators, SEQ operator defines the occurrence order of the events, for example SEQ(A, B, C) [1 h] defines a sequence of events of types A, B and C occur in order ABC within an hour. Here [1 h] is a sliding window. Sliding window defines the lifespan of an event existing during event processing. Sliding window can be considered as a temporal context in CEP.

3.2   Event Specification Language

In this paper, we use NEEL [10] as event specification language. To support context definition in pattern specification, we extend the NEEL language with HAVING clause. The language has the following overall structure:

```
[PATTERN <event pattern>]
[WHERE <qualification>]
```

```
[HAVING <pattern filtering condition, context list>]
[WITHIN <window>]
```

in which, the PATTERN clause contains a sequence construct in particular order, whose components are the occurrences or non-occurrences of primitive events; the WHERE clause filters events through predicate constraints which involve attribute value comparison; the HAVING clause specifies context definition; the WITHIN clause specifies the sliding window during which the whole sequence of events should occur.

3.3   Context Model

Within event processing languages, contexts may be explicit, implicit, or partially explicit. Explicit context means that context primitives are first class primitives in the language [21]. For example, in NEEL, the sliding window is a temporal context which is defined explicitly in a patter definition. Some languages do not support any notion of context, and some support partial notion of contexts. A survey of contexts in various languages can be found in [21]. [21] shows the context dimensions: temporal, spatial, state oriented, and segmentation oriented as shown in Fig. 1.

In this paper, we mainly focus on temporal and spatial context dimensions, other dimensions are our future work. Some temporal and spatial contexts can be

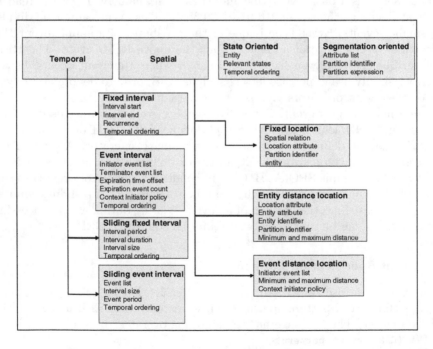

Fig. 1. Context dimensions [21]

explicitly found in event query specification in the form of event operator (SEQ) and clauses (where) indicating the time and spatial order of sub events in NEEL. But as shown in Fig. 1, not all the temporal and spatial context dimensions can be defined in NEEL without introducing the HAVING clause. Generally, there are four types of temporal contexts: fixed interval, event interval, sliding fixed interval and sliding event interval [20, 21]. Fixed temporal interval context utilizes one or more temporal intervals which are defined as event timestamp constants. This context can be a one time interval: [June 6 2015 14:20, June 6 2015 17:00); This context also can be stated as [June 6 2015 14:20, + 1.5 h), where Te (end of the interval) is an offset relative to Ts (start of the interval) [20, 21].

Event interval is a temporal interval which specifies the opened or closed time when one or more events occur. Note the meaning of "event occurs" is determined by the temporal ordering parameter and can be interpreted either as the detection time or the occurrence time as defined in pattern specification. The collection of events that open such an interval are called initiator (trigger) events, and the collection of events that close the temporal interval is called terminator events. An interval may also expire after a certain time offset is reached. For example, in RFID enabled hospital monitoring, temporal interval is initiated when an equipment is admitted to a hospital sanitation process and ends at the end of the sanitation process.

For sliding fixed interval context, windows are defined in NEEL using the WITHIN clause. New windows are sliding at regular intervals. Each window has a fixed size, defined either as a time interval (for example, 1 h) or a count of event instances (for example 1000 events).

For spatial context there are three classic types of spatial contexts: fixed location, entity distance location and event distance location [20, 21]. Fixed location context partitions the location of a reference entity into geo-fence. An event instance is classified into a context partition if its location's attribute correlates with the spatial entity. Entity distance location context defines one or more context partitions, based on the distance between the event's location attribute and some other entity. Note that distance means the shortest distance between two spatial entities. Event distance location context specifies an event type and a matching expression predicate. For example, detecting the shopping activity of an item within a 10 h after detection of expiration of that batch of items.

3.4   Motivation Example

In this section, we present a motivation example to illustrate context aware nested CEP query. The scenario used in this paper is: suppose in an RFID enabled supply chain monitoring system, objects with an RFID tag move from one reading point to another point following pre-defined business flow, and the traces of different kinds of objects need to be monitored continuously in real time as shown in Fig. 2. Suppose the monitored query is shown as Query 1 in Fig. 2. Query 1 tends to detect complex events $SEQ(A; !AND(D, E); B; C)$ over RFID streams. This event query is a nested definition. In NEEL [10–14], there does not exist the "HAVING" clause, so evaluation of $SEQ(A; !AND(D, E); B; C)$ would

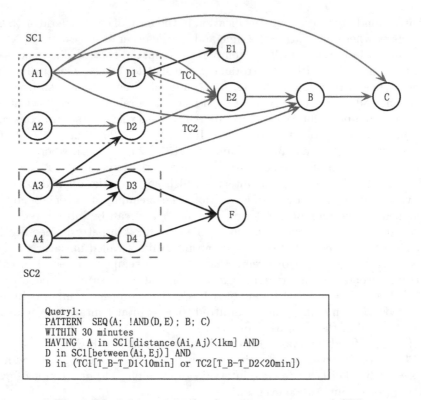

Fig. 2. Motivation example of context aware nested CEP

be: first detect pattern $SEQ(A; B; C)$, and to verify if there exists complex pattern $AND(D, E)$, if there is a corresponding $AND(D, E)$, the complex event $SEQ(A; B; C)$ is deleted (not fulfilling the constraint), else output a complex event. Note that, during CEP, many partial matches need to be maintained until the end of the sliding window. In our motivation example, any event of type A would trigger a instance of the subexpression $SEQ(A; B; C)$ and any event of type D or C would also trigger instances of $AND(D, E)$. So there would exist many potential instances during evaluation of nested RFID event queries which would incur great system overhead in a high speed realtime decision making monitoring system. In NEEL [10–14], although authors propose optimization methods such as caching and partition to reduce partial matches. However, they does not consider context information during event detection, which, we think, would be critical in nested CEP evaluation. Note that in this scenario, we use A_i and D_i to represent different reading points of the same event types, respectively. Spatial and temporal contexts in this scenario are denoted as SC_i and TC_i. Funtions $distance()$ and $between()$ are spatial context operators. T_B and T_{D1} are used to denote arriving time of a specified event at reading point B and D1. Due to space limitation, some details are omitted.

4   Context Aware Nested Complex Event Detection Model

4.1   NFA Model of Context Aware Nested CEP

As a nested complex event is defined using declarative languages NEEL, NFA is a natural execution model which is widely used in many CEP engines. Figure 3 shows NFA of Query1.

As shown in Fig. 3, a nested CEP query is parsed into corresponding NFA. Event types are transformed into NFA states, sliding window constraint is transformed into time discrepancy between different event types. Negation nested subexpression is transformed into small group pattern. Note that spatial and temporal contexts are evaluated with the help of distance matrix and betweeness lists which we can predefine in an RFID scenario.

Generally, NFA-based CEP evaluation model is implemented into stack-based execution model. In this model, nested query evaluation suffers from several inefficiencies. First, partial results of SEQ(A;B;C)generated may be discarded later. Another potential overhead is that complete matches for the negation event AND(D;E) are constructed. Iterative execution method in NEEL does not solve these problems [10]. To overcome these problem, we utilize rewriting techniques to optimize context aware nested CEP queries.

4.2   Rewriting Rules for Context Aware CEP

As we introduce context into evaluation of nested event queries, we exploit rewriting context to reduce instance numbers and earlier partial matches pruning. Rewriting rules for nested event queries are described as follows.

Spatial-Share Rule. This rule transforms sequence operator with other operators in a single sequence operator if they have the same spatial context. Transformation follows the operators priorities. For Query1, $SEQ(A; !AND(D, E); B; C)$

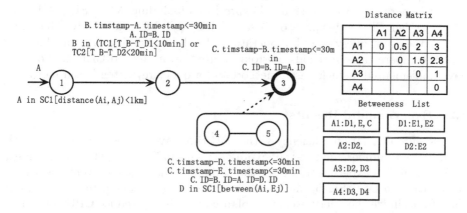

Fig. 3. NFA of the motivation example

is rewritten as $SEQ(SEQ(A;!D));!E;B;C)$. With this rewritten rule, instances that match SEQ(A;D) would be deleted earlier during event detection.

Temporal-Share Rule. This rule transform event types into the same temporal context when evaluation begins. For Query1, $SEQ(A;!AND(D,E);B;C)$, as we define temporal context over events of type B and D, so we transform Query1 into $SEQ(SEQ(A;!D;B[10\,min]));!E;C)$. This transformation means if events time discrepancy between A and B is less than 10 min, we would omit the evaluate of !AND(D,E); otherwise, we still need to wait for the verification of !AND(D,E) before output of complex event. Due to space limitation, we omit formal semantics of these rewriting rules.

5   Experimental Evaluation

To evaluate the proposed method, we have implemented a prototype CEP engine using C++. We compared our methods with the corresponding methods of NEEL. Comparisons are made between the iterative processing technique, the alternative caching techniques of NEEL on query execution time. Experimental event streams used in this paper is explained in the next section.

5.1   Experiment Data Description

As we do not have real RFID scenario event streams, we generated event stream in our motivation example. We simulate objects's moving route in an RFID enabled supply chain. Locations are denoted by (A,B,C,D,E,F). An object's moving route is simulated by a walk on the work flow graph consisting of these nodes. The elapsed time of moving between two locations are set with normal distribution within an interval $[t_1, t_2]$. In a supply chain, spatial and temporal context of some reading points are pre-defined and kept as matrix or list in the CEP engine. Nested event queries over RFID streams are generated with spatial and temporal context number changed with number 2, 4, and 6. We have utilized different operators in the nested subexpressions including AND, SEQ, OR and Negation, length of subexpression is set with 2, 4 and 6 respectively. The outer level of the event query is a SEQ operator. The sliding window is set at different size to compare these methods. For our context aware nested CEP method, we have implemented a stack-based method and a query rewritten based method.

5.2   Experimental Results

We measure evaluation efficiency with CPU time. We first generate event with fixed context number and fixed subexpression length. Experiment result is shown in Fig. 4. In this experiment, we set the sliding window as 100 events per slide. As we can see from Fig. 5, our method utilized context based query rewritten rules to optimize the query execution plan which greatly reduce CPU time. The NEEL-caching method performs better than our stack based method because

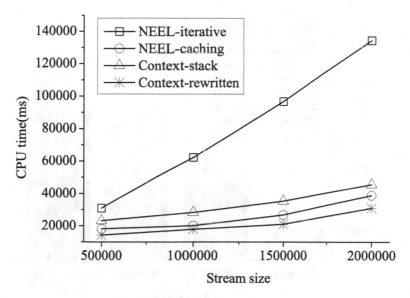

Fig. 4. Comparison with methods in NEEL (sliding window size 100, context number: 2, length of nested subexpression: 2)

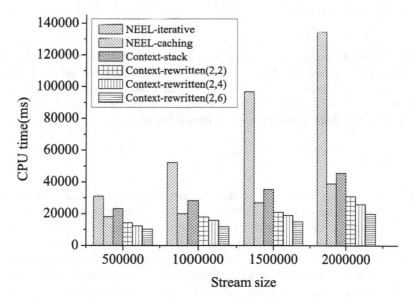

Fig. 5. Comparison with methods in NEEL (sliding window size 100, context number: 2, 4, 6, length of nested subexpression: 2)

caching utilizes effective partial match caching and discarding mechanism than our stack based method. We also tuned the context length and subexpression length parameters and generated different sizes of streams to evaluate these methods. Figures 5 and 6 are the performance comparisons with different para-

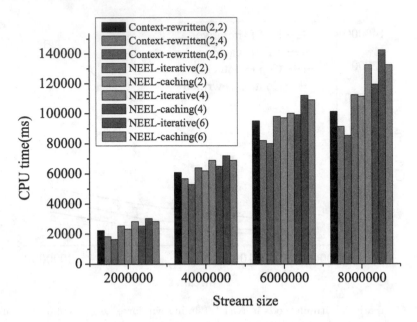

Fig. 6. Comparison with methods in NEEL (sliding window size 100, context number: 2, length of nested subexpression: 2, 4, 6)

meters. As we can see from Figs. 5 and 6, when context numbers and subexpression become larger, context based methods outperform other method because as context number increases, the instances generated and deleted during event detection are reduced greatly. When subexpression length become larger, partial matches that fulfill the subexpression pattern become less.

6   Conclusion

In this paper, we try to exploit context information to optimize evaluation of nested RFID complex event processing. We extend context semantic into NEEL language, and we propose transformation rules to incorporate context into the nested query evaluation model. Experimental analysis compared with methods proposed in NEEL verifies effectiveness of our proposed context aware method. However, some our work is not formalized in this paper, and we also need to experiment our method over some industry scenario and data sets.

References

1. Luckham, D.C.: The Power of Events: An Introduction to Complex Event Processing in Distributed Enterprise Systems. Addison-Wesley, Boston (2002)
2. Wu, E., Diao, Y.L., Rizvi, S.: High-performance complex event processing over streams. In: SIGMOD, pp. 407–418 (2006)

3. Zhang, H., Diao, Y., Immerman, N.: Recognizing patterns in streams with imprecise timestamps. PVLDB **3**(1), 244–255 (2010)
4. Nie, Y., Cocci, R., Cao, Z., Diao, Y., Shenoy, P.J.: SPIRE: efficient data inference and compression over RFID streams. IEEE Trans. Knowl. Data Eng. **24**(1), 141–155 (2012)
5. Zhang, H., Diao, Y., Immerman, N.: On complexity and optimization of expensive queries in complex event processing. In: International Conference on Management of Data, SIGMOD 2014, Snowbird, UT, USA, 22–27 June 2014, pp. 217–228 (2014)
6. Mei, Y., Madden, S.: Zstream: a cost-based query processor for adaptively detecting composite events. In: SIGMOD (2009)
7. Brenna, L., Demers, A., Gehrke, J., et al.: Cayuga: a high-performance event processing engine (demo). In: SIGMOD (2007)
8. Demers, A., Gehrke, J., Hong, M., et al.: Cayuga: a general purpose event monitoring system. In: CIDR (2007)
9. Barga, R.S., Goldstein, J., Ali, M.H., Hong, M.: Consistent streaming through time: a vision for event stream processing. In: CIDR 2007, Third Biennial Conference on Innovative Data Systems Research, Asilomar, CA, USA, 7–10 January 2007, Online Proceedings, pp. 363–374 (2007)
10. Liu, M., Rundensteiner, E.A., Dougherty, D.J., Gupta, C., Wang, S., Ari, I., Mehta, A.: High-performance nested CEP query processing over event streams. In: Proceedings of the 27th International Conference on Data Engineering, ICDE, 11–16 April 2011, Hannover, Germany, pp. 123–134 (2011)
11. Liu, M., Ray, M., Rundensteiner, E.A., Dougherty, D.J., Gupta, C., Wang, S., Ari, I., Mehta, A.: Processing nested complex sequence pattern queries over event streams. In: Proceedings of the Seventh International Workshop on Data Management for Sensor Networks, DMSN 2010, pp. 14–19. ACM, New York (2010)
12. Ray, M., Liu, M., Rundensteiner, E.A., Dougherty, D.J., Gupta, C., Wang, S., Mehta, A., Ari, I.: Optimizing complex sequence pattern extraction using caching. In: Workshops Proceedings of the 27th International Conference on Data Engineering, ICDE, 11–16 April 2011, Hannover, Germany, pp. 243–248 (2011)
13. Liu, M., Ray, M., Zhang, D., Rundensteiner, E.A., Dougherty, D.J., Gupta, C., Wang, S., Ari, I.: Realtime healthcare services via nested complex event processing technology. In: 15th International Conference on Extending Database Technology, EDBT 2012, Berlin, Germany, 27–30 March 2012, Proceedings, pp. 622–625 (2012)
14. Ray, M., Rundensteiner, E.A., Liu, M., Gupta, C., Wang, S., Ari, I.: High-performance complex event processing using continuous sliding views. In: Joint 2013 EDBT/ICDT Conferences, EDBT 2013 Proceedings, Genoa, Italy, 18–22 March 2013, pp. 525–536 (2013)
15. Chakravarthy, S., Krishnaprasad, V., Anwar, E., Kim, S.: Composite events for active databases: semantics, contexts and detection. In: VLDB, pp. 606–617 (1994)
16. Gatsiu, S., Dittrich, K.R.: Events in an active object-oriented database system. In: International Conference on Rules in Database Systems, pp. 23–39 (1993)
17. Hirzel, M.: Partition and compose: parallel complex event processing. In: DEBS, pp. 191–200. Citeseer (2012)
18. Wu, S., Kumar, V., Wu, K.L., Ooi, B.C.: Parallelizing stateful operators in a distributed stream processing system: how, should you and how much? In: Proceedings of the 6th ACM International Conference on Distributed Event-Based Systems, pp. 278–289. ACM (2012)
19. Schneider, S., Hirzel, M., Gedik, B., Wu, K.L.: Auto-parallelizing stateful distributed streaming applications. In: Proceedings of the 21st International Conference on Parallel Architectures and Compilation Techniques, pp. 53–64. ACM (2012)

20. Etzion, O., Magid, Y., Rabinovich, E., Skarbovsky, I., Zolotorevsky, N.: Context-based event processing systems. In: Helmer, S., Poulovassilis, A., Xhafa, F. (eds.) Reasoning in Event-Based Distributed Systems. SCI, vol. 347, pp. 257–278. Springer, Heidelberg (2011)
21. Etzion, O., Niblett, P.: Event Processing in Action, 1st edn. Manning Publications Co., Greenwich (2010)
22. Taylor, K., Leidinger, L.: Ontology-driven complex event processing in heterogeneous sensor networks. In: Antoniou, G., Grobelnik, M., Simperl, E., Parsia, B., Plexousakis, D., De Leenheer, P., Pan, J. (eds.) ESWC 2011, Part II. LNCS, vol. 6644, pp. 285–299. Springer, Heidelberg (2011)
23. Teymourian, K., Paschke, A.: Enabling knowledge-based complex event processing. In: Proceedings of the 2010 EDBT/ICDT Workshops, EDBT 2010, pp. 37:1–37:7. ACM, New York (2010)
24. Cao, K., Wang, Y., Wang, F.: Context-aware distributed complex event processing method for event cloud in internet of things. Adv. Inf. Sci. Serv. Sci. 5(8), 1212 (2013)

Using Convex Combination Kernel Function to Extract Entity Relation in Specific Field

Qi Shang[1], Jianyi Guo[1,2]($\boxtimes$), Yantuan Xian[1,2], Zhengtao Yu[1], and Yonghua Wen[1]

[1] The School of Information Engineering and Automation,
Kunming University of Science and Technology, Kunming 650500, China
gjade86@hotmail.com
[2] Computer Technology Application Key Laboratory of Yunnan Province,
The Institute of Intelligent Information Processing, Kunming 650500, China

Abstract. Kernel method has been proven to be effective in measuring the similarity of two complex relation patterns. Aim at the optimization problem of compound kernel functions, this paper presents a method of finding the optimal convex combination kernel function, which is comprised of multiple kernel functions and needs to be optimized. After preprocessing the corpus and selecting features including lexical information, phrases syntax information and dependency information, the feature matrix was constructed by using these features. The optimal kernel function can be found in the process of mapping the feature matrix to different high-dimensional matrix, and the different classification models can be obtained. The experiments are conducted on the domain dataset from Web and the experimental results show that our approach outperforms state-of-the-art learning models such as ME or Convolution tree kernel.

Keywords: Entity relation extraction · Compound kernel functions · Optimization · Convex combination of kernel functions

1 Introduction

Entity relation extraction refers to the automatic identification the associated relation between two entities expressed with the natural language (e.g. in the sentence "Yunnan produces a lots of wild mushrooms", the relationship between entity "Yunnan" and "mushroom" is "production" relations). Entity relation extraction has made an important role in information extraction, automatic question answering systems, machine translation and knowledge-base construction [1].

The current methods for relation extraction mainly include: knowledge-based [2], the pattern matching [3] and machine learning-based [4–12]. Knowledge-based method requires experts build different knowledge-bases on different specific areas, this method is time-consuming and labor-intensive and has many limitations. If an entity does not exist in the knowledge base, the relationship extraction cannot be carried out. Some common knowledge repositories have now been built up [15–18], while specific areas knowledge-bases are less. Pattern matching method is based on the words located

© Springer International Publishing AG 2016
S. Song and Y. Tong (Eds.): WAIM 2016 Workshops, LNCS 9998, pp. 137–148, 2016.
DOI: 10.1007/978-3-319-47121-1_12

before or after the given entity in a sentence, matching these words or the syntax format to the target pattern. But in the process of matching, due to the position of the words, which were usually disaccord with the given pattern that may result in a very low similarity. Especially for Chinese, because of the complexity of the language, accuracy and recall rate of this method is very low.

Machine learning methods for entity relation extraction can be divided into several ways: features-based method [4, 5], bootstrapping-based method [6, 7], deep learning method [8], and kernel methods [9–14]. The key of features-based method is the effectiveness of the selected features. For example, Dong (2007) [5] selected a syntax tree, lexical features, physical features and other features as well as their combination, using CRF model, performed relation extraction between inclusion and non-inclusion. However, it is difficult for this method to adapt to another relation extraction system, and its disadvantage lies in the difficultness of discovery of new plane features to improve the extraction performance. The bootstrapping is used to learn relation patterns. For example, Ye (2014) [6] took the bootstrapping to extract entity relation. However, Komachi' analysis in his paper [7] showed that semantic drift is an inherent property of iterative bootstrapping algorithms and, therefore, poses a fundamental problem. As a new machine learning algorithms, deep learning has made remarkable achievements in speech recognition and image processing. Recently it has been widely studied and applied on natural language processing field. Zeng (2014) in his paper [8] proposed to have used convolution method on DBN to predict the relationship between the two marks word. But the limitation of deep learning requires a lot of corpus for the experiment.

2   Related Work

Kernel methods [9–14] have been proven to be effective in measuring the similarity of two complex relation patterns.

For extracting entity relationship in the specific field, different kernel functions for different relation extraction have differences, the usage of a single kernel cannot solve the universal problem but the compound kernel methods, which may reduce the risk caused by the loss of important features by combining different single kernel function to be a combinatorial function. However, the composite kernel functions are not optimalizing. This paper proposes a way of optimizing the compound kernel to improve extracting performance by finding the optimal kernel function. The target kernel function is formed by convex combination of each single kernel function [13, 14], and the lexical and sentence information such as context words, part of speech, phrase syntactic and dependency syntax information extracted from the pretreated corpus are all used to form the feature matrix, and then the feature matrix is mapped to the different convex functions which are made of the combination among radial basis function (RBF), linear kernel function (LKF) or the polynomial kernel. Function (PKF) The classification model can be obtained by machine-learning method which supporting the kernel function, enumerating the classified result by making use of the testing corpus to find the optimal model. At last the entity relation in specific areas can be identified with the optimal model of convex combination kernel.

The task of entity relation extraction in this paper is divided into four parts: pre-processing the corpus, extracting feature to form feature matrix, mapping kernel function to high dimensional matrix, learning different classification model to find the optimal classification model (e.g. in tourism field). As shown in Fig. 1.

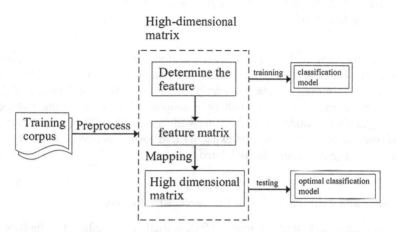

Fig. 1. General framework for entity relation extraction in specific fields

2.1 Preprocess

Preprocess includes segmentation, speech tagging, named entity recognition, sentence segmentation, relationship candidate generation and label candidate instances.

First, this paper employs ICTCLAS [19] tool from Chinese Academy of Sciences for the segmentation and speech tagging of the input text; Secondly, artificial tagging of named entities, which will be conducted on the corpus that has completed segmentation and speech tagging. In compliance with rules of conditional random fields (CRFs), the training of named entity recognition model should be performed, so as to achieve automatic recognition of named entities. Meanwhile, it requires pre-defined categories of entities, specifically: sights, location, snacks, specialties, hotels, numeric expressions, dates and festivals. Then, based on punctuation and contextual features, tagged corpus should be cut into separate sentences. It should be noted that types of entity relationships also need to be pre-defined, (for example, located, apart, adjacent, these words belong to location relationship, or ticket prices, altitudes, customs, these words belong to attribute relationship). Take the sentence "In 2004, Huangshan was first selected as a World geological park, making itself a tourist destination which won honors of World Cultural Heritage and Natural Heritage and World Geological Park at the same time." Its segmentation result is: In 2004/m, /w Huangshan /ns was first /m selected /v as a World Geological Park /n, /w making /v itself a tourist destination /n which /w won /v honors of /n World Cultural /n and /c Natural Heritage /n and/c World Geological Park /n at the same time/c. Among them, entities are "Huangshan", "World Geological Park", "World Cultural Heritage", "Natural Heritage", "tourist destination", whose relationship is shown in Table 1.

Table 1. Entity relationship

Entity 1	Entity 2	Relationship
Huangshan	World Geological Park	Listed (in 2004)/whole-part relation
Huangshan	tourist destination	whole-part relation
Huangshan	World Cultural Heritage	whole-part relation
Huangshan	Natural Heritage	whole-part relation

After enumerating and finding all possible combinations of entities in each sentence, each combination has become a candidate example, whose artificial tagging will be carried out according to whether relationships exist. With 500 pieces of data as artificial tagging examples, there will be examples of relations being marked out. Employing CRFs for model training, along with a test on 1000 pieces of data in the model, instances of relations will be extracted and form a desired corpus. By way of previous steps, pre-processing is completed.

2.2   Extracting Features to Form Characteristic Matrix

In this paper, the selection of features will follow methods introduced in the References [4, 16]. In the field of Chinese tourism, it focuses on features of syntax and dependency information of phrases, with an expectation to optimize the performance of relation extraction. After completing the selection of features, according to which features matrix can be formed, such as the formula (1), where $X_i (i = 1, 2, \ldots m)$ is the vector of each example after characteristic extraction.

$$K = (X_1, X_2, \ldots X_m)^T \tag{1}$$

2.2.1   Lexical Information

(1) Entity Information. Entity information is the basic vocabulary information, including the first entity category, the first entity subcategory, the first entity's syntactic functions, the second entity's category, the second entity's subcategory and the second entity syntactic functions.

(2) Local Contextual Information of Words. Referred literature [16] has verified that the window of lexical features should not be too large, in order to prevent excessive noise. Usually, windows of 2-2 are selected. This paper chooses 2-3-2 mode, that is, select two words on entity one's left, two words on entity two's right and three words between the entities as features.

(3) Inclusion Information. Inclusion information mainly reflects the lexical information and inclusion relations between entities. This paper select the number of inter-entity vocabulary, number of entities and whether the entity is one with inclusion relationship as nested information.

2.2.2 Phrases Syntax Information

The syntax tree of phrases reflects the grammatical structure of sentences and expresses the semantic information over long distances. In the sentence "In 2004, Huangshan was first selected as a World geological park, making itself a tourist destination which won honors of World Cultural Heritage and Natural Heritage and World Geological Park at the same time.", its minimal complete syntax tree is shown in Fig. 2.

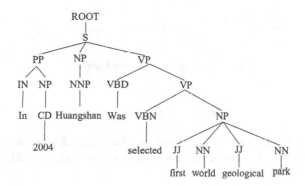

Fig. 2. Minimal complete syntax tree of an entity

Minimal complete syntax tree refers to one whose root node is the nearest public root node between two entities. Since the minimal complete syntax tree contains certain contextual information and reduces noise to some degree, this paper uses the minimal complete syntax tree to extract features. On account of the fact that paths between two entities are too specific in the syntax tree, the problem of sparse data is likely to occur. In order to avoid it, this paper regards the number of nodes in two entities' paths and the types of two entities' root nodes as features. Due to rather specific structural information contained in the syntax tree, low recall rate are to follow in addition to data sparseness.

2.2.3 Dependency Information

Dependency tree reveals long-distance dependencies of sentences, avoids the noise occurring in unstructured features, and provides more useful information for relationship extraction. For the sentence "In 2004, Huangshan was first selected as a World geological park, making itself a tourist destination which won honors of World Cultural Heritage and Natural Heritage and World Geological Park at the same time.", the dependency tree is shown in Fig. 3. Likewise, the distribution of structural information in the dependency tree is very specific, and it may generate the problem of low recall rate.

2.3 Improved Radial Base of Training Matrixes

RBF kernel function is a kernel function invariant in translation, and its concrete expression is shown as the formula (2)

Fig. 3. Example of a dependency tree

$$rbf(x) = \exp(-a \cdot x + b) \tag{2}$$

When the coefficient a = 1, b = 0, the radial basis function is shown in Fig. 4. As can be seen from Fig. 4, when x increases, the function value quickly reaches extremely close to zero.

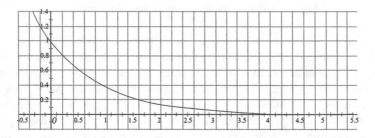

Fig. 4. Curve of radial basis functions

For two random feature vectors X_i, X_j, i, j•1,2,…, m, their maps form the j-th feature of i-th vectors in the new map space. Due to the property of radial basis functions, each character approaches very close to 0, making it very detrimental to classification. For example, there are three two-dimensional vectors a, b, c, respectively: a = (2, 3); b = (4, 10); c = (3, 5); the training matrixes of their corresponding space that they mapped to the radial basis functions are:

$$K_{a,b,c} = \begin{bmatrix} 0 & 9.602 \times 10^{-24} & 6.738 \times 10^{-3} \\ 9.602 \times 10^{-24} & 0 & 5.109 \times 10^{-12} \\ 6.738 \times 10^{-3} & 5.109 \times 10^{-12} & 0 \end{bmatrix} \tag{3}$$

As it can be seen, in the new feature space, three new vectors are very close to the origin, but data like these are quite unfavorable for classification. In order to solve problems mentioned above, this paper employs a method that limits each feature of the training matrix within an appropriate range to facilitate classification. For example, the range is limited to between 0.2 and 1. The standardized method of training matrix is:

(1) Calculate $Feature_{ij} = (Xi - Xj) \cdot (Xi - Xj)^T$, among them, $i,j \in 1,2,..., m$;
(2) Enumerate each $Feature_{ij}$, find the maximum value, denoted as $Feature_{max}$;
(3) The matrix is multiplied by a constant $\delta = (-ln_{0.2})/Feature_{max}$;
(4) The training matrix is standardized as:

$$K_{new} = [k(Feature_{ij} \times \delta)]_{m \times m} \tag{4}$$

Among them, $i,j \in 1,2,...,m$。

If three vectors mentioned previously, a, b, c, have been standardized, the training matrix will be:

$$K_{new} = \begin{bmatrix} 0 & 0.2000 & 0.8593 \\ 0.2000 & 0 & 0.4541 \\ 0.8593 & 0.4541 & 0 \end{bmatrix} \tag{5}$$

It can be perceived that three standardized vectors are more distant than that without standardizing in spacing.

2.4 Getting the Optimal Convex Combination Kernel Function

In terms of varied Chinese relation extraction, there are differences in effects for different kernel functions. In order to make kernel functions perform good adaptability on characteristic represented by different pieces of information, this paper will make convex combination of different single-core functions, expecting that the optimal convex combination of kernel functions also possess good adaptability. According to different expressions of kernel functions, they can be divided into translational invariant kernel functions and inner product kernel functions, and their expressions are: $k(x,y) = f(x - z)$ and $k(x,y) = f(<x,z>)$, respectively.

In order to fuse these two kernel functions' features in the relationship extraction, this paper focuses on three single-core functions, which include radical kernel functions, polynomial kernel functions and linear kernel functions, as shown in formulas (6), (7) and (8), where x, y are two vectors with arbitrary dimension.

$$K_{RBF}(x,y) = \exp(-|x - y|^2) \tag{6}$$

$$K_{polynomial}(x,y) = (x \cdot y^T)^3 \tag{7}$$

$$K_{linear}(x,y) = x \cdot y^T \tag{8}$$

The compound of various kernel functions offers a solution to the problem that the single-core functions do not have universal properties. For convex combination kernel functions, determining parameters of the combination ratio is of great importance. Suppose there are n kinds of kernel functions $k1,..., kn$, parameters options of combination ratios may have m kinds; then enumerate all the computational complexity of the kernel functions as $O(m^n)$. And the n kinds of kernel functions of convex combination kernel functions are shown in formula (9), where a_i is called convex combination parameter, $\sum_{i=1}^{n} a_i = 1$.

$$CCK = \sum_{i=1}^{n} a_i k_i \tag{9}$$

This paper presents a strategy that by constantly looking for the optimal convex combination kernel function of these two kernel functions based on some principles to substitute these two kernel functions, so as to find out the optimal convex combination kernel function of various kernel functions. This strategy can reduce computational complexity to $O(m*(n-1))$. The algorithm to get the optimal convex combination of kernel functions is shown in Fig. 5.

Generation Algorithm of Optimal Convex Combined Kernel Functions

Algorithm Input: n kinds kernel functions PPI extracted performance collection F

Algorithm Output: Optimal Convex Combined Kernel Functions (OCCK)

Algorithm Steps:

While(number(F)>1)

candidate1 ←min(F)

F=F-min(F)

candidate2←min(F)

F=F-min(F)

F=F+Optimal(candidate1,candidate2)

if number(F)=1

then

Return Optimal(candidate1,candidate2)

Δnumber(F) is the number of F_i in collection F, min(F) is the minimum in collection F;

ΔOptimal (k1,k2) is the optimal kernel function of kernel functions k1,k2

Fig. 5. Generation algorithm of optimal convex combination of kernel functions

After the kernel function is determined, feature matrix can be mapped to high-dimensional matrix by formula (9).

2.5   Finding the Optimal Classification Model

Via Sect. 2.4, a high-dimensional matrix which is mapped by different convex combinations of the kernel functions can be acquired. It is not probable to tell classification model obtained from high dimensional matrix training possess better extraction per-formance in entity relations, that is, which convex combination of kernel functions has better adaptability, by direct observation of high dimensional matrixes. In order to get the optimal convex combination of kernel function, it is a top priority to train each high-dimensional matrix and to obtain the corresponding training model. By running tests on the corpus, a subsequent step is to test out a training model with the best performance in extraction of entity relations. The corresponding kernel function of this training model is the optimal convex combination of kernel function.

3 Experimental Results and Discussion

3.1 The Experimental Dataset

The corpus used in this paper has been manually acquired by us from the Web and relevant literature, more than 1000 Chinese tourism texts. Among various machine learning algorithms applicable to kernel functions in the field of entity relation extraction, support vector machines (SVMs) have the best performance. So this paper chooses SVM as the machine learning algorithms, taking LIBSVM developed by Lin Zhiren from Taiwan University as SVM tools [20]. In the analysis of phrases, Stanford Parser [21] is used to conduct phrases parsing and dependency study, selecting prob-ability context-free grammars. CCprocessed dependency expression is used to form the dependency information. As for training, a method of 10-fold cross-validation is employed to maximize the use of data. The performance is measured by precision, recall, and F-measure.

$$F = \frac{2 \times P \times R}{P + R} \times 100\% \tag{10}$$

3.2 Experiments

Experiment 1, on the lexical information basis of word, speech, etc. in contexts of entities, the syntactic information and dependency syntactic information of phrases are added as features, to find the impacts on the extraction performance of entity relation in Chinese. In the test, different features are added, with a purpose to find the optimal feature combination. Parameters in formula (9), $\alpha_1 = 1$, $\alpha_2 = 0$, $\alpha_3 = 0$ are chosen as the weighting parameter of convex combined kernel functions.

Experiment 2, comparing convex combined kernel functions composed of three single-core functions with single kernel functions and combinations of two kernel functions in relation extraction performance, In this test, coefficients of the convex combination in formula (9) are set, with the upper and lower limits from 0 to1, steps of 0.1, 11 options in total. By means of enumeration, the best coefficients can be deter-mined. For n kinds of single-core functions, it requires a calculation of $11 \times (n - 1)$ times to find the best convex combination.

Experiment 3, in order to verify the effectiveness of the system proposed in this paper, a comparison with other methods has been conducted under the same corpus used in this paper.

3.3   Results and Discussion

3.3.1   Performance of Different Features in Relation Extraction

As it can be seen from Table 2, the entity relation extraction performance is unsatisfactory when only lexical information is selected as features in the field of Chinese tourism. After syntax and dependency information of phrases are added, the extraction performance improves, particularly with an increase in recall rate. It indicates that it is an effective method, which increases the performance of entity relation extraction in the field of Chinese tourism, to include syntax and dependency syntactic information of phrases to features.

Table 2. Performance under different features

Features	Accuracy	Recall Rate	F value
Lexical information	61.3 %	48.4 %	53.7 %
+ Syntactic information of phrases	62.6 %	52.7 %	57.1 %
+ Dependency of syntactic information	62.9 %	53.5 %	57.8 %

3.3.2   Comparison of Relation Extraction Performance Between Three Single-Core Functions and Convex Combined Kernel Functions

As Table 3 shows, in the entity relation extraction performance of single kernel function concerning Chinese tourism, polynomial kernel functions(PKF) have highest accuracy and F value; Radial kernel functions(RBF) have the highest recall rate; the performance of linear kernel function(LKF) is rather poor. These have also verified that single-core functions have differences in extraction performance in regard to the same feature matrix. In combinations of two convex combinations of kernel functions, the convex combination kernel functions, consisted of radial basis kernel functions and polynomial kernel functions, have the best performance, proving that the performance of convex combination bases has a positive impact on the performance related to convex combination of kernel functions. The best extraction performance is the convex combination of three single-core functions. And the performance of convex combinations of two single-core functions is superior to that of single-core functions. It can be described that convex combinations of single-core functions can solve the problem that single-core functions do not possess universality property.

3.3.3   Comparison with Other Methods

Table 4 suggests that the use of maximum entropy method in the field of Chinese tourism for relation extraction generates the highest accuracy. Using the shortest path tree in convolution tree kernel functions in relation extraction can get the best recall rate.

Table 3. Performance of single-core functions and convex combination kernel function

Categories of kernel functions	Accuracy	Recall Rate	F value
RBF	62.7 %	53.4 %	57.65 %
PKF	70.1 %	49.5 %	57.78 %
LKF	65.8 %	51.9 %	58.1 %
RBF + LKF	63.7 %	53.9 %	58.5 %
RBF + PKF	72.6 %	55.5 %	62.9 %
LKF + PKF	70.5 %	49.7 %	58.3 %
RBF + LKF + PKF	72.6 %	56.1 %	63.4 %

The method of optimal convex combination of kernel functions proposed in this paper has better recall rate and accuracy and the best F value, which fully verifies the effectiveness of the proposed optimal convex combination method.

Table 4. Comparison with other methods

Methods	Accuracy	Recall Rate	F value
Maximum entropy	74.1 %	48.9 %	58.9 %
Convolution tree kernel	61.3 %	58.9 %	60.1 %
Proposed method	72.6 %	56.1 %	63.4 %

4 Conclusions

Based on the lexical information of word, speech, etc. in entity contexts, this paper includes syntactic and dependency information of phrases as features. With different convex combinations of radial kernel functions, polynomial kernel functions and linear kernel functions, the feature matrix is mapped to a different high-dimensional matrix. To obtain different classification models, training support vector machine is used; and the optimal performance of the classification model is found by enumerating. Finally, the classification model is used in the field of Chinese entities' relation extraction. In the field of tourism, the optimal convex combination kernel extraction system proposed in this paper has achieved an F value of 63.4. In the further studies, efforts will be made to unveil other useful information as features, to find the deep relationship between kernel functions and the corpus, and to further improve the performance of relation extraction concerning entities in Chinese.

Acknowledgments. This work was supported in part by the National Natural Science Foundation of China (Grant Nos. 61262041, 61472168 and 61562052) and the key project of National Natural Science Foundation of Yunnan province (Grant No. 2013FA030).

References

1. Zhao, J., Liu, K., Zhou, G.Y.: Open information extraction. J. Chin. Inf. Process. **25**(6), 98–110 (2011)
2. Aone, C., Ramos-Santacruz, M.: Rees: A large-scale relation and event extraction system. In: Proceedings of the 6th Applied Natural Language Processing Conference, pp. 76–83. ACM Press, New York (2000)
3. Califf, M.E., Mooney, J.: Bottom-up relational learning of pattern matching rules for information extraction. J. Mach. Learn. Res. **4**, 177–210 (2003)
4. Zhou, G., Su, J., Zhang, J.: Exploring various knowledge in relation extraction. In: ACL, June 2005, pp. 427–434 (2005)
5. Dong, J., Sun, L., Feng, Y.Y.: Chinese automatic entity relation extraction. J, Chin. Inf. Process. **21**(4), 80–85 (2007)
6. Ye, F., Shi, H., Wu, S.: Research on pattern representation method in semi-supervised semantic relation extraction based on bootstrapping. In: 2014 Seventh International Symposium on Computational Intelligence and Design (ISCID). IEEE(2014)
7. Komachi, M., Kudo, T., Shimbo, M., Matsumoto, Y.: Graph-based analysis of semantic drift in espresso-like bootstrapping algorithms. In: Proceedings of the Conference on Empirical Methods in Natural Language Processing. Association for Computational Linguistics, pp. 1011–1020 (2008)
8. Zeng, D., Liu, K., Lai, S.: Relation classification via convolutional deep neural network. In: Proceedings of COLING (2014)
9. Liu, K.B., Li, F., Liu, L., Han, Y.: Implementation of a kernel-based Chinese relation extraction system. J. Comput. Res. Dev. **44**(8), 1406–1411 (2007)
10. Zhuang, C.L., Qian, L.H., Zhou, G.D.: Research on tree kernel-based entity semantic relation extraction. J. Chin. Inf. Process. **23**(1), 3–9 (2009). ISSN: 1003-0077
11. Yang, Z., Tang, N., Zhang, X., et al.: Multiple kernel learning in protein–protein interaction extraction from biomedical literature. J. Artif. Intell. Med. **51**(3), 163–173 (2011)
12. Peng, C., Gu, J., Qian, L.: Research on tree kernel-based personal relation extraction. In: Zhou, M., Zhou, G., Zhao, D., Liu, Q., Zou, L. (eds.) NLPCC 2012. CCIS, vol. 333, pp. 225–236. Springer, Heidelberg (2012). doi:10.1007/978-3-642-34456-5_21
13. Arenas-García, J., Martínez-Ramón, M., Gómez-Verdejo, V., Figueiras-Vidal, A.R.: Multiple plant identifier via adaptive LMS convex combination. In: Proceedings of the IEEE International Symposium on Intelligent Signal Processing, Budapest, Hungary, pp. 137–142 (2003)
14. Arenas-García, J., Figueiras-Vidal, A.R., Sayed, A.H.: Mean-square performance of a convex combination of two adaptive filters. IEEE Trans. Signal Process. **54**(3), 1078–1090 (2006)
15. Knowloge- base, CYC. http://www.cyc.com/2008
16. Miller, G.: Introduction to wordnet: an on-line lexical database. Int. J. Lexicograhy **3**(4), 235–3244 (1990)
17. Dong, Z.D., Dong, Q.: National Knowledge Infrastructure (2005)
18. Suchanek, F.M., Kasneci, G., Weikum, G.: Yago: a core of semantic knowledge. In: 16th International World Wide Web Conference (WWW2007). ACM Press, New York (2007)
19. ICTCLAS tool from Chinese Academy of Sciences. http://ictclas.nlpir.org/downloads
20. LIBSVM developed by Lin, Z.R from Taiwan University. http://www.csie.ntu.edu.tw/~cjlin/libsvm
21. Stanford Parser. http://nlp.stanford.edu/software/lexparser.shtml

A Novel Method of Influence Ranking via Node Degree and H-index for Community Detection

Qiang Liu[1(✉)], Lu Deng[1], Junxing Zhu[1], Fenglan Li[1],
Bin Zhou[1,2], and Peng Zou[1]

[1] College of Computer, National University of Defense Technology,
Changsha 410073, Hunan, People's Republic of China
liuqiang1981@nudt.edu.cn
[2] State Key Laboratory of High Performance Computing,
National University of Defense Technology,
Changsha 410073, Hunan, People's Republic of China

Abstract. Identifying influential nodes is critical to have a better understanding of the network function and the process of information diffusion. Traditional methods of evaluating influential nodes such as degree centrality ignore the location of a node and its neighbors' influence in networks, while this plays an important role in revealing the node's local influence in spreading information. In this paper, we propose a novel method, named DH-index (node Degree and H-index), to measure a node' importance by considering its and neighbors' influence simultaneously. Meanwhile, we put forward a node DH-index based label propagation algorithm (DH_LPA) for community detection. We demonstrate its validity and feasibility on a set of real-world and synthetic networks for our new proposed community detection method.

Keywords: Influence ranking · Information diffusion · Community detection · Label propagation

1 Introduction

In recent years, community detection problem has attracted widespread research from many scholars around the world, and a great number of methods have been proposed. A detailed survey of community detection can be found in [1]. One general problem concerning community detection is that there is still no well-established precise definition of community. In general, a community in a network is described as a group of nodes with dense connections within groups and sparse connections with others. Discovering communities plays an important role in revealing the structure and function characteristics of networks. For instance, in virtual social network such as twitter, it's necessary to detect possible communities of terrorists or reactionary organization so as to avoid any criminal behaviors in real life, which may bring tremendous damage to a country or its people.

Among all the proposed community detection algorithms [2–7], the label propagation algorithm (LPA), proposed by Raghavan [7], has greatly received attention for its near linear time complexity in finding communities in large scale networks.

© Springer International Publishing AG 2016
S. Song and Y. Tong (Eds.): WAIM 2016 Workshops, LNCS 9998, pp. 149–160, 2016.
DOI: 10.1007/978-3-319-47121-1_13

The LPA utilizes the diffusion of label information of each node to detect communities and does not need any prior knowledge of community structure, such as the number of communities. Nevertheless, the random update order of label information lead to the poor robustness of community detection results. Then, a lot of improved LPA methods were proposed. Barber [8] et al. reformulated the LPA as an equivalent optimization problem, and put forward an improved LPA based on modularity constraints. Leung et al. [9] found that the original LPA may produce large communities due to the fact that some labels can plague a large amount of nodes during the process of label propagation. Then, they proposed an improved LPA based on hop attenuation and node preference so as to avoid finding monster communities. Subelj et al. [10] also presented an improved LPA that combines two unique strategies of community formation, namely, defensive preservation and offensive expansion of communities. Besides, in view of the disadvantage of randomly selecting initial nodes problem of traditional LPA, other methods based on how to select initial nodes for LPA are also proposed. He et al. [11] utilized PageRank to measure node centrality and put forward a node importance based LPA. Sun et al. [12] proposed a centrality-based LPA with specific update order and node preference to uncover communities. However, the PageRank method of [11] is degenerated into degree centrality and does not consider the importance of the node to its neighbors; the centrality-based LPA also uses the degree centrality to computer local density for selecting initial nodes for expansion.

In fact, the initial nodes selection problem for community detection can also be seen as an influence ranking problem. That is because the formation of communities in a network is decided by its important nodes. These nodes are more influential than other nodes, and then other nodes around the influential nodes form communities. Therefore, how to specify a quantitatively exact influence measure is crucial. By far, there are six widely used methods to measure a node's influence, which are degree, closeness, betweenness, eigenvector, katz and core centrality. The disadvantage of the former five methods had been illustrated in Ref. [13]. The utilization of core to measure a node's influence is proposed by Kitsak et al. [14], they deemed that a node's location is more important than the number of its linked neighbors. According to core theory, a node with more linked neighbor nodes on the edge of a network may not be influential compared to a node in the center of a network. Therefore, they advised that the coreness can better measure a node's influence for spreading information than degree centrality. However, calculating coreness needs global topological information of the network, while obtain this information is difficult, especially for the dynamic network whose network structure changes with time passing. Then, the coreness can not better measure a node's influence. Lately, Lü et al. [15] extended the concept of the H-index, which was originally used to measure the citation impact of a scholar or a journal, to qualify how important a node is to its network, and showed the H-index can better measure a node's influence in several cases compared to traditional centrality measurements mentioned above. Nonetheless, the H-index only takes into account of the influence of prominent neighbor nodes of a node to measure its influence in the light of the idea that a node is prominent if many other prominent nodes are around it. Then, the H-index ignores the influence of the node itself. Therefore, the H-index calculation of each node in the network will not reflect the node's local influence fairly.

In order to single out influential nodes for community detection, a better measurement of a node's influence is important. In the light of the advantage and disadvantage of H-index in judging influential nodes, we define a DH-index function, which not only consider the prominent neighbor nodes' influence, but also take into account of the node's influence to less prominent neighbors, to measure nodes' influence and ranking them according to this function. Then, a community detection algorithm is proposed based on spreading influential labels of ranking order according to DH-index, named DH-LPA.

The rest of paper is organized as follows. In Sect. 2, we explain and define some fundamental concepts. The Sect. 3 shows the proposed algorithm DH-LPA. In Sect. 4, we give some applications of the DH-LPA algorithm to some synthetic and real-world networks. Section 5 concludes this paper.

2 Preliminaries

2.1 Label Propagation Algorithm (LPA)

Label propagation algorithm [7] is an efficient algorithm for its nearly linear time complexity in detecting communities. According to the theory of the LPA, each node is initialized with a unique label and then let the label spread throughout the network. During the process of propagating label, each node will choose the label which is owned by most of its neighbors. Then, densely connected modules of nodes will reach a consensus on a unique label, and nodes with the same label form a community. The rule of updating community labels can be expressed as follows:

$$C_n = \arg\max_l |N^l(n)| \tag{1}$$

In (1), the $|N^l(n)|$ shows the neighbors of node n which has the label l. If there exists multiple most frequent neighbor labels, a random label will be selected among them. The course of separating label will be iterative until each node does not change its label and has a label that most of their neighbors have. As far as the efficiency of the LPA is concerned, due to its simple computation process and low time complexity, it is very fit for community detection for very large networks. However, random update order leads to the unstable detected results, which hampers its robustness and stability.

2.2 H-index

Due to the disadvantage of traditional centrality methods in measuring the influence of nodes in networks, Lü et al. [15] introduced the H-index concept to quantify how important a node is to its network in 2016. The H-index of a node is defined to be the maximum value h such that there exists at least h neighbors of degree no less than h.

For instance, the Fig. 1 is an example network consisted of 23 nodes and 40 edges [16], the degree of node 1 is 8. However, the H-index of it is 2. Because there exists at least 2 neighbors of degree no less than 2. That is to say, if the Fig. 1 is a citation network, the citation impact of the scholar (node 1) is 2.

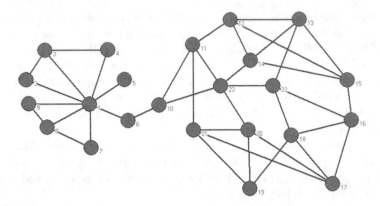

Fig. 1. An example network consisted of 23 nodes and 40 edges [16].

2.3 DH-Index

According to the definition of the H-index above, we can see that it only measures the influence of neighbor nodes, and ignores the influence of the node itself. Therefore, we combine the H-index and node degree to take into account of the influence of the node and its neighbors, and defined a function, named DH-index, to better measure a node influence in networks.

$$DHindex(n) = Hindex(n) \times Degree(n) \tag{2}$$

In (2), the $Hindex(n)$ shows the H-index of a node n, the $Degree(n)$ presents the degree of a node n. According to the DH-index function, we can see that a node's influence is not only related to the node degree (itself influence), but also is associated with the node H-index (neighbor nodes' influence). Then, the DH-index is reasonable in measuring the local influence around the node.

For instance, the Table 1 shows values of degree, H-index and DH-index for each node in the network of Fig. 1. From Table 1, we can see that the node 1 has largest degree among all 23 nodes. Then, if we want to choose a node to spread information faster and most broadly over the network, is node 1 a better choice for its biggest degree? Lǚ et al. [15] pointed out a node's influence should consider its prominent neighbors' influence, and introduced the H-index to describe a node's influence. They regarded the bigger the H-index value of a node, the more influential of it. However, the H-index only considers the prominent neighbors' influence, and does not take into the node's influence of itself. Take the node 22 and 23 for example. The H-index of node 22 and 23 are 4, and then it is difficult to distinguish which node is more influential. Therefore, we regard that we should simultaneously consider the influence of node itself and its neighbors, and defined the DH-index to measure a node's influence. Although the node 22 and 23 have the same H-index, we can see that the DH-index of node 23 is bigger than 22, which shows that the node 23 is more influential than node 22.

Table 1. Nodes information in the network

Node	Degree	H-index	DH-index
1	8	2	16
2	2	2	4
3	3	2	6
4	2	2	4
5	1	1	1
6	2	2	4
7	2	2	4
8	3	2	6
9	2	2	4
10	3	2	6
11	4	3	12
12	4	4	16
13	4	4	16
14	4	4	16
15	4	4	16
16	4	4	16
17	4	4	16
18	4	3	12
19	3	3	9
20	4	3	12
21	4	3	12
22	4	4	16
23	5	4	20

3 DH-Index Based Label Propagation Algorithm (DH-LPA)

In order to resolve the limit of the traditional LPA, we put forward a novel DH-index based label propagation algorithm for community detection. Our community detection algorithm DH-LPA includes two phases. In the first phase, we measure and quality the importance of each node. Specially speaking, we rank each node according to the DH-index value in a descending order. Then, the ranking results can better show the importance of each node in networks. The former nodes in the ranking results are more influential compared to the latter ones. In the second phase, based on the obtained update order from the first phase, the nodes in the ranking results will separate their labels to neighbors one by one. Due to the fact that the former nodes have a higher DH-index value than the latter ones, the labels of latter ones will be updated by former nodes. During the course of label propagation, there exists nodes that belong to neighbors of more than one node, and they have been updated by former nodes. For this condition, we should consider the neighbor nodes' influence of this node. If there are more nodes with the same label having high influence, we should change its label with the current ones. Eventually, theses nodes with the same label will form a community.

The details of the DH-index based label propagation algorithm are shown in Algorithm 1.

Algorithm 1 the DH-index based label propagation algorithm(DH-LPA)

Input: Network $G = (V, E)$

Output: Set of Communities C

1: Calculating the DH-index value of each node;
2: Ranking the nodes in a descending order according to the DH-index value,
   and getting the sorted nodes list L
3:**for** node in L **do**
4:     **for** neighbor in neighbors(node) **do**
5:         **if** neighbor.update==False **then**
6:             neighbor.label=node.label
7:         **else**
8:             neighbor.label=choose the most frequent and influential node's label
9:         **end if**
10:    **end for**
11: **end for**
12: Saving nodes with the same label into the same community c and add it to C

In Algorithm 1, the step 1 to step 2 are the first phase, and the step 3 to step 12 are the seconding phase. The main idea of the Algorithm 1 is that we first select these nodes with higher DH-index values to spreading their labels to neighbor nodes. That is because these nodes own high local influence compared to their neighbor nodes. For the latter nodes with lower DH-index values, the labels of them has been updated by the former nodes, then it's necessary to taking into account of the neighbor conditions of this node to again update this node's label according to influential nodes around it.

Let's consider the computational complexity of the DH-LPA. Suppose n be the number of nodes and m be the number of edges. According to Algorithm 1, there is two phases: (1) Calculating the DH-index values of each node and sorting them in a descending order; (2) Spreading labels according to nodes' influence. For the first phase, we need to compute the degree and the H-index of each node so as to get the DH-index value, then the time complexity is $O(3n)$. For the second phase, the label propagation process for each node has a time complexity of $O(n)$. Therefore, the total time complexity of the DH-LPA is $O(4n)$. After omitting the constant, the time complexity is $O(n)$.

4 Experiments

In this section, we conduct some experiments on several real world networks and synthetic networks so as to evaluate the performance of our proposed algorithm DH-LPA. Meanwhile, we also compare our algorithm with other well-known algorithms on benchmark network [17, 18] with known community structure. Our algorithm is implemented in Python 2.7. All the experiments were conducted on windows 7 with Intel(R) Core(TM) i5-2520 M processor, 2.5 GHz, 4G RAM.

4.1 Evaluation Metrics: Normalized Mutual Information (NMI) and Modularity

A great many of methods have been proposed for community detection, but it is not clear which method is reliable. In other words, when community partitions are found by an algorithm, a reasonable evaluation criterion should be used to evaluate how accurately the detection algorithm has performed. At present, there are two widely used evaluation methods for testing the efficiency of community detection algorithm. One is the Normalized Mutual Information (NMI), and the other is modularity. For these networks, the real community partitions of which are known, we can use NMI to test the performance of algorithm. If we do not know the real partitions of corresponding network, such as real-world networks, we can use the modularity to check the performance of community detection method. The bigger of NMI and modularity, the better of the partition results are, which can illustrate the efficiency of community detection algorithm.

The NMI, proven by Danon et al. [19], is a reliable criterion in evaluating community partitions. It can evaluate the similarity between the real partitions and the detected ones. Given two partitions A and B of a network in communities. Let C be the confusion matrix whose element $C_{i,j}$ is the number of nodes of community i of the partition A that is also in the community j of the partition B. The normalized mutual information $I(A, B)$ is defined as follows:

$$I(A, B) = \frac{-2 \sum_{i=1}^{C_A} \sum_{j=1}^{C_B} C_{ij} \log\left(\frac{C_{ij}N}{C_{i.}C_{.j}}\right)}{\sum_{i=1}^{C_A} C_{i.} \log\left(\frac{C_{i.}}{N}\right) + \sum_{j=1}^{C_B} C_{.j} \log\left(\frac{C_{.j}}{N}\right)} \tag{3}$$

Where C_A (C_B) is the number of groups in the partition A (B), $C_{i.}$ $(C_{.j})$ is the sum of the elements of C in row i (column j), and N is the number of nodes. If $A = B$, $I(A, B) = 1$; if A and B are completely different, then $I(A, B) = 0$.

Modularity [17] is also a most widely used function for testing efficiency of partitioning communities for a community detection algorithm. Consider an unsigned network denoted as $G = (V, E)$, where V is the vertex set with the number of it is n; and E is the edge set with the number of it is e. The adjacent matrix of G is A. If V_1 and V_2 are two disjoint subsets of V, then we define $L(V_1, V_2) = \sum_{i \in V_1, j \in V_2} A_{ij}$, $L(V_1, V_1) = \sum_{i \in V_1, j \in V_1} A_{ij}$, and $L(V_1, \overline{V_1}) = \sum_{i \in V_1, j \notin V_1} A_{ij}$, where $\overline{V_1} = V - V_1$. Meanwhile, we also define a partition of a network $G, G_1(V_1, E_1), G_2(V_2, E_2), \ldots, G_m(V_m, E_m)$, where V_i and E_i are the aggregation of vertices and edges of G_i for $i = 1, 2, \ldots, m$, the modularity Q can be defined as follows:

$$Q = \sum_{i=1}^{m} \left[\frac{L(V_i, V_i)}{L(V, V)} - \left(\frac{L(V_i, V)}{L(V, V)}\right)^2 \right] \tag{4}$$

According to the above function Q, we can see that the main idea of modularity comes from a comparison between real community partitions structure and network partitions allocated without any regard to the underlying structure. Then, sum over all the partitions differences of this two kinds of network structure.

4.2   Test on Real-World Networks

Test on Zachary's Karate Club Network. Zachary's karate club network [20] was generated by Zachary, who studied the friendship of 34 members of a karate club over a period of 2 years. In the course of research, he found a disagreement developed between the administrator and the instructor of karate club. Eventually, the club was divided into two groups almost of the same size. This network consists of 34 nodes and 78 edges.

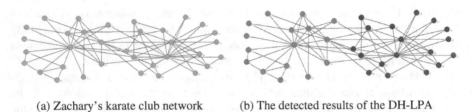

(a) Zachary's karate club network   (b) The detected results of the DH-LPA

Fig. 2. Zachary's karate club network and the detected results of the DH-LPA

Figure 2(b) shows the detected community partitions of the DH-LPA. We can see that two partitions are found, which is equal to the real partitions of the network. In Fig. 2(b), The value of modularity is 0.3715, and the NMI is 1, which illustrates the efficiency of DH-LPA on this network.

Test on Bottlenose Dolphin Network. Bottlenose dolphin network [21] describes a network of 62 bottlenose dolphins living in Doubtful Sound, New Zealand, was compiled by Lusseau after studying their behavior for 7 years. A tie between two dolphins was established by their statistically significant frequent association. The network split naturally into two large groups where the number of ties was 159.

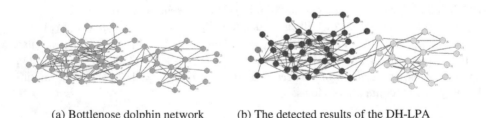

(a) Bottlenose dolphin network   (b) The detected results of the DH-LPA

Fig. 3. Bottlenose dolphin network and the detected results of the DH-LPA

From Fig. 3(b), we can see that the DH-LPA found 3 communities, which is a little different from the real partitions of the network. The value of modularity is 0.3749, and the NMI is 0.8069, which is very close to real partitions. During the course of label propagation, we found that the former updated labels may be again relabeled by other influential and frequent nodes' labels, such as the node 52. At the previous label propagation process, the node 52, 5 and 12 own the same label, but latter the label of the node 52 was relabeled by influential nodes, so we see this condition in Fig. 3(b).

Test on NetScience Network. NetScience network contains a co-authorship network of scientists working on network theory and experiment, as compiled by M. Newman in May 2006. The network was compiled from the bibliographies of two review articles on networks [22, 23], with a few additional references added by hand, which contains 1461 nodes and 2742 edges in total. This network is weighted, but we handle it as an unweighted one in our experiments.

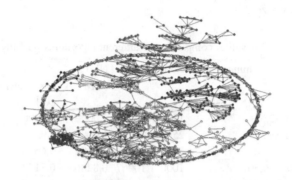

(a) The complete detection partitions

(b) Bigger communities A and B found by the DH-LPA

Fig. 4. The detected results of the DH-LPA on NetScience network

As far as the NetScience network is concerned, the DH-LPA is also competent in detecting better communties. On this network, our algorithm obtains 277 community partitions with a big modularity value of 0.9541, which illustrates that our algorithm has found communities with strong structures. The complete detection results of the DH-LPA can be seen in Fig. 4(a). Figure 4(b) presents two bigger community partitions found by the DH-LPA.

4.3   Test on Synthetic Networks

In this section, we first test accuracy of the DH-LPA on synthetic benchmark network with a known community structure, so as to illustrate that our proposed algorithms can figure out the real community partitions. We use the Lancichinetti-Fortunato-Radicchi (LFR) benchmark networks proposed by Lancichinetti et al. [24] to evaluate the performance of the DH-LPA. By tuning the parameters of the networks, different benchmark network can be generated. This kind of generated networks is defined as $LFR(N, k, \max k, mu, \min c, \max c)$. Where N is the number of nodes in network, k is the average degree of nodes, $\max k$ is the maximum degree of the nodes, mu is the mixing parameter, $\min c$ is the minimum for the community sizes, and $\max c$ is the maximum for the community sizes.

Table 2. The NMI values comparison of different algorithms on LFR Networks

No	$LFR(N, k, \max k, mu, \min c, \max c)$	Nodes	Edges	NMI		
				DH-LPA	Fast Newman [17]	Danon [18]
1	(128,16,16,0.1,32,32)	128	1024	1.0000	1.0000	1.0000
2	(256,16,16,0.1,32,32)	256	2048	1.0000	0.9549	0.9549
3	(512,10,16,0.2,10,50)	512	2532	0.9182	0.6518	0.8262
4	(1000,20,50,0.2,40,50)	1000	10067	0.8286	0.8153	0.6153
5	(2000,20,50,0.2,50,60)	2000	20660	0.8828	0.7864	0.8566
6	(4000,30,50,0.2,50,100)	4000	59679	0.9042	0.7731	0.9257

We generated 6 different LFR benchmark networks according to Ref. [24]. In order to better illustrate the performance of different algorithms, we set different parameters for benchmark networks. The number of edges for each networks is presented according to LFR code, which contains bilateral edges. In Table 2, from network 1 to 6, the number of nodes is increased, and other parameters are also changed to improve the complexity of the benchmark network. From Table 2, we can see that DH-LPA obtains better results compared to Fast Newman and Danon algorithm in most conditions. For the network 6, the Danon method gets a higher NMI value than DH-LPA and Fast Newman. However, in our experiments, we find the DH-LPA can quickly generate community partition results for its nearly linear time complexity. In larger networks, such as the network 6, the Fast Newman and Danon algorithm run for nearly 50 min and then generated results, which illustrates their high time complexity.

5   Conclusion

In this paper, we put forward a community detection algorithm DH-LPA based on DH-index, and test the effectiveness of DH-LPA on 3 real world networks and the artificial benchmark networks. Meanwhile, we compare it with other algorithms in these networks. The experiment results show that the DH-LPA is very effective in community detection problems. In our future work, we will pay attention to improve the efficiency of DH-LPA so as to make it work on dynamic networks, which is a very interesting work.

Acknowledgments. This work was supported by 973 Program of China (Grant No. 2013CB329601, 2013CB329604, 2013CB329606).

References

1. Fortunato, S.: Community detection in graphs. Phys. Rep. **486**, 75–174 (2010)
2. Girvan, M., Newman, M.E.J.: Community structure in social and biological networks. Proc. Natl. Acad. Sci. U.S.A. **99**(12), 7821–7826 (2002)
3. Clauset, A., Newman, M.E.J., Moore, C.: Finding community structure in very large networks. Phys. Rev. E **70**, 066111 (2004)
4. Xu, B., Deng, L., Jia, Y., Zhou, B., Han, Y.: Overlapping community detection on dynamic social network. In: Proceedings of the 6th International Symposium on Computational Intelligence and Design, Hangzhou, China, vol. 2, pp. 321–326 (2013)
5. Liu, Q., Zhou, B., Li, S., Li, A.P., Zou, P., Jia, Y.: Community detection utilizing a novel multi-swarm fruit fly optimization algorithm with hill-climbing strategy. Arab. J. Sci. Eng. **41**, 807–828 (2016)
6. Lin, W., Kong, X., Yu, P.S., Wu, Q., Jia, Y., Li, C.: Community detection in incomplete information networks. In: 21st International Conference on World Wide Web, pp. 341–350. ACM Press (2012)
7. Raghavan, U.N., Albert, R., Kumara, S.: Near linear time algorithm to detect community structures in large-scale networks. Phys. Rev. E **76**, 036106 (2007)
8. Barber, M.J., Clark, J.W.: Detecting network communities by propagating labels under constraints. Phys. Rev. E **80**, 026129 (2009)
9. Leung, I.X., Hui, P., Lio, P., Crowcroft, J.: Towards real-time community detection in large networks. Phys. Rev. E **79**, 066107 (2009)
10. Subelj, L., Bajec, M.: Unfolding communities in large complex networks: combining defensive and offensive label propagation for core extraction. Phys. Rev. E **83**, 036103 (2011)
11. He, M., Leng, M., Li, F., Yao, Y., Chen, X.: A node importance based label propagation approach for community detection. In: Sun, F., Li, T., Li, H. (eds.) ISKE2012. Advances in Intelligent Systems and Computing, vol. 214, pp. 249–257. Springer, Heidelberg (2014)
12. Sun, H., Liu, J., Huang, J., Wang, G., Yang, Z., Song, Q., Jia, X.: CenLP: a centrality-based label propagation algorithm for community detection in networks. Phys. A Stat. Mech. Appl. **436**, 767–780 (2015)
13. Wang, W., Street, W.N.: A novel algorithm for community detection and influence ranking in social networks. In: 2014 IEEE/ACM International Conference on Advances in Social Networks Analysis and Mining, pp. 555–560. IEEE Press, New York, August 2014
14. Kitsak, M., Gallos, L.K., Havlin, S., Liljeros, F., Muchnik, L., Stanley, H.E., Makse, H.A.: Identification of influential spreaders in complex networks. Nat. phys. **6**, 888–893 (2010)
15. Lü, L., Zhou, T., Zhang, Q.M., Stanley, H.E.: The H-index of a network node and its relation to degree and coreness. Nat. Commun. **7**, 10168 (2016)
16. Chen, D., Lü, L., Shang, M.S., Zhang, Y.C., Zhou, T.: Identifying influential nodes in complex networks. Phys. A Stat. Mech. Appl. **391**, 1777–1787 (2012)

17. Newman, M.E.J.: Fast algorithm for detecting community structure in networks. Phys. Rev. E **69**(6), 066133 (2004)
18. Danon, L., Díaz-Guilera, A., Arenas, A.: The effect of size heterogeneity on community identification in complex networks. J. Stat. Mech. Theory Exp. **11**, P11010 (2006)
19. Danon, L., Diaz-Guilera, A., Duch, J., Arenas, A.: Comparing community structure identification. J. Stat. Mech. Theory Exp. **9**, P09008 (2005)
20. Zachary, W.W.: An information flow model for conflict and fission in small groups. J. Anthropol. Res. **33**, 452–473 (1977)
21. Lusseau, D., Schneider, K., Boisseau, O.J., Haase, P., Slooten, E., Dawson, S.M.: The bottlenose dolphin community of doubtful sound features a large proportion of long-lasting associations. Behav. Ecol. Sociobiol. **54**, 396–405 (2003)
22. Newman, M.E.J.: The structure and function of complex networks. SIAM Rev. **45**, 167–256 (2003)
23. Boccaletti, S., Latora, V., Moreno, Y., Chavez, M., Hwang, D.U.: Complex networks: Structure and dynamics. Phys. Rep. **424**, 175–308 (2006)
24. Lancichinetti, A., Fortunato, S., Radicchi, F.: Benchmark graphs for testing community detection algorithms. Phys. Rev. E **78**, 046110 (2008)

Efficient and Load Balancing Strategy for Task Scheduling in Spatial Crowdsourcing

Dezhi Sun[1(✉)], Yong Gao[2], and Dan Yu[3]

[1] State Key Laboratory of Software Development Environment,
Beihang University, Beijing, People's Republic of China
csusun12@gmail.com

[2] Key Laboratory of Yunnan Province Universities of the Diversity
and Ecological Adaptive Evolution for Animals and Plants on YunGui Plateau,
Qujing Normal University, Qujing, Yunnan, People's Republic of China
yonggaoqj@gmail.com

[3] China Standard Software Co., Ltd., Beijing, People's Republic of China
danyu@cs2c.com.cn

Abstract. With the wide use of mobile devices, spatial crowdsourcing platforms are becoming popular. An important problem of spatial crowdsourcing is assigning a set of spatial tasks tagged with location and time for workers according to their location. In most cases, existing approaches usually take the matching algorithm as a fundamental step to solve this problem which aims to maximize the number of completed tasks. However, in the present of many spatial crowdsourcing platforms, how to assign the tasks at high efficiency and make a relatively fair schedule for multiple workers is a new challenge. In this paper, we study the problem of load balancing based task scheduling for multiple workers. We present fast and effective approximate algorithms for task scheduling problem. With both real and synthetic datasets, we verify the effectiveness of our proposed methods.

1 Introduction

With the popularity of online to offline(O2O) and location-based services(LBS), crowdsourcing is becoming more and more popular nowadays [2,16,17]. In particular, many new spatial crowdsourcing platforms are emerging, involving food delivery [1], ride sharing [7] and so on. In general, all these tasks referring to tasks tagged with location and time can be treated as spatial tasks. Spatial task assignment is a basic problem of spatial crowdsourcing, referring to assign a set of workers to a specific location of one task. In most cases, spatial task assignment mainly concerns the number of completed tasks, while ignoring other important objective, like the workload of workers and the maximum travel cost for one worker. Consider the scenario of take-out, some users who order the take-out would have to wait for a long time, even though there are a significant number of orders reached. The key to reduce waiting time of user is improving the task schedule method. That is, for these time-aware applications, we should propose more effective strategy to reach balance between workers so as to minimize the

© Springer International Publishing AG 2016
S. Song and Y. Tong (Eds.): WAIM 2016 Workshops, LNCS 9998, pp. 161–173, 2016.
DOI: 10.1007/978-3-319-47121-1_14

time that the users wait. Additionally, we should also concern the number of completed spatial tasks.

2   Related Work

There are various studies on spatial assignment problem [5, 6, 10, 12–14, 19–21]. Different from previous research of personalized task recommendation [8, 9], which usually recommends frequent items to potential users according to users' preferences and spatial information [18], the issue of task assignment in spatial crowdsourcing is to find a global optimal assignment for all tasks and workers. According to the taxonomy defined in [10], we consider the server assigned tasks mode. Kazemi et al. propose three approaches to solve the spatial task assignment problem which is to maximize the number of completed tasks. The first is greedy strategy, which is reducible to the maximum flow problem. The second is least location entropy priority strategy, which consider the distribution of tasks, and give higher priority to the tasks with low entropy. The third is nearest neighbor priority, which performs fast but have low efficiency sometimes. In [4], Deng et al. propose the problem of worker selected tasks mode and give solutions to find a schedule for one worker that maximizes the number of performed tasks. They first give exact algorithms which contain dynamic programming and branch-and-bound strategy. However, these two methods will grow exponentially when the number of tasks grows. Later they propose fours approximation algorithms, which is least expiration time heuristic, nearest neighbor heuristic, most promising heuristic and progressive algorithms. In [5], Deng et al. consider the problem which combines task matching and task scheduling, they first propose a 3-phase approach GALS which can generates high quality results but slow, so they propose a Bisection-based method which divides the workers and tasks into balanced partitions and performs faster. The main idea of GALS is iteratively running the procedure of maximum matching and task schedule, until there are no more tasks match the worker. In [12–14], She et al. study how spatio-temporal conflicts affect spatial task assignment. In [11], Liu et al. propose the problem of cost minimization and social fairness for spatial crowdsourcing tasks. They want to minimize the total cost incurred by workers. In the meanwhile, they find solutions that are socially fair. Two distribution aware algorithms are proposed which use the technical of cluster. In particular, different from the assumption that the aforementioned studies focus on static scenario, where all spatio-temporal information of tasks and workers is known in advance, Tong et al. propose the online task assignment over real-time dynamic scenarios in [19, 20].

Howerver, as mentioned above, most existing approaches only consider the objective of the number of completed tasks and do not provide an intelligent and global arrangement for workers, which balances the workload between workers and will be helpful to accelerate the completion of some tasks. In the same time, in practical tasks also need time to finish which means workers overload will spend more time to finish tasks. Note that the workload in this paper presents the total length of task schedule plan assigned to a worker, which is different from previous work [5] that may indicates the number of workers or tasks per partition or the cluster based partition in [11]. A possible goal may be to maximize the average value of the workload for all the works, but obviously, it still can lead to imbalance arrangement between workers. So a reasonable goal is to minimize the maximum workload of workers which can increase utilization of worker resources.

In this paper, we explore the load balancing based spatial task scheduling problem(LBSTS), which, given a set of spatial tasks and workers, we want to find an arrangement that minimize the maximum workload of all workers. In the meanwhile, the arrangement should satisfy the spatial-temporal constrain and take into account the completion time of tasks, which we also consider the capacity of each task t and the cost produced by completing task.

3   Problem Statement

We use t to denote a spatial task with location l_t, and a worker is represented with $w = <l_w, c_w, b_w>$ which l_w is the location of worker w, c_w is the capacity of w, and b_w is the travel budget of w. We first introduce some basic concepts and then formulate the problem of load balancing based spatial task scheduling(LBSTS).

Definition 1 (Task Schedule Plan). Given a set of workers W, and a set of tasks T, the feasible schedule for each $w \in W$, denoted by $P_w(t_1, t_2, ..., t_m)$, that not exceed the travel budget of w. The size of P_w, denoted by $|P_w|$, is the number of tasks in P_w.

Definition 2 (Workload). The sum of travel distance between tasks in task schedule plan P_w and the cost to finish each task, denoted by $|L_w|$, is defined as the workload of one worker w. The cost to finish one task t is represented as $\gamma(t)$. The cost of task schedule plan can be computed by the following:

$$L(w) = cost(l_w, t1) + \sum_{i=1}^{m-1} cost(t_i, t_{i+1}) + \sum_{i=1}^{m} \gamma(t_m)$$

Definition 3 (LBSTS Problem). Given a set of workers W, each worker tagged with location l_w, capacity c_w, travel budget b_w, a set of tasks T, each t of which is associated with location l_t, capacity c_t, the LBSTS problem is to find a workload balanced arrangement P for all task schedule plan $Pw \in P$ that minimize $max_{w \in W} L\{w\}$, and satisfy all the spatial-temporal constrain C. We use D(w) as travel distance of worker w. The travel distance of worker w D(w) can be computed as follows:

$$D(w) = cost(l_w, t1) + \sum_{i=1}^{m-1} cost(t_i, t_{i+1})$$

We use C(w) as the number of tasks assigned to worker w, and C(t) as the number of workers assigned to task t. The spatial-temporal constrain C can be described as follows:

$$D(w) < b_w$$
$$C(w) < c_w$$
$$C(t) < c_t$$

4   Solutions for LBSTS

In this section, we present solutions for the LBSTS problem. We first present a baseline algorithm. We then propose efficient one load balancing based heuristic algorithm. The first algorithm is based on an equal division strategy while the second considers three factors related to the workload balance. But there are some drawbacks of this baseline algorithm which should improve. LBA algorithms considers three factors that influences the result. For the improvement of efficiency of LBA, we then propose the divide-and-conquer method, which partitions the set of tasks and workers first and then use LBA for each partition. We discuss them as follows:

4.1   Baseline Method

The baseline algorithm tries to reduce the imbalance between workers through equally sharing the tasks. At the beginning, we should check the capacity and travel budget constraint. That is, for a fair allocation between workers, we should find an allocation that satisfy the constraint of worker and the validity of inserting task t into task schedule plan of worker w. At each step, we aims at inserting task t that minimize the task schedule plan cost.

The baseline method aims at keeping balance between workers by give higher priority to workers with smaller workload. More specifically, for each task t, the baseline method checks all the workers that can finish this task and satisfy the capacity constraint of workers. The worker with the minimum workload will be assigned to complete task t. After each allocation, we will remove task t from task set T and update the capacity of worker w. Although the baseline algorithm aims at arrange all the workers equally, there still be some drawbacks. One is the order of assignment may influence overall effect of arrangement. Even though the workload between workers may be fair at each step, we should try to use geographic information of workers that can effectively prevent local optimization.

4.2   Load Balancing Strategy

In this section, we present a load balancing Algorithm(LBA). Instead of visiting each task in order as the baseline algorithm does, we discuss three main factors that influence the maximum cost of completing tasks and workload balancing as follows:

Algorithm 1. Baseline Algorithm

Input: worker set t, task set T, spatial-temporal constrain C
Output: arrangement P for all workers
1: **while** $W \neq \phi$ **do**
2:    **for** each $t \in T$ **do**
3:       **for** each $w \in W$ **do**
4:          **if** $c_w \neq \phi$ **then**
5:             check the validity of w according to the travel budget b_w
6:             check the validity of $P_{(t,w)}$
7:             $w \leftarrow argmin_{t|P_{(t,w)} \ is \ feasible} L(t, w)$
8:             insert a new task t to the schedule P_w of worker w
9:          **end if**
10:       **end for**
11:       remove the new assigned task t from T
12:       update the capacity of worker w
13:       **if** $c_w = \phi$ **then**
14:          remove w from W
15:       **end if**
16:    **end for**
17: **end while**

Imbalance Ratio: We introduce the concept of imbalance ratio. When we assign one task to a worker, we want to make the task schedule plan of each worker more fair, which means reduces the difference between the task schedule plan of workers. We use the variance of task schedule plan of workers as imbalance ratio. We use $\overline{L(w)}$ as the average value of task schedule plan cost of all workers. The imbalance ratio can be computed as follows:

$$imbalance\_ratio(t, w) = \sum_{i=1}^{m} \left(L(w_i) - \overline{L(w)}\right)^2$$

With the definition of imbalance ratio, we can use it to evaluate the workload banlace of task schedule plans caused by inserting task t for worker w. That is, for each unassigned task t, we compute all the possibilities that task t can be inserted to task schedule plan. Compared with the baseline algorithm, we dont simply assign this task to the worker with minimum workload. We verify the workload balance after inserting task t into P_w, and choose the one with minimum value. However, imbalance ratio only focus on the balance of workers, we should consider the geographic information of workers to make an allocation more feasible. Because all tasks are tagged with location, we should make use of their geographic information and optimize the solution. So we introduce the concept of task location entropy next.

Task Location Entropy: [3] introduces the concept of location entropy, and [10] proposed least location entropy priority strategy as one method to solve the task assignment problem. Location entropy can be treated as the probability of one task's being visited by workers according to the location of task. Specifically, one task will have a low location entropy if there are only a few workers can visit it. Consider the load balancing based task scheduling problem, we should give higher priority to tasks which have smaller location entropy. That is, when there are many unassigned tasks have the same imbalance ratio for worker w, we should insert the task with the smallest locatioin entropy to worker w so as to keep fair allocation between tasks relatively. Inspired by this idea, we use P(t) as the probability of one task t being visited by the workers that t satisfies their travel budget. The task location entropy can be computed as follows:

$$task\_location\_entropy(t) = -\sum_t P(t) log P(t)$$

Workload: Although the task location entropy can be used to assignment tasks for workers more fairly, we still need balance the workload between workers. There are two reasons for us to consider workload between workers. First, with the above two strategy, some workers may still be overload. And the second, we consider the cost of completing one task which represented as $\gamma(t)$, so the length of task schedule plan of each worker has to be controlled.

Now we define the ratio between worker w and task t as follows:

$$ratio(t, w) = \alpha/imbalance\_ratio(t, w) + (1 - \alpha)/task\_location\_entropy(t)$$

Based on the discussions above, we propose our load balancing algorithm(LBA). The main idea of LBA is that we use the two strategies, which named imbalance ratio and location entropy, to find a feasible task schedule plan in a greedy way. More specifically, given the task set T and worker set W, we should try to find an allocation that keep workload balance between workers. As mentioned above, we should try to find the task t with smaller imbalance value and task location entropy. At the same time, we want to maximum the utility of workload of each worker. So we handle the (t,w) pairs with high ratio which is the ratio(t,w) divided by workload of worker w. We maintain the (t,w) pairs and fetch the one with the largest value among remaining elements iteratively. The algorithm terminates when the worker set or task set is empty. We maintain the (t,w) pairs and fetch the one with the largest value among remaining elements iteratively. The algorithm terminates when the worker set or task set is empty.

The detailed method of LBA is as follows. At the beginning, the arrangement P for all workers is empty. We use maintains the unassigned task-worker sets. As our algorithm is based on the workload balancing, we need to initialize the assignment at first. That is, we compute the ratio for (w,t) pairs and iteratively extract one with the largest ratio value, then insert t into the task schedule plan of worker w, and update the workload of worker w. After initialization, we insert (t,w) pairs into E in the same way. While task set T and worker set W is not empty, we iteratively find the pair(t,w) with highest ratio, which is the ratio between worker w and task t divided by the workload of worker w. The intuition of this step is that we want to maximize the utility of workload of workers. Then we check the feasibility of P(t,w) and insert task t if it satisfies the condition. At the same time, if the count of workers participate in task t is more than the capacity of task t, we remove task t from task set T. In a similar way, we update the workload of each worker as well. The algorithm will terminate when the task set or worker set is empty, and at that time, we have find an arrangement for all workers.

The description of LBA Algorithm is as follows:

Algorithm 2. Load Balancing Algorithm

Input: worker set t,task setT, spatial-temporal constrain C
Output: arrangement P for all workers
1: **for each** $w \in W$ **do**
2:     $t \leftarrow argmax_{t|P_{(t,w)} \ is \ feasible} ratio(t,w)$
3:     insert t into Pw
4:     update workload(w)
5: **end for**
6: **for each** $t \subset T$ **do**
7:     **for each** $w \in W$ **do**
8:         Find (t,w) pair with the maximum $\theta = \frac{ratio(t,w)}{workload(w)}$
9:         **if** P(t,w) is feasible **then**
10:             insert t into Pw
11:             count[t] $\leftarrow$ count[t]+1
12:             update workload(w)
13:         **end if**
14:         **if** count[t]= c_t **then**
15:             remove task t from task set T
16:         **end if**
17:         **if** workload(w)= c_w **then**
18:             remove worker w from worker set W
19:         **end if**
20:     **end for**
21: **end for**

4.3 Divide-and-Conquer Algorithm

Even though the LBA algorithm considers the overall efficiency, it may still encounters the bottleneck due to the construction of worker-task pairs. We consider the divide-and-conquer algorithm that partitions the tasks and their relevant workers first, and then use the LBA algorithm in each partition. The intuition of the partition should consider the workload balance between different partitions, which will be helpful to the results of LBA algorithm.

There have been many partitioning techniques [5] proposed for balancing prtitioin is spatial crowdsourcing. In this paper, we propse the partition methods based on task location entropy. We first sort all the tasks according to task location entropy, and choose the one with minimum value, we add it to the first partition, in the next we find the worker nearest to this task. and add this worker to this partition as well. For every task satisfies the travel budget, we add each task into the partition with probability in the inverse ratio of its location entropy. Then we iteratively continue this process until the number of tasks in this partition satisfies a threshold value. At last we get the partitions which is nearly balanced, and then with each partition, we can handle it with LBA algorithm. And after one execution of this process, there may exists some tasks and workers still not be assigned. So we use a process to merge the remaining results, and iteratively continue this processs. The method will terminates when there are no more tasks/workers can be assigend. That is the basic idea of our divide-and-conquer algorithm(DCA). We use R_t to represent the number of unassigned tasks, and R_w to represent the number of unassigned workers. The divide-and-conquer algorithm is as follows:

Algorithm 3. Divide-and-Conquer Algorithm

Input: worker set t, task setT, spatial-temporal constrain C
Output: arrangement P for all workers
1: Initiallize
2: $R_t \leftarrow$ the number of task T
3: $R_w \leftarrow$ the number of workers W
4: **for** *task t ∈ remaining task set T* **do**
5:     $t \leftarrow argmin\ task\_location\_entropy(t)$
6:     add t into partition P
7:     find the nearest worker w, add w into partition P
8:     add task wt with probability in the inverse ratio of its location entropy
9:     until partition P is full
10: **end for**
11: **while** $R_t \neq \phi$ *and* $R_w \neq \phi$ **do**
12:     **for** each *partiioin ∈ allpartitionsset* **do**
13:         LBA
14:         Record the unassigned workers and tasks
15:     **end for**
16:     merge the unassigned workers and tasks
17:     $R_t \leftarrow$ the remaining number of tasks.
18:     $R_w \leftarrow$ the remaining number of workers.
19: **end while**

---

5   Experimental Study

Datasets: Although the application of spatial task assignment is very wide, it not easy to get real data in practical. We adopt the method of [5], which uses the Gowalla check-in dataset for simulation. For the synthetic, we use some data

Table 1. Experiment parameters

$|T|$	$|W|$	$max(c_t)$	$max(c_w)$	$max(b_w)$	$max(\gamma)$
1 K	1 K	1	10	200	10
2 k	2 K	2	20	400	20
3 K	3 K	3	30	600	30
4 k	4 K	4	40	800	40
5 K	5 K	5	50	1000	50

generating algorithms of [15] to produce the synthetic dataset we want to test. More specifically, we varied both the number of tasks and workers from 1K-5K. The travel cost between two locations is their Euclidean distances. We generated synthetic data with uniform(UNI) and mixture of uniform and gaussian(MIX) distributions in a grid with size 500*500 (Table 1).

Configuration and Measures: We evaluated the effeciency of our algorithms by varying different values of our parameters. There are many parameters that influence the result of our algorithms. For synthetic data, we try to simulate the scenario of take-out and evaluate the maximum cost of completing one task and load balance. Specifically, we variy the number of tasks from 1 K to 5 K, and the number of workers variy from 1 K to 5 K too. The distribution of tasks/worers follow the uniform(UNI) and mixture of uniform and gaussian(MIX) distributions. And both the tasks and worker are generated on a grid with 500*500. So we set the maximum budget of a worker from 200–1000, whose range is around the length of side of the grid. That means, a worker can move locally in a small region, or move for a long distance which is larger than the length of side. Moreover, we set the capacity of a worker which means the maximum number of tasks one work can accept. The maximum capacity value of a worker varied from 10 to 60, which is suitable for completing tasks. In the meanwhile, our algorithm can handle the tasks with capacity, so we set the capacity of tasks from 1 to 5, which may be similar to the case of ridesharing. In the last, we consider the parameter of cost of completing one task, with the maximum value from 10 to 60. Because in many real applications, it is an important factor that influences the wait time of one task too. The default setting of our experiment is the value in the middle of each parameter list, that is, the number of tasks is 3 k, the number of workers is 3 K, the maximum capacity of all tasks is 3, the maximum capacity of all workers is 30, the maximum travel budget of all workers is 600 and the maximum time of completing one task is 30.

Algorithms Evaluating: In this paper we present three methods, which is baseline Algorithm(BLA), the load balancing algorithm(LBA), and the divide and conquer algorithm(DCA). While existing methods usually consider the number of completed tasks, our methods concerns the maximum wait time of one task, and the workload for each worker. So we compare our three methods and the

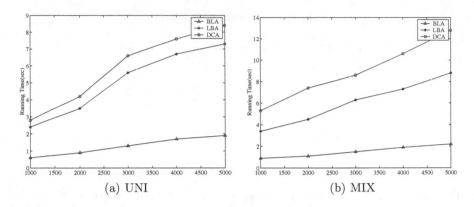

(a) UNI   (b) MIX

Fig. 1. Running time of varying number of tasks $|T|$.

results verifies the effectiveness of our algorithms. Specifically, our major concern is the maximum length of one task schedule plan, the second is the average workload of all workers. In the meanwhile, we check results of the number of completed tasks of the algorithms we proposed.

Running Time of Varying $|T|$**:** We first evaluate the proposed algorithms by varying the number of tasks under different distribution. That is, with different distribution of uniform(UNI) and mixture of uniform and gaussian(MIX), we vary the number of workers from 1 K to 5 k. From Fig. 1 we can see that Baseline algorithm runs faster than LBA and DCA, and due to the preprocess of DCA, LBA runs faster on both datasets than DCA. Them maximum cost incurred by LBA and DCA is relatively small than the size of the grid.

Running Time of Varying of $|W|$**:** We evaluate the proposed algorithms by varying the number of workers under different distribution. Same with the evaluation of the change of tasks, we vary the number of workers from 1 K to 5 k under different distribution of uniform(UNI) and mixture of uniform and gaussian(MIX). In fact, the effect of changing the number of workers is similar to change the number of tasks, so we omit it here.

Maximum Cost of Varying of $|T|$**:** We check the maximum cost by varying the number of workers under different distribution. Maximum cost is related to the completion time of tasks, so our objective is to reduce the maximum cost. As metioned obove, we consider three factors that influence the workload of workers in the LBA algorithm, which is imbalance ratio, task location entrop and workload. Same with the evaluation of running time, we vary the number of workers from 1 K to 5 k under different distributions of uniform(UNI) and mixture of uniform and gaussian(MIX). In fact, the effect of changing the number of workers is similar to change the number of tasks, so we omit it here. And we run

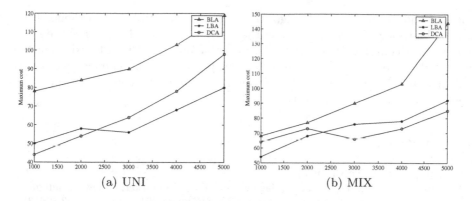

Fig. 2. Maxiumum cost of varying number of tasks $|T|$.

Table 2. Experiment results on Gowalla

	No. of completed tasks	Avg(travel cost)
BLA	8318	7.8
LBA	9604	5.5
DCA	9523	6.1

every instance with 50 times, and compute the average maximum cost of all the instances. The experimental results are showed in Fig. 2, which shows that both LBA and DCA get better results than BLA. In the meanwhile, DCA performs better than LBA in mixture of uniform and gaussian(MIX) distribution. The results show that the workoad balacing strategy which considers three factors is effective, and is useful to make an balanced assignment for workers, which would reduce the completion time of tasks.

Experiment Results on Gowalla: We also test the algorithms on Gowalla, which is showed in Table 2, the result shows that the LBA algorithm is also effective in reality. And we can see that LBA algorithm performs better than both BLA and DCA algorithms and reaches the minimum average travel cost.

Experimental Environment: We implement our algorithms in C++, and the experiments were performed on a machine with Intel i5-5200 CPU @ 2.2 GHZ and 8 GB memory.

6   Conclusion

In this paper, we study the problem of load balancing based spatial task scheduling problem, which is different from existing methods. We devise a baseline method and two approximate algorithms to solve this problem. In the load balancing based algorithm, we consider the three factors which influence the maximum cost of completing one task, which is imbalance ratio, task location entropy and workload of workers. We further develop a divide and conquer method to enhance the efficiency. We verify the efficiency and effectiveness of the proposed methods through experimental results.

References

1. https://favordelivery.com/apply/
2. Cao, C.C., Tong, Y., Chen, L., Jagadish, H.V.: Wisemarket: a new paradigm for managing wisdom of online social users. In: SIGKDD 2013, pp. 455–463 (2013)
3. Cranshaw, J., Toch, E., Hong, J., Kittur, A., Sadeh, N.: Bridging the gap between physical location and online social networks. In: Ubicomp 2010, pp. 119–128 (2010)
4. Deng, D., Shahabi, C., Demiryurek, U.: Maximizing the number of worker's self-selected tasks in spatial crowdsourcing. In: GIS 2013, pp. 324–333 (2013)
5. Deng, D., Shahabi, C., Zhu, L.: Task matching and scheduling for multiple workers in spatial crowdsourcing (2015)
6. Gao, D., Tong, Y., She, J., Song, T., Chen, L., Xu, K.: Top-k team recommendation in spatial crowdsourcing. In: Cui, B., Zhang, N., Xu, J., Lian, X., Liu, D. (eds.) WAIM 2016. LNCS, vol. 9658, pp. 191–204. Springer, Heidelberg (2016). doi:10. 1007/978-3-319-39937-9_15
7. Huang, Y., Bastani, F., Jin, R., Wang, X.S.: Large scale real-time ridesharing with service guarantee on road networks. Proc. VLDB Endow. **7**(14), 2017–2028 (2014)
8. Jiang, D., Leung, K.W.T., Ng, W.: Fast topic discovery from web search streams. WWW **2014**, 949–960 (2014)
9. Jiang, D., Leung, K.W.T., Vosecky, J., Ng, W.: Personalized query suggestion with diversity awareness. In: ICDE 2014, pp. 400–411 (2014)
10. Kazemi, L., Shahabi, C.: Geocrowd: enabling query answering with spatial crowdsourcing. In: GIS 2012, pp. 189–198 (2012)
11. Liu, Q., Abdessalem, T., Wu, H., Yuan, Z., Bressan, S.: Cost minimization and social fairness for spatial crowdsourcing tasks. In: Navathe, S.B., et al. (eds.) DASFAA 2016. LNCS, vol. 9642, pp. 3–17. Springer, Heidelberg (2016). doi:10.1007/978-3-319-32025-0_1
12. She, J., Tong, Y., Chen, L.: Utility-aware social event-participant planning. In: SIGMOD 2015, pp. 1629–1643 (2015)
13. She, J., Tong, Y., Chen, L., Cao, C.C.: Conflict-aware event-participant arrangement. In: ICDE 2015, pp. 735–746 (2015)
14. She, J., Tong, Y., Chen, L., Cao, C.C.: Conflict-aware event-participant arrangement and its variant for online setting. IEEE Trans. Knowl. Data Eng. **28**, 2281–2295 (2016). doi:10.1109/TKDE.2016.2565468
15. To, H., Asghari, M., Deng, D., Shahabi, C.: Scawg: a toolbox for generating synthetic workload for spatial crowdsourcing (2016)
16. Tong, Y., Cao, C.C., Chen, L.: TCS: efficient topic discovery over crowd-oriented service data. In: SIGKDD 2014, pp. 861–870 (2014)
17. Tong, Y., Cao, C.C., Zhang, C.J., Li, Y., Chen, L.: Crowdcleaner: data cleaning for multi-version data on the web via crowdsourcing. In: ICDE 2014, pp. 1182–1185 (2014)

18. Tong, Y., Chen, L., Cheng, Y., Yu, P.S.: Mining frequent itemsets over uncertain databases. Proc. VLDB Endow. **5**(11), 1650–1661 (2012)
19. Tong, Y., She, J., Ding, B., Chen, L., Wo, T., Xu, K.: Online minimum matching in real-time spatial data: experiments and analysis. Proc. VLDB Endow. **9**, 1053–1064 (2016)
20. Tong, Y., She, J., Ding, B., Wang, L., Chen, L.: Online mobile micro-task allocation in spatial crowdsourcing. In: ICDE 2016, pp. 49–60 (2016)
21. Tong, Y., She, J., Meng, R.: Bottleneck-aware arrangement over event-based social networks: the max-min approach. World Wide Web J. **19**(6), 1151–1177 (2016)

How Surfing Habits Affect Academic Performance: An Experimental Study

Xing Xu[1,2]($\boxtimes$), Jianzhong Wang[1], and Haoran Wang[3]

[1] School of Humanities and Social Sciences, Beihang University, Beijing, China
{xuxing66,wjz}@buaa.edu.cn
[2] Network Information Center, Beihang University, Beijing, China
[3] School of Software, Beihang University, Beijing, China
18811399416@163.com

Abstract. The issues regarding relationships between surfing habits and academic performance of university students have attracted much attention of all kinds of research communities. The approaches of statistical analysis and data mining with imperfect datasets are used in previous works. In this paper, an experimental study about the relationships between surfing habits and academic performance is conducted. Particularly, we observe a surprising results on extensive datasets, which contains the information of students, e.g. basic profiles, Internet using logs and course scores. First, several statistical methods are used to find the associations between students' surfing habits and academic performance. Then, a learning algorithm is devised to cluster the students according to their different surfing habits. Furthermore, we develop a BP neural network to predict the rate of failing a test of a student based on his/her basic information, surfing habits and the clustering information. According to the aforementioned approaches, we find an interesting result that the academic performance of students in universities is quite possibly enhanced if the network connections are always terminated at 24:00 p.m.

Keywords: Internet surfing habits · Course score · Exam-failure rate · Cluster analysis · Neural network analysis

1 Introduction

According to the Statistics of Chinese Ministry of Education in 2016 [1], there are 2,852 universities and colleges in China, and the number of university students reaches 3.7 billion, which is the most around the world. With the development the Internet, it is convenient for university students to use it in almost everywhere in the campus through PCs, Laptops, PADs, Smart phones and so on. The Internet affects every aspect of the life of students including learning, entertainment, social activity, etc. Students are encouraged to communicate with teachers, instructors and classmates through the Internet and search information on it.

© Springer International Publishing AG 2016
S. Song and Y. Tong (Eds.): WAIM 2016 Workshops, LNCS 9998, pp. 174–185, 2016.
DOI: 10.1007/978-3-319-47121-1_15

Some researches indicate that academic performance can be promoted with the help of the Internet. Rashid and Asgharb suggests that the use of the Internet has a direct and positive influence on students' engagement and self-directed learning [9]. Dr. Suhail and Bargees's investigation finds that a great number of students demonstrate positive effects of Internet using on educational aspect [13]. In the U.S., 68 % of the parents, 69 % of the students, and 69 % of the teachers say that they have authenticated the improvement of students' grades through the use of the Internet on an online survey reports [2]. In addition, the Internet contributes significantly to the academic performance of the students in university of Botswana, Gaborone, possibly because the students can learn at anytime and anywhere with the Internet [14]. The broad use of the Internet in universities can make students have better perceptions of learning [4].

However, there are also some researches indicating that the Internet using has negative effects. General Internet surfing habits of students show that some people may experience psychological problems such as social isolation, depression, loneliness, and time mismanagement because of their Internet using [7]. Bad academic performance has been confirmed to be correlated highly with heavier recreational Internet using. Loneliness, staying up late, tiredness, and missing classes are also correlative to self-reports of Internet-caused impairment [8]. Heavy Internet usage may have bad relationships with teachers, bad academic grades and low learning satisfaction than general users. And they are apt to be depressed, physically ill, lonely, and introverted [3]. Especially, with the increasing popularity of smart phone, more people start relying on it. The data from the Pew Research Center in 2015 shows that 46 % of smart phone owners in the U.S. say that their smart phone is something "they could not live without". Smart phones are used all day for quite a lot of reasons, including communication, learning, entertainment, information seeking, social networking and gaming [17]. In particular, recent research shows that online social communications based on smart phones significantly affect offline behavior of users [10–12,18–20]. University students who are at a high risk of smart phone addiction may be less likely to have good GPAs [5]. In addition, excessive use of the social network systems results in negative effects. For example, multitasking with Facebook has been found to be significantly negatively predictive of GPAs for freshmen and sophomores [6].

The correlation between students' Internet using and their academic performance is a complex issue. Average daily connection time, as a general used metric to determine Internet addiction, may not be accurate because students may study using the Internet for a long time other than being addicted to online entertainment. In fact, Internet addiction of students in universities is reflected not only by Internet using time but also by other features such as online and offline time, volume of Internet traffic and other potential factors. The relationships between these factors and the performance of students are very complex. In addition, students' background features, including gender, major, hometown, etc., differ from one another, which may lead to different online behaviors. All these features should be taken into consideration when evaluating the impact of

Internet using habits on academic performance of university students. We make the following contributions.

- We reconfirm that no obvious relationship between the Internet using habits and the academic performance can be found using linear regression.
- We construct a model to predict whether a student will fail a test using the Internet using and other features.
- Based on the prediction model, we make a suggestion that the network connections should be terminated at 0:00 to improve students' academic performance.

The rest of the paper is organized as follows. In Sect. 2, we introduce the datasets and the methods used in this paper. In Sect. 3, we give the analysis results. We conclude the paper in Sect. 4.

2 Preliminaries

2.1 Datasets

In order to get accurate results, the data employed in this study are obtained from several source providers, e.g., the Internet using data are obtained from ITS (IT Service Department) of the university, the student academic performance data are obtained from the Academic Affairs office, and the students' basic data (background features) are obtained from the Admission department. There are four grades of students, ranging from freshman to senior college students. Students' basic dataset has 23,175 records, Internet using dataset has 321,651,912 items, and academic performance dataset has 510,261 courses information records. Each record of the students' basic dataset contains the following attributes: ID number, gender, grade, school, major and source province. Each record of the Internet using dataset contains the following attributes: ID number, online time, offline time, Internet traffic volume (bit) and connection time. Each record of the academic performance dataset contained the following attributes: ID number, course number, course grade, year and semester.

2.2 Data Preparation and Statistics

A record of in the students' basic dataset is represented as $(d_1, p_1, b_1, m_1, e_1)$, where d_1 is the key of the student's basic information table, $p_1 \in P$ is the province from which the student comes, b_1 is the birthday, m_1 is the major and $e_1 \in E$ is the nationality. P and E are the collections of provinces and nationalities in China, respectively.

A record in the Internet using habit dataset is represented as (d_2, l_2, f_2, b_2), where d_2 is the key of the table, l_2 and f_2 is the online time and the offline time respectively and b_2 is the Internet traffic volume. Two important features included to reflect the Internet using habit of the students are the online time and the Internet traffic volume.

A record in the dataset of student grade is represented as (d_3, c_3, g_3), where d_3 is the unique ID of the student in this table, c_3 is the course ID, and g_3 is the examine score of the corresponding course.

2.2.1 Internet Traffic Volume per Week of Students

We use the Internet using habit data for statistics, and get the weekly Internet traffic volume of the university students. The average traffic volume per week of a student is about 9 Gb. The average weekly traffic volume is divided into seven categories: (1) less than 5 Gb, (2) 5–10 Gb (excluding 10 Gb), (3) 10–15 Gb (excluding 15 Gb), (4) 15–20 Gb (excluding 20 Gb), (5) 20–25 Gb (excluding 25 Gb), (6) 25–30 Gb (excluding 30 Gb), (7) no less than 30 Gb. 36.5 % of the students use 10–15 Gb Internet traffic volume weekly, taking the majority. 24.3 % of the students use 15–20 Gb Internet traffic volume, which is the second most, and 5–10 Gb followed as the third. Only 2.2 % of the students use no less than 30 Gb per week. More than 60 % of the students' weekly Internet traffic volume is between 10 Gb and 20 Gb. Figure 1 shows the distribution of Internet traffic volume per week.

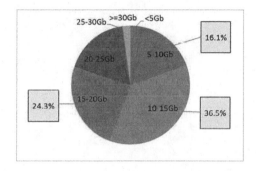

Fig. 1. The distribution of Internet traffic volume per week

2.2.2 Internet Connection Time of Students

We use average Internet connection time per week of a student as a factor. The average value of all students is 32 hours. Figure 2 shows the distribution of average Internet connection time per week. similarly, we divide the time into seven intervals: (1) less than 10 h, (2)10–20 h (excluding 20 h), (3) 20–30 h (excluding 30 h), (4) 30–40 h (excluding 40 h), (5) 40–50 h (excluding 50 h), (6) 50–60 h (excluding 60 h), (7) no less than 60 h. From the figure it can be observed that most students keep online for 20–50 h per week. The top-3 intervals with the most students are 30–40 h, 20–30 h and 40–50 h, taking the percentage of 37.3 %, 25.6 % and 19.1 % respectively. More than 1/3 of the students keep online for 30–40 hours per week. It seems that the students are most likely to connect Internet for 20 to 50 h every week, which accounted for about 82 % in the general situation. The distribution of Internet using time in one-day (24 h) is depicted as Fig. 3. 11:00 a.m., 20:00 p.m. and 24:00 p.m. takes the leading places, and the accumulative quantities of the Internet using time at 11:00, 12:00, 13:00, 20:00, 21:00, 23:00, 24:00 are more than 50 % of the total. After 24:00 p.m., about

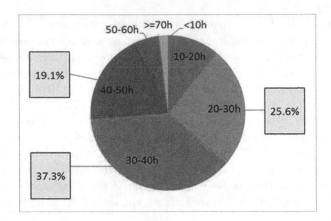

Fig. 2. The distribution of average weekly Internet connection time

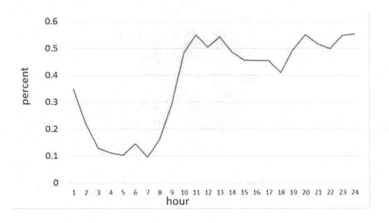

Fig. 3. The distribution of average weekly Internet connection time

20 % of the students disconnect the Internet for sleeping, but there are still more than 20 % of the students keeping online. From 3:00 to 7:00 a.m., the Internet connection quantities reach the lowest level.

2.2.3   Academic Performance of the Students

Figure 4 shows the distribution of the academic performance. According to the scores the students obtain, grades are divided into five parts: (1) less than 60, (2) 60–70 (excluding 70), (3) 70–80 (excluding 80), (4) 80–90 (excluding 90), (5) no less than 90. The maximum score on all the tests is 100. The scores of more than 1/3 of all students range from 80 to 90. The rate of failing an exam is 13.7 %.

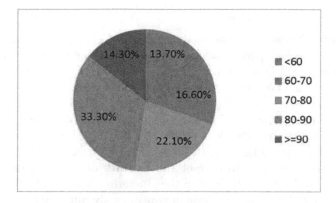

Fig. 4. Academic performance of students

2.2.4 Relationship Between Connection Time and Academic Performance

Will academic performance be determined by Internet using? Do the 1.9 % of the students who connect the Internet more than 60 h per week always fail the exams? The distribution of the academic performance based on the connection time is showed in Fig. 5. The outliers should be neglected, and it seems that there is no obvious characteristic of their relationship. The students with different connection time have similar distribution of scores. In order to do further analysis, the linear regression is used to analyze the relationship between the Internet connection time and the academic performance. We get the correlation coefficient of 0.016321. This result suggests that they no significant correlationship.

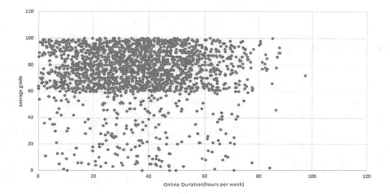

Fig. 5. The distribution of academic result based on Internet connection time

2.3   The System Framework

However, due to the data complexity, the relationship between the Internet using habit and the academic performance cannot be analyzed simply by calculating the correlation between the exam performance and the Internet connection time. More features should be taken into consideration. Therefore, multivariate fitting is introduced to get a higher degree of fitting relationship. In order to get more accurate relationship, the techniques of clustering and neural network are used. The workflow of the framework is showed in Fig. 6. Three types of datasets including the Internet using data, the academic performance data and the students' basic data, are used to extract the features of students for clustering. Then we use neural network to fit with the clustering results. The recommended disconnection time is calculated through continuously automatic fitting.

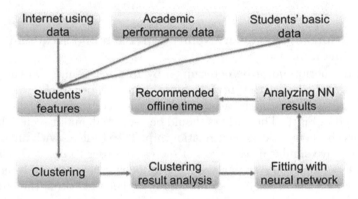

Fig. 6. Process of the framework

2.4   The Data Models

The three datasets which are the Internet using data, the academic performance data and the student basic data, are employed to extract the following features: Internet using habit, source province, gender, major, connection time per week and Internet traffic volume per week.

Internet using habit is defined as the connection time series of one-day. 24 elements $(t_0, t_1, t_2, t_3, t_4, \ldots, t_{23})$ are contained, where, $t_i (i = 0, 1, 2, \ldots, 23)$ represents the number of days, (d_i) that a student is online during a certain hour of day (i.e. the i th hour), T is the total days of sampled days(a year, 365 days):

$$t_i = d_i/T \tag{1}$$

Every 4 continuous hours in a day is used to be a period as the factors to run clustering algorithms. The items are expressed as $(f_1, f_2, f_3, f_4, f_5, f_6)$, and

$$f_j = \frac{\sum_{l=4(j-1)}^{4j-1} t_l}{4} \tag{2}$$

A student's performance in courses is represented by a boolean value where 0 represents passing the course and 1 represents failing. P is the score of course. C is the collection of all the courses studied in the sample semesters

$$P = \begin{cases} 0, \forall p \in C, p \geq 60 \\ 1, \exists p \in C.p < 60 \end{cases} \qquad (3)$$

2.5   Cleansing Sensitive Data

To get accurate result, the noise and uncertain data should be eliminated firstly. For example, the records of very short connection time or very low Internet traffic volume should be removed. Since these sensitive data are infrequent, we use the uncertain frequent itemset mining techniques [15,16,21] to discover infrequent records. Although some data reach the selection standard of the Internet connection time threshold and the Internet traffic volume threshold, their periods (number of days) are relatively low, possibly less than one month during the two-year period. These discontinuous data should not be used to represent long-term behavior, and is removed as well.

2.6   Clustering Analysis

We use DB-SCAN as the clustering algorithm. A 6-dimensional vector is defined as the input, the smallest class cluster is limited within 1,500, and the scanning radius is 0.15623. Finally, 15,062 results data and 11,023 noise data are obtained during a year. The result containing six clusters is depicted in Fig. 7, which shows the line chart of the Internet habit of the six clusters. The before-dawn online period (from 1 am to 6 am) of clusters 4, 5 and 6 are nearly 0. For cluster 1, 2, and 3, the data may refer to the Internet online records of several students. Comparing to other periods, the Internet connections of this period in these clusters are also low, although, the percentage of the average connection is approximately 5 %. 20:00 p.m. is the peak of the average online time. At this time point, the online proportions are the highest among most clustered groups. For each cluster, the Internet using habits of morning and afternoon are different. Some fluctuates largely, while the others are relatively steady.

2.7   Classification Analysis

These three datasets in that year are used for classification analysis. There are 15,062 records of the students' basic information and the Internet using habit data, among which, 12,000 records of data are used as training set, and 2,500 used as testing set for accuracy verification with cross validation. Classification analysis is carried out using BP Neural Network. BP Neural Network is composed of input layer, hidden layers and output layer. The structure of the BP Neural Network is showed in Fig. 8. In this study, 7 features of each student are used in a BP neural network training, and the output layer node represents the exam failure. The used 7 features are: the student's Internet habit, source province,

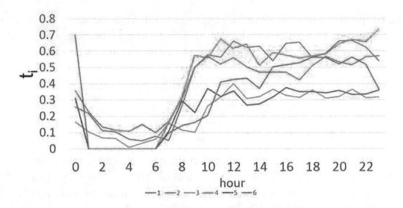

Fig. 7. Internet habit of 6 clusters

gender, major, weekly Internet connection time, weekly Internet traffic volume, and the cluster ID which the student belongs to. The offline in evening may be the best time for improving exam score, so we make the hypothesis that adjusting the offline time in evening will help reducing the exam failure rate. The student's Internet habit is represent with nighttime Internet using habits $(t_{21}, t_{22}, t_{23}, t_0, t_1, t_2, t_3, t_4, t_5)$. Thus, The input layer is divided into 15 nodes: province, major, gender, age, Internet traffic volume, Internet connection time, and Internet using habit which comprises $(t_{21}, t_{22}, t_{23}, t_0, t_1, t_2, t_3, t_4, t_5)$. The output layer is exam failure, where, 1 represents that the student has failed at least one test of the course, and 0 indicates that the student has no exam failure.

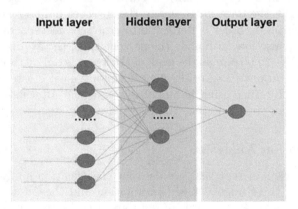

Fig. 8. The structure of the BP Neural Network

With 15 input layer nodes, and 1 output layer node, according to the empirical formula, the quantity of the hidden layer nodes ranges from 5 to 14. Making the largest study iteration number as 5,000, and under the accuracy condition of 0.1, the results indicates that with 11 hidden layer nodes, the optimal effect reaches 91.625 %. So, accuracy of the hidden layer increases when the nodes added from 5 to 11, while decreases with nodes from 11 to 14.

After the establishment of neural network modeling, the model is used to predict the effect of the disconnection time on students' exam failure rate. Five groups of students are selected randomly from cluster 4, 5, 6. Each group has 500 students, and their actual failure rates are as shown in Table 1.

Table 1. Failure rate of each group

Group number	Failure rate
1	0.132
2	0.126
3	0.112
4	0.122
5	0.126

3   The Experimental Results

3.1   Disconnection Time v.s. Exam Failure

By changing $t_{21}, t_{22}, t_{23}, t_0, t_1$ to 0 respectively, the simulation data is obtained and shown in Fig. 9. The failure rates of the five groups is declining smoothly till 24:00 p.m., and reach the lowest at 24:00 p.m. It indicates that there may be some students use Internet for course learning before 24:00 p.m, and hence 24:00 p.m. should be the time to disconnect the Internet. If the students keep online till 1:00 a.m., the failure rates increase sharply. So taking the 24:00 p.m. as the disconnection time would improve students' academic performance in this university.

3.2   The Influence of Other Factors

In addition to online time, there are other elements being used for neural network modeling to predict the students' academic performance. Source provinces and academic majors have some influence on academic performance. Hebei province, Hunan province and Hubei province where students come from are the top 3 provinces of reduction of exam failure rate in response to adjusted offline time (12:00 p.m.). Similarly, students who major in mechanics, transportation, and Liberal arts show a similar trend in their academic performance in response to the adjusted time. However, the difference between students of Hans and other nationalities is not obvious. That is to say the nationality of students has nothing to do with the Internet connection limitations which would lead to decline of the exam-failure rate.

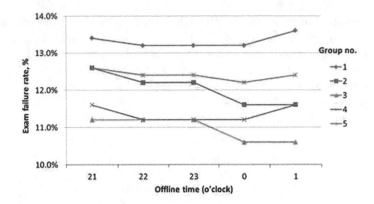

Fig. 9. The failure rate of the five group

3.3 Incorrect Prediction Explanation

About 9 % of the incorrect predictions mainly come from the insufficient data acquisition and condition prediction. Even though neural network has high multivariate fitting effect, the analysis of students' academic performance has high complexity, and factors with the students' attributes of this study are not enough.

4 Conclusion

With the popularity of campus network service, it is meaningful to research the relationship between the Internet using habits and the academic performance of university students. A comprehensive dataset of the past five years from an university in China are analyzed in this paper. Firstly, statistical methods are used, and we reconfirm that no obvious relationship between the Internet using habits and the academic performance can be found with linear regression. Then, data mining methods are applied. The clustering algorithm DB-SCAN is used to classify the online users, and a BP neural network is trained for deep mining. Lastly, the best moment of 24:00 p.m. is gotten by simulating the disconnection time with the trained BP neural network model. Therefore, in order to improve students' academic performance, we suggest that the Internet should be disconnected at 24:00 p.m.

References

1. http://www.moe.gov.cn
2. http://www.att.com/press/0998/980902.csa.html
3. Chen, Y., Peng, S.: University students' internet use and its relationships with academic performance, interpersonal relationships, psychosocial adjustment, and self-evaluation. CyberPsychol. Behav. **11**(4), 467–469 (2008)

4. Cheung, W., Huang, W.: Proposing a framework to assess internet usage in university education: an empirical investigation from a student's perspective. Br. J. Educ. Technol. **36**(2), 237–253 (2005)
5. Hawi, N., Samaha, M.: To excel or not to excel: Strong evidence on the adverse effect of smartphone addiction on academic performance. Comput. Educ. **98**, 81–89 (2016)
6. Junco, R.: Student class standing, facebook use, and academic performance. J. Appl. Dev. Psychol. **36**, 18–29 (2015)
7. Kraut, R., Patterson, M., Lundmark, V., Kiesler, S., Mukophadhyay, T., Scherlis, W.: Internet paradox: A social technology that reduces social involvement and psychological well-being? Am. Psychol. **53**(9), 1017 (1998)
8. Kubey, R., Lavin, M., Barrows, J.: Internet use and collegiate academic performance decrements: Early findings. J. Commun. **51**(2), 366–382 (2001)
9. Rashid, T., Asghar, H.: Technology use, self-directed learning, student engagement and academic performance: Examining the interrelations. Comput. Hum. Behav. **63**, 604–612 (2016)
10. She, J., Tong, Y., Chen, L.: Utility-aware social event-participant planning. In: SIGMOD 2015, pp. 1629–1643 (2015)
11. She, J., Tong, Y., Chen, L., Cao, C.C.: Conflict-aware event-participant arrangement. In: ICDE 2015, pp. 735–746 (2015)
12. She, J., Tong, Y., Chen, L., Cao, C.C.: Conflict-aware event-participant arrangement and its variant for online setting. IEEE Trans. Knowl. Data Eng. **28**(9), 2281–2295 (2016)
13. Suhail, K., Bargees, Z.: Effects of excessive internet use on undergraduate students in pakistan. CyberPsychol. Behav. **9**(3), 297–307 (2006)
14. Tella, A.: Undergraduates uses of the internet: implications on academic performance. J. Educ. Media Libr. Sci. **45**(2), 161–185 (2007)
15. Tong, Y., Chen, L., Cheng, Y., Yu, P.S.: Mining frequent itemsets over uncertain databases. Proc. VLDB Endow. **5**(11), 1650–1661 (2012)
16. Tong, Y., Chen, L., Ding, B.: Discovering threshold-based frequent closed itemsets over probabilistic data. In: ICDE 2012, pp. 270–281 (2012)
17. Tong, Y., She, J., Chen, L.: Towards better understanding of app functions. J. Comput. Sci. Technol. **30**(5), 1130–1140 (2015)
18. Tong, Y., She, J., Ding, B., Chen, L., Wo, T., Xu, K.: Online minimum matching in real-time spatial data: experiments and analysis. Proc. VLDB Endow. **9**(12), 1053–1064 (2016)
19. Tong, Y., She, J., Ding, B., Wang, L., Chen, L.: Online mobile micro-task allocation in spatial crowdsourcing. In: ICDE 2016, pp. 49–60 (2016)
20. Tong, Y., She, J., Meng, R.: Bottleneck-aware arrangement over event-based social networks: the max-min approach. World Wide Web J. **19**(6), 1151–1177 (2016)
21. Tong, Y., Zhang, X., Chen, L.: Tracking frequent items over distributed probabilistic data. World Wide Web J. **19**(4), 579–604 (2016)

Preference-Aware Top-k Spatio-Textual Queries

Yunpeng Gao[1](✉), Yao Wang[2], and Shengwei Yi[3]

[1] State Key Laboratory of Software Development Environment,
Beihang University, Beijing, China
setfire1@163.com
[2] Yellow River Engineering Consulting Co., Ltd., Zhengzhou, China
[3] China Information Technology Security Evaluation Center, Beijing, China
yisw@itsec.gov.cn

Abstract. A novel query type named preference-aware top-k spatial-textual query is proposed in this paper. Unlike the common measure, this query returns spatial-textual objects ranked according to the warefare of the facilities in conditional range. Suppose a tourist that looks for motels that have nearby a highly rated Japanese restaurant that serves sushi. The proposed query considers not only the spatial location and textual description of spatial-textual objects (such as motels and restaurants), but also additional information such as ratings that describe their quality. Furthermore, spatial-textual objects (i.e., motels) are ranked according to the features of facilities (i.e., restaurants) in their neighborhood. However, it is time-consuming to compute the score of data object. To address this issue, we propose an efficient algorithm for processing this query. Last but not least, we conduct extensive experiments for evaluating the performance of the proposed methods.

1 Introduction

An increasing number of applications support location-based queries, which retrieve the most interesting spatial objects based on their geographic location. Recently, spatial-textual queries are gaining in prominence since such queries integrate location-based retrieval with textual information. Most of the existing queries only consider retrieving objects that satisfy a spatial constraint ranked by their spatial-textual similarity to the query point. Actually people are quite often interested in spatial objects according to the quality of other facilities (feature objects) that are located in their vicinity. Feature objects are typically described by non-spatial attributes such as quality or rating, in addition to the textual description. In this paper, we propose a novel query type, preference-aware top-k spatial-textual query, for spatial-textual objects ranked by the warefare of the facilities in conditional range.

Suppose a tourist that looks for motels that have nearby a well rated Japanese restaurant that serves sushi. Figure 1 depicts a spatial area containing motels (data objects) and restaurants (feature objects). The quality of the restaurants based on existing reviews is depicted next to the restaurant. The tourist also specifies a spatial constraint (in the figure depicted as a range around each

© Springer International Publishing AG 2016
S. Song and Y. Tong (Eds.): WAIM 2016 Workshops, LNCS 9998, pp. 186–197, 2016.
DOI: 10.1007/978-3-319-47121-1_16

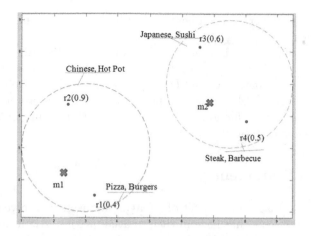

Fig. 1. A motivating example

motel) to restrict the distance of the restaurant to the motel. Obviously, the motel m_2 is the best option for a tourist that poses the aforementioned query. Even though spatial preference queries have been studied before [9,18], their definition ignores the available textual information. In our example, the spatial preference query would correspond to a tourist that searches for motels that are nearby a qualified restaurant and the motel m_1 would always be retrieved, irrespective of the textual information.

In this paper, we define preference-aware top-k spatial-textual queries and provide efficient algorithms for processing this query type. Compared to traditional spatial preference queries [9,18], a main challenge is that the score of a data object changes depending on the query keywords, which makes methods that rely on materialization such as [9] not suitable. Moreover, processing spatial preference queries is costly in terms of both I/O and execution time [18]. Intuitively, processing preference-aware top-k spatial-textual queries is to compute the spatial-textual preference score for each object and then return the k objects with the highest score. We call this approach $spatial - textual\ Data\ Ordering\ (STDO)$ and take it as a baseline. Moreover, an efficient and scalable algorithm, called $Preference - Aware\ Spatial - Textual\ search$, is proposed for processing the queries. $PASTS$ follows a different strategy than $STDO$, as it retrieves highly ranked feature objects first, and then searches data objects in their conditional range. To summarize the contribution of this paper are:

(a). We propose a novel query type, called preference-aware top-k spatial-textual query that ranks the data objects due to the quality and textual relevance of feature objects located in their vicinity.

(b). We present two algorithms for processing spatial-textual preference queries, namely $spatial - textual\ Data\ Ordering$ and $Preference - Aware\ spatial - textual\ Search$.

(c). We conduct several experiment for studying the performance of our proposed algorithms.

The rest of this paper is organized as follows: In Sect. 2, we define the preference-aware top-k spatial-textual query. In Sect. 3 we describe our baseline algorithm, called spatial-textual Data Ordering ($STDO$). Preference-Aware spatial-textual Search ($PASTS$), is proposed in Sect. 4. We present the experimental evaluation in Sect. 5 and overview the relevant literature in Sect. 6. Last, we make a conclusion in Sect. 7.

2   Problem Statement

Given an object dataset O and a set of c feature datasets, we solve the problem of finding k data objects that have in their spatial proximity well ranked feature objects that are relevant to the given query keywords. Each data object has a spatial location. Similarly, each feature object is associated with a spatial location but also with a non-spatial score that indicates the quality and its value domain is $[0, 1]$. Table 1 provides an overview of the symbols used in this paper.

Definition 1. The preference score $s(t)$ of feature object t based on a user-specified set of keywords W is defined as $s(t) = (1-\lambda) \cdot t.s + \lambda \cdot sim(t, W)$, where $\lambda \in [0, 1]$ and $sim()$ is a textual similarity function.

For example, consider the restaurants depicted in Table 2. Given a set of keywords $W = \{italian, pizza\}$ and $\lambda = 0.5$, the restaurant with the highest preference score is $Ontario's\ Pizza$ with a preference score $s(r4) = 0.9$, while the score of $Beijing Restaurant$ is $s(r1) = 0.3$.

Definition 2. The preference score $\tau_i(p)$ of data object p based on the feature set F_i is defined as $\tau_i(p) = \max\{s(t)|t \in F_i : dist(p, t) \leq r, sim(t, W_i) > 0\}$,

Table 1. Overview of symbols.

Symbol	Description
O	Set of data objects
p	Data object, $p \in O$
c	Number of feature sets
F_i	Feature sets, $i \in [1, c]$
t	Feature object, $t \in F_i$
$t.s$	Non-spatial score of t
$t.W$	Set of keywords of t
$sim(t, W)$	Textual similarity between t and W
$s(t)$	Preference score of t
$\tau_i(p)$	Preference score of p based on F_i
$\tau(p)$	Spatial-textual preference score of p

Table 2. Feature objects (Restaurants).

	Name	Rating	x	y	Description
r1	Beijing Restaurant	0.6	1	2	Chinese, Asian
r2	Japanese Restaurant	0.5	4	1	Japan, Sushi
r3	Espanol Restaurant	0.8	5	8	Italian, European
r4	Ontario's Pizza	0.8	7	6	Pizza, Italian

where the $dist(p, t)$ denotes the spatial distance between data object p and feature object t and we employ the Euclidean distance function.

Definition 3. The overall spatial-textual preference score $\tau(p)$ of data object is defined as $\tau(p) = \sum_{i=1}^{c} \tau_i(p)$

PROBLEM. Preference-Aware Top-k Spatial-Textual Queries (PATSTQ): Given a query Q, defined by an integer k, a radius r and c sets of keywords W, find the k data objects $p \in O$ with the highest spatial-textual score $\tau(p)$.

3   Spatial-Textual Data Ordering (STDO)

Our baseline approach, called spatial-textual data ordering ($STDO$), computes the spatial-textual score $\tau(p)$ of each data object $p \in O$ and then returns the k data objects with the highest score.

Algorithm 1 shows the process of STDO. In more detail, for a data object p, its score $\tau_i(p)$ for every feature set F_i is computed (lines 3–5). The details on this computation for range queries are described that will be presented in the sequel. Interestingly, for some data objects p we can avoid computing $\tau_i(p)$ for some feature sets. This is feasible because we can determine early that some data objects cannot be in the result set R. To achieve this goal, we define a threshold τ which is the k-th highest score of any data object processed so far. In addition, we define an upper bound τ' for the spatial-textual preference score $\tau(p)$ of p, which does not require knowledge of the preference scores $\tau_i(p)$ for all feature sets F_i. The algorithm tests the upper bound τ' based on the already computed $\tau_i(p)$ against the current threshold (line 6). If it is smaller than the current threshold, the remaining score computations are avoided. After computing the score of p, we test whether it belongs to R (line 6). If this is case, the result set R is updated (line 7), by adding p to it and removing the data object with the lowest score (in case that $|R| > k$). Finally, if at least k data objects have already been added to R, we update the threshold based on the k-th highest score (line 9).

Performance improvements. For sake of simplicity, we omit the implementation details, even though we use this improved modification in our experimental evaluation. The remaining challenge is to compute efficiently the score based on the spatial-textual information of the feature objects. The goal is to reduce the number of disk accesses for retrieving feature objects that are necessary for computing the score of each element $p \in O$.

Algorithm 1. spatial-textual Data Ordering (STDO)

    Input: Query $Q = (k, r, W_i)$
    Output: Result set R sorted based on $\tau(p)$

1: $R = \emptyset; \tau = -1;$
2: **for** each $p \in O$ **do**
3:    **for** each $i = 1 \ldots c$ **do**
4:      **if** $\tau' > \tau$ **then**
5:        $\tau_i(p) = F_i.computeScore(Q, p)$
6:      **end if**
7:    **end for**
8:    **if** $\tau(p) > \tau$ **then**
9:      $update(R);$
10:     **if** $|R| > k$ **then**
11:       $\tau = k^{th} score;$
12:     **end if**
13:    **end if**
14: **end for**

---

4   Preference-Aware Spatial-Textual Search (PASTS)

In this section we propose a novel and efficient algorithm, called Preference-Aware spatial-textual Search ($PASTS$), for processing preference-aware top-k spatial-textual queries. $PASTS$ follows a different strategy than $STDO$, as it involves two major steps, namely finding highly ranked feature objects first, and then retrieving data objects in conditional range. Intuitively, if we find a neighborhood in which highly ranked feature objects exist, then the neighboring data objects are naturally highly ranked as well.

4.1   Combination of Feature Objects

In a nutshell, the goal is to find sets of feature $C = \{t_1, t_2, \ldots, t_c\}$ objects where $t_i \in F_i$ $(1 \leq i \leq c)$, such that the spatial-textual preference score of each t_i is as high as possible and the feature objects are located in conditional range.

    In the general case, a data object may be highly ranked even in the case where a certain kind of feature object does not exist in its neighborhood, though feature objects of other kinds might compensate for this. For example, consider the extreme case where all data objects have only one type of feature object in their spatial neighborhood. This feature object is used for presenting unified definitions for the case where the spatial-textual score of the top-k data objects is defined based on less than c feature objects.

4.2   PASTS Overview

Algorithm 2 provides an insight to PASTS. At each iteration, the following steps are followed: (i) a special iterator (line 3) returns successively the combinations of feature objects sorted based on their score, (ii) up to k data points in the conditional range of these features are retrieved (line 5). Data objects that have already been previously retrieved are discarded, while the remaining data object p have a score $\tau(p) = s(C)$ and can be returned to the user incrementally. If k data objects have been returned to the user (line 2), the algorithm terminates without retrieving the remaining combinations of feature objects. Differently to the $STDO$, $PASTS$ retrieves only the data objects that most certainly belong to the result set.

Algorithm 2. Preference-Aware spatial-textual Search (PASTS)

    Input: Query $Q = (k, r, W_i)$
    Output: Result set R sorted based on $\tau(p)$
1: $C = \emptyset$;
2: **while** ($|R| \leq k$) **do**
3:    **while** ($\exists i : not\ heap_i.isEmpty()$) **do**
4:      $i = \text{nextFeatureSet}()$;
5:      $e_i = heap_i.pop()$;
6:      **while** ($not\ e_i\ is\ a\ data\ object$) **do**
7:        **for** $childEntry$ in $e_i.childNodes$ **do**
8:          $heap_i.push(childEntry)$,
9:        **end for**
10:       $e_i = heap_i.pop()$;
11:      **end while**
12:      $D_i = D_i \bigcup e_i$;
13:      $heap.push(validCombinations(D_1, \ldots, e_i, \ldots, D_c))$;
14:      $min_i = s(e_i)$;
15:      $\tau = \max(max_1 + \ldots + min_j + \ldots + max_c)$;
16:      $C = heap.top()$;
17:      **if** ($score(C) \geq \tau$) **then**
18:        $heap.pop()$;
19:      **end if**
20:    **end while**
21:    $R = R \bigcup getDataObjects(C)$;
22: **end while**
23: **return** R;

---

5   Experimental Evaluation

In this section, we evaluate the performance of our algorithms $STDO$ and $PASTS$, presented previously in Sects. 3 and 4 respectively, for processing preference-aware top-k spatial-textual queries over large data. All experiments implemented in C++, and are performed on an Intel i5-4590 processor equipped with 8 GB RAM.

Table 3. Experimental parameters.

Parameters	Range
Cardinality of dataset	50 K, **100 K**, 500 K, 1 M
Cardinality of features sets	50 K, **100 K**, 500 K, 1 M
Number of feature sets C	**2**, 3, 4, 5
Indexed keywords	64, **128**, 192, 256
Radius r	0.005, **0.01**, 0.02, 0.04, 0.08
k	5, **10**, 20, 40, 80
Smoothing parameter	0.1, 0.3, **0.5**, 0.7, 0.9
Queried keywords	1, **3**, 5, 7, 9

5.1 Experimental Setup

In our experimental evaluation, we vary four important parameters of the datasets in order to study the scalability of the proposed techniques. These parameters are: the cardinality of the feature sets $|F_i|$, the cardinality of the set of data objects $|O|$, the number of feature sets $|C|$, and number of distinct keyword. Moreover, we study four different query parameters to study how the characteristics of the query influence the performance of the algorithms. We vary the query radius r, the number k of retrieved data objects, the smoothing parameter λ between textual similarity and non-spatial score, and the number of keywords of the query for each feature set. Tested ranges for all parameters are shown in Table 3. The default values are denoted as bold. When we vary one parameter, all others are set to their default values.

For evaluating our algorithms, we use both real and synthetic datasets. The real dataset, which was obtained from yelp.com, describes hotels ($\simeq 25$ K objects) and restaurants ($\simeq 79$ K objects). In more details we collected restaurant and hotels that are annotated with their location. Moreover, for the collected restaurants we extracted their rating and their textual description of the served food. The number of distinct values of keywords is around 130 and each restaurant description may contain one or more keywords. Our datasets contain collected hotels and restaurants are sufficient data. In addition, we created synthetic clustered datasets of varying size, number of keywords and number of feature sets. Approximately 10000 clusters constitute each synthetic dataset.

5.2 Scalability Analysis

In this section, we evaluate the impact of varying different parameters on the efficiency of our algorithms. In order to perform a scalability analysis, we employ the synthetic dataset for this set of experiments.

Figure 2 illustrates the results for the experiment for $STDO$ and $PASTS$. In summary, the results clearly demonstrate that two proposed algorithms scale

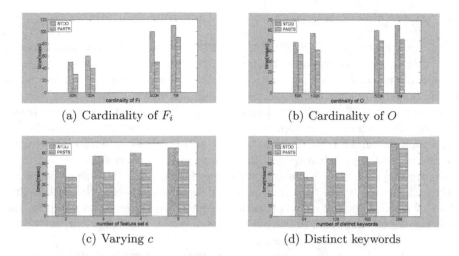

(a) Cardinality of F_i (b) Cardinality of O

(c) Varying c (d) Distinct keywords

Fig. 2. Scalability for synthetic dataset

with all parameters and in all cases, the $PASTS$ algorithm exhibits high performance, as witnessed by the low execution time, which stems from its ability to quickly identify qualified feature combinations. Consequently, the significant gains in processing time (orders of magnitude compared to $STDO$) are mostly due to the effective design of the algorithm.

Figure 2(a) shows the execution time when increasing the cardinality of the feature sets. $PASTS$ scales well since the execution time increases only when increasing the dataset by one order of magnitude. This increase is due to the increased size of the data structures and the additional processing required to traverse a bigger data structure and find valid combinations of high score.

Figure 2(b) shows the obtained results when increasing the number of data objects. Again, $PASTS$ scales well. In fact, it is even better than in the previous $STDO$. Obviously, a larger dataset of data objects does not affect the performance so much as larger feature sets.

In Fig. 2(c), we increase the number of feature categories. As expected, this has a stronger effect on performance, since the cost required to retrieve the highest ranked combinations increases with the number of possible combinations. Still, the performance of $PASTS$ is not severely affected.

In Fig. 2(d), we illustrate how the performance is affected by the number of distinct keywords in the dataset. Apparently, a higher number of keywords cause higher execution times. The reason is twofold. First, as the number of distinct keywords increases, it is less probable to find feature objects that are described by all queried keywords, thus more feature objects need to be retrieved in order to ensure that no other combination has a higher score. Secondly, the node capacity of the index structures drops, thus the height of the index structures may increase, thus causing more IOs.

5.3   Varying Query Parameters

In Fig. 3, we study the effect of varying query parameters for the real dataset. First, in Fig. 3(a), we evaluate the impact of increasing the query radius on the performance of $PASTS$. We notice that for smaller values of r the execution time increases. For small radius, access to more qualified combinations of feature objects is required, since only few data objects are located in their neighborhood. Therefore, for both indexing approaches the execution time increases mainly due to the increase of the IOs. Since a high percentage of the feature objects need to be retrieved for each feature set, the gain of indexing is small.

Figure 3(b) illustrates the execution time when varying the size of result set k. Overall, the execution time increases as k increases. Specifically, with higher values of k more combinations of feature objects are retrieved to compose the result set, which again lead to more IOs to retrieve the qualifying feature objects that constitute valid combinations.

In Fig. 3(c), we vary the smoothing parameter λ. In general, both approaches exhibit relatively stable performance for varying values of λ. The performance of the methods not affected by the smoothing parameter, since the feature objects are not grouped into blocks based on the non-spatial score nor based on their textual similarity. We note for that objects with similar textual descriptions are stored throughout the index, regardless of their non-spatial score; unlike the index where they are clustered together in the same block. As a result, a significant overhead is evident when searching for relevant objects all over the index. On the other hand, the index is built by taking into account non-spatial score, the textual information and the spatial location. Thus, $PASTS$ that uses index is consistently more efficient regardless of the value of the smoothing parameter.

In Fig. 3(d), we vary the number of queried keywords per feature set from 1 to 9. The number of queried keywords has little impact on performance, except

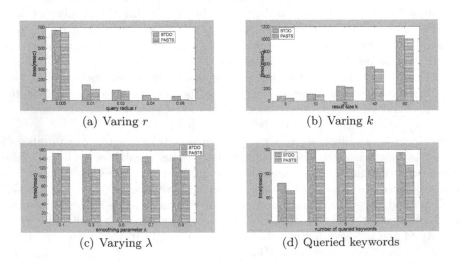

(a) Varing r (b) Varing k

(c) Varying λ (d) Queried keywords

Fig. 3. Scalability for synthetic dataset

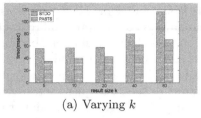 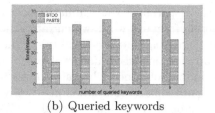

(a) Varying k (b) Queried keywords

Fig. 4. Query parameters for synthetic dataset

for the special case where one keyword is queried for each feature set. This is because both of the indexing techniques aggregate in the non-leaf nodes the textual information of the leaf nodes, which makes it much easier to find objects that contain one keyword, rather than finding objects that are described with more keywords.

Figure 4 depicts results obtained from the synthetic dataset, when varying different query parameters. We notice the same tendency as in the case of the real dataset. In general, we observed that range queries are costlier for the real dataset. This is due to the data distribution: our real dataset, which was extracted from yelp.com, consists of restaurants and hotels forming just a few clusters. On the other hand, our synthetic dataset is substantially larger and contains a few thousands of clusters. Hence, the data from the latter dataset are more dispersed compared to the former.

6   Related Work

In this section, we review related works about spatial queries from two categories, spatial keyword queries and spatial assignment.

Spatial Keyword Queries. Recently several approaches have been proposed for spatial keyword search. In [6], the problem of distance-first top-k spatial keyword search is studied. To this end, the authors propose an indexing structure (IR^2-$Tree$) that is a combination of an R-Tree and signature files. The IR-Tree was proposed in another conspicuous work [5,7], which is a spatial-textual indexing approach that employs a hybrid index that augments the nodes of an R-Tree with inverted indices. The inverted index at each node refers to a pseudo document that represents all the objects under the node. During query processing, the index is exploited to retrieve the top-k data objects, defined as the k objects that have the highest spatial-textual similarity to a given data location and a set of keywords. Moreover, in [8] the Spatial Inverted Index (S2I) was proposed for processing top-k spatial keyword queries. The S2I index maps each keyword to a distinct aggregated R-Tree or to a block file that stores the objects with the given term. All these approaches focus on ranking the data objects based on their spatial-textual similarity to a query point and some keywords. This is different from our work, which ranks the data objects based on textual relevance and a non-spatial score (quality) of the facilities in their spatial neighborhood. [4] provides an all-around evaluation of spatial-textual indices and reports on the findings obtained when applying a benchmark to the indices. Spatial-textual similarity joins were studied in [1]. Given two data sets, the query retrieves all pairs of objects that have spatial distance smaller than a given value and at the same time a textual similarity that is larger than a given value. This differs from the top-k spatial-textual preferences query, because the spatial-textual similarity join does not rank the data objects and some data objects may appear more than once in the result set. Prestige-based spatio textual retrieval was studied in [3]. The proposed query takes into account both location proximity and prestige based text relevance.

Spatial Assignment. Recently, with the popularity of mobile Internet, another type spatial query [15], called spatial assignment, is becoming more and more popular. The query of spatial assignment aims to find a global optimal assignment result based on the location distance and textual similarity between users and eventsIn [17]. In [10–12], the spatio-temporal conflict between different events is avoided in the global optimal assignment. Moreover, in [15,16], the real-time dynamic spatial assignment is also studied due to the wide range of crowdsourcing applications [2,13,14]. Although the aforementioned studies solve the many variants of spatial assignment, they ignore the influence of both location proximity and prestige based text relevance.

7   Conclusions

Recently, the database research community has lavished attention on spatial-textual queries that retrieve the objects with the highest spatial-textual similarity to a given query. Differently, in this paper, we address the problem of ranking data objects based on the facilities (feature objects) that are located in their vicinity. A spatial-textual preference score is defined for each feature object that takes into account a non-spatial score and the textual similarity to user-specified keywords, while the score of a data object is defined based on the scores of feature objects located in its neighborhood. Towards this end, we proposed a novel query type called preference-aware top-k spatial-textual query and present two query processing algorithms. Spatial-textual Data Ordering ($STDO$) first retrieves a data object and then computes its score, whereas Preference-Aware spatial-textual Search ($PASTS$) first retrieves highly ranked feature objects and then searches for data objects nearby those feature objects. In our experimental evaluation, we put all methods under scrutiny to verify the efficiency and the scalability of our method for processing preference-aware top-k spatial-textual query.

References

1. Bouros, P., Ge, S., Mamoulis, N.: Spatio-textual similarity joins. Proc. VLDB Endowment **6**(1), 1–12 (2012)
2. Cao, C.C., Tong, Y., Chen, L., Jagadish, H.V.: Wisemarket: a new paradigm for managing wisdom of online social users. In: SIGKDD 2013, pp. 455–463 (2013)
3. Cao, X., Cong, G., Jensen, C.S.: Retrieving top-k prestige-based relevant spatial web objects. Proc. VLDB Endowment **3**(1–2), 373–384 (2010)
4. Chen, L., Cong, G., Jensen, C.S., Wu, D.: Spatial keyword query processing: an experimental evaluation. Proc. VLDB Endowment. **6**, 217–228 (2013)
5. Cong, G., Jensen, C.S., Wu, D.: Efficient retrieval of the top-k most relevant spatial web objects. Proc. VLDB Endowment **2**(1), 337–348 (2009)
6. De Felipe, I., Hristidis, V., Rishe, N.: Keyword search on spatial databases. In: ICDE 2008, pp. 656–665 (2008)
7. Li, Z., Lee, K.C., Zheng, B., Lee, W.C., Lee, D., Wang, X.: Ir-tree: an efficient index for geographic document search. IEEE Trans. Knowl. Data Eng. **23**(4), 585–599 (2011)
8. Rocha-Junior, J.B., Gkorgkas, O., Jonassen, S., Nørvåg, K.: Efficient processing of top-k spatial keyword queries. In: Pfoser, D., Tao, Y., Mouratidis, K., Nascimento, M.A., Mokbel, M., Shekhar, S., Huang, Y. (eds.) SSTD 2011. LNCS, vol. 6849, pp. 205–222. Springer, Heidelberg (2011)
9. Rocha-Junior, J.B., Vlachou, A., Doulkeridis, C., Nørvåg, K.: Efficient processing of top-k spatial preference queries. Proc. VLDB Endowment **4**(2), 93–104 (2010)
10. She, J., Tong, Y., Chen, L.: Utility-aware social event-participant planning. In: SIGMOD 2015, pp. 1629–1643 (2015)
11. She, J., Tong, Y., Chen, L., Cao, C.C.: Conflict-aware event-participant arrangement. In: ICDE 2015, pp. 735–746 (2015)
12. She, J., Tong, Y., Chen, L., Cao, C.C.: Conflict-aware event-participant arrangement and its variant for online setting. IEEE Trans. Knowl. Data Eng. **28**(9), 2281–2295 (2016)
13. Tong, Y., Cao, C.C., Chen, L.: TCS: efficient topic discovery over crowd-oriented service data. In: SIGKDD 2014, pp. 861–870 (2014)
14. Tong, Y., Cao, C.C., Zhang, C.J., Li, Y., Chen, L.: Crowdcleaner: data cleaning for multi-version data on the web via crowdsourcing. In: ICDE 2014, pp. 1182–1185 (2014)
15. Tong, Y., She, J., Ding, B., Chen, L., Wo, T., Xu, K.: Online minimum matching in real-time spatial data: experiments and analysis. Proc. VLDB Endowment **9**(12), 1053–1064 (2016)
16. Tong, Y., She, J., Ding, B., Wang, L., Chen, L.: Online mobile micro-task allocation in spatial crowdsourcing. In: ICDE 2016, pp. 49–60 (2016)
17. Tong, Y., She, J., Meng, R.: Bottleneck-aware arrangement over event-based social networks: the max-min approach. World Wide Web J. **19**(6), 1151–1177 (2016)
18. Yiu, M.L., Dai, X., Mamoulis, N., Vaitis, M.: Top-k spatial preference queries. In: ICDE 2007, pp. 1076–1085 (2007)

Result Diversification in Event-Based Social Networks

Yuan Liang[1]([⊠]), Haogang Zhu[1], and Xiao Chen[2]

[1] State Key Laboratory of Software Development Environment,
Beihang University, Beijing, China
liangyuan040@gmail.com, haogangzhu@buaa.edu.cn
[2] School of Computer Science and Technology, Beijing University of Posts
and Telecommunications, Beijing, China
xiaochen@bupt.edu.cn

Abstract. Result diversification is an important aspect in query events, web-based search, facility location and other applications. To satisfy more users in event-based social networks (EBSNs), search result diversification in an event that covers as many user intents as possible. Most existing result diversification algorithms recognize an user may search for information by issuing the different query as much as possible. In this paper, we leverage many different users in a same event such that satisfy the maximum benefit of users, where users want to participate in an event that s/he did not know any users, for example, blind date, Greek and other activities. To solve this problem, we devise an effective greedy heuristic method and integrate simulated annealing techniques to optimize the algorithm performance. In particular, the *Greedy* algorithm is more effective but less efficient than *Integrate Simulated Annealing* in most cases. Finally, we conduct extensive experiments on real and synthetic datasets which verify the efficiency and effectiveness of our proposed algorithms.

1 Introduction

In recent years, various event-based social networks (EBSNs) are more and more popular and practical in real world, many websites are produced in this scenario, such as Meetup, Plancast and Eventbrite. These websites are mainly provide online events which users can resigned and manage offline social activities, such as fellowship activities, gathering, sports activities and others events, and sends such information to users.

However, most existing EBSNs only provide a set of events that users can participate in them [9], where consider that the diversity of users in an event are absent. Image the following scenario, Bob has no special partner, he want to meet his true love through bind dating. And he did not want to with his familiar friends in the dating occasions, so he choose the event that there are no any his friends to take part in. Though Bob desired to join a fraternity, he have to consider his friends that they whether in the fellowship hall, as he did not want to let other friends who know he go to a blind date. In fact, many users usually

S. Song and Y. Tong (Eds.): WAIM 2016 Workshops, LNCS 9998, pp. 198–210, 2016.
DOI: 10.1007/978-3-319-47121-1_17

encounter the same problem: they want to attend an event that there are no any familiar friends such that other users did not know his/her information.

Besides resolving result diversification, it is appealing to have an user-event arrangement strategy that optimizes the benefits of both event organizers and users, e.g. for organizing fellowship or bind date. Particularly, [2,8,12–14,19] are recent studied the arrangement of users to events such that all users obtain maximum satisfaction. However, the motivation of [2,8,12–14,19] have no considered that result diversification of users in a same event.

In addition to the aforementioned scenario, Bob did not want to take the far of distance from his home, so he will choose a place that organizes activities and he could accept the distance from his home to the activities. Therefore, the problem of result diversification in EBSNs should consider two factors as follows: the location influence between events/activities and users and the social friendship among users. Therefore, a new arrangement strategy not only consider the aforementioned two factors, but also guarantee all events satisfy its capacity. In the following, we illustrate a toy example to explain our motivation in details.

Example 1. Suppose we have five users ($u_1 - u_5$) and two events ($v_1 - v_2$) (each location corresponds to an event) in an EBSN. We illustrate this example in Fig. 1, the edge weights indicate the strength of social connections, the Euclidean distance $||u_i, v_j||$ between users $u \in U$ and events $v \in V$ are showed in the table of Fig. 1. The larger the distance, the higher arrangement cost will be taken for users. Furthermore, each event includes a capacity, which is the maximum number of users. In this example, the capacities of v_1 and v_2 are 4 and 3, respectively. If we assume that the cost function between a pair of users and events is the linear of normalized factors between their location and social friendship. The optimal goal is to minimum the sum of the spatial distance and the social friendship to obtain result diversification. A feasible arrangement is showed in following. The arrangement of the feasible arrangement is $\langle u_1, v_2 \rangle, \langle u_2, v_2 \rangle, \langle u_3, v_1 \rangle, \langle u_4, v_1 \rangle, \langle u_5, v_1 \rangle$ and the total cost of current arrangement is 1.20.

As discussed in the motivation example, a novel EBSNs is introduced, which mainly devises result diversification of all users in a same event/activity. There are some researches relevant with result diversification, [11] introduces users did not clearly specified from the initial query, and users want to find a result

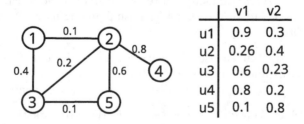

	v1	v2
u1	0.9	0.3
u2	0.26	0.4
u3	0.6	0.23
u4	0.8	0.2
u5	0.1	0.8

Fig. 1. Running example

consistent with their intents. However, our work aim at the user have known her/his interest of events, and s/he want to attend the event that her/his friends can not recognize her/him such that her/his information is secret. Specifically, given a set of users and a set of events, each user has a distance from event and each user has friendship of others. In our paper, we find an arrangement such that the minimum the sum of distance between users and events and social friendship of each users. In particular, we make the following contributions.

(a). We introduce a new social event arrangement problem and propose the formal definitions of result diversification in EBSNs.
(b). For the result diversification in EBSNs, we devise a baseline algorithm and a optimization algorithm, *Greedy* and *Integrate Simulated Annealing*. In particular, the *Greedy* algorithm is more effective but less efficient than *Integrate Simulated Annealing* in most cases.
(c). We conduct extensive experiments on real and synthetic datasets which verify the efficiency and effectiveness of our proposed algorithms.

The rest of the paper is organized as follows. In Sect. 2, we formally formulate the result diversification in EBSNs. In Sect. 3, we propose a baseline algorithm, *Greedy* algorithm. Moreover, a optimization algorithm (*Integrate Simulated Annealing*) is proposed in Sect. 4. Section 5 shows extensive experiments on both synthetic and real datesets. The related works are presented in Sect. 6. We finally conclude this paper in Sect. 7.

2   Problem Statement

We first introduce several concepts and then formally define the result diversification in EBSNs.

Definition 1 (User). A user is defined as $u(x_u, y_u)$, x_u presents the longitude of the user, y_u presents the latitude of the user. x_u and y_u presents the location of the user.

Definition 2 (Event). An event is defined as $v(x_v, y_v, \delta_v)$, x_v presents the longitude of the event, y_v presents the latitude of the event, δ_v presents the capacity of event v, x_v and y_v presents the location of the event.

Basically, we consider two factors of result diversification in EBSNs, the spatial distance of users and events, social friendship among users. And we have the following definition.

Definition 3 (Distance). As users have its location and events have its location, we use Euclidean distance to compute the distance of users and events. We denote $d = \sqrt{(x_u - x_v)^2 + (y_u - y_v)^2}$ as the distance of users and events.

Definition 4 (Social Graph). Let $G = (U, E, W)$ presents the social graph, where U is the set of users, E is the set of edges (i.e., social connections), and W is the set of edge weights (denoting the strength of social connections).

Given a set of users U, each u of which with longitude x_u and latitude y_u. a set of events V, each v of which with capacity δ_v, longitude x_v and latitude y_v, Euclidean distance formula and a social graph, find an arrangement between users and events to minimize the total cost $Cost(U, V, \alpha)$ such that

(1) Each event is not exceed its capacity δ_v.
(2) $d(u, v) \geq 0$
(3) $Cost(U, V, \alpha) = \alpha \cdot \sum_{i \in v_i} d(i, v_i) + (1 - \alpha) \cdot \sum_{e=(u_i, u_j) \in v, v_{u_i} = v_{u_j}} w(u_i, u_j)$

where the preference parameter $\alpha \in (0, 1)$ adjusts the relative importance of the two factors. If $\alpha > 0.5$, the result diversification should aim more at minimizing the arrangement cost of the distance. In this formula, the first term is the sum of all distances between each user and its event, and the second term is the social friendship of users.

3   The Baseline of EBSNs

In this section, we propose a greedy algorithm as baseline to solve the result diversification in EBSNs, in which we minimize the spatial distance between users and events and the social friendship of users when they participate in a same event. Since an user would like to attend en event that it is near from her/his home, and s/he did not want to see too familiar friends in some activities, such as bind date, fellowship club and other events/activities.

The baseline algorithm is thinking about the following. There are a set of users and a set of events, and find an arrangement such that each user obtain the maximum satisfaction. We first assign pairs of users and events into H, taking into account the spatial distance. H contains a tuple $\langle u, v, g \rangle$ representing potential arrangement of pairs of users and events. The H of g is ordered by non-increasing according to potential gain, defined as $g(u, v|\emptyset) = d(u, v)$. Then we extract the pair with the smallest $g(u, v|\emptyset)$ from heap H which stores tuple containing user u, event v and the potential gain g. If the event is not up to its capacity, we will assign the user u to the event v. Let $M(u)$ denotes the arrangement of user u, (u_i, v_j) presents the arrangement of u_i is v_j, then if the neighbours of user u (i.e., u') is not assigned, we update $g(u', v|S_v) = \alpha \cdot d(u', v) + (1 - \alpha) \cdot \sum_{v_{u'} = v_u} w(u, u')$ based on the heap H. Finally, we extract the smallest the potential gain of user u', and assign u' to event v which satiated $|S_v| < \delta_v$. This process can be repeated as needed. The post-processing stops when either all users are assigned or there are not enough available events/activities. More details are show in Algorithm 1.

We illustrate the procedure in Algorithm 1. In line 1, we first initialize the heap H, and let $M(u) \leftarrow \emptyset$ denotes user u is not assigned, S_v presents the set of users are assigned to event v. In lines 2–4, We compute the potential gain according to pairs of users and events, and put them into the heap H. In lines 6–15, We first extract the pair with the smallest potential gain which contains user u, event v and gain value g from H, and then assign user u to event v.

Algorithm 1. Baseline

1: Initialize heap H, $M(u) \leftarrow \emptyset$ for all u, $S_v \leftarrow \emptyset$ for all v.
2: **for all** $(u, v) \in U \times V$ s.t. $d(u, v) > 0$ **do**
3:     Insert $\{u, v, g(u, v|\emptyset)\}$ into H
4: **end for**
5: Heapify H
6: **while** $H \neq \emptyset$ **do**
7:     Extract the pair with the smallest function $g(u, v|S_v)$ from H
8:     **if** $|S_v| < \delta_v$ and $M(u) \leftarrow \emptyset$ **then**
9:         $S_v \leftarrow S_v \cup \{u\}$
10:        **for all** u': $w(u, u') > 0$ and $M(u') = \emptyset$ **do**
11:            Update $\{u', v, g(u', v|S_v)\}$ into H
12:        **end for**
13:    **end if**
14:    Heapify H
15: **end while**
16: **return** the final arrangement and the total cost.

Then compute potential gain of the neighbours of user u, and update H. Finally, find the minimum potential gain which contains the neighbours of user u, event v and the gain utility, then assign the neighbours of user u to the event v.

Example 2. Running our algorithm in Example 1. We first make an initialization pair of users and events according to the spatial distance. We construct a heap H that stores the spatial distance of users and events. There are 10 potential gains and 10 pairs of users and events in H, with which $\langle u_1, v_1, 0.9 \rangle$, $\langle u_1, v_2, 0.3 \rangle$, $\langle u_2, v_1, 0.26 \rangle$, $\langle u_2, v_2, 0.4 \rangle$, $\langle u_3, v_1, 0.6 \rangle$, $\langle u_3, v_2, 0.23 \rangle$, $\langle u_4, v_1, 0.8 \rangle$, $\langle u_4, v_2, 0.2 \rangle$, $\langle u_5, v_1, 0.1 \rangle$, $\langle u_5, v_2, 0.8 \rangle$, respectively. From above computation, we found 0.1 is the smallest number of all potential gain, the capacity of v_1 is 4, so user u_5 can be assigned to event v_1. We found that u_5 has two friends u_2 and u_3, respectively. if assign u_2 to v_1, s/he will obtain the potential gain is 0.86, and assign u_2 to v_2, s/he will obtain the potential gain is 0.4. Then if assign u_3 to v_1, s/he will obtain the arrangement cost is 0.23, and assign u_3 to v_2, s/he will obtain the arrangement cost is 0.7. In this step, 0.23 is the smallest potential gain, so assign u_3 to v_1. Then assign user u_4 to v_1, assign u_1 to v_2. Finally, the final arrangement is $\langle u_1, v_2 \rangle$, $\langle u_2, v_2 \rangle$, $\langle u_3, v_1 \rangle$, $\langle u_4, v_1 \rangle$, $\langle u_5, v_1 \rangle$ and the final total cost is 1.2.

Complexity Analysis. In the arrangement of pair of users and events, There are $|u||v| \cdot \log(|u||v|)$ iterations and it takes at most $O(|u||v|)$ time to compute potential gain. Thus, then we exact the smallest potential gain, this step takes $O(\log(|u||v|))$. Then we visited each user u, we will visit his friends at most the number of edges $|E|$ times. Therefore, we can visit the event which has many users are assigned to, this step spend the maximum degree of the graph, it takes $O(d)$ time. In overall, the worst-case time complexity of the baseline algorithm is $O(|u||v| \cdot \log(|u||v|) + |E| \cdot d \cdot \log(|u||v|))$.

4   The Optimization of EBSNs

Simple greedy algorithm is easily fallen into local optimum. In this section, in order to solve the limitation, we propose a hybrid heuristics to optimize EBSNs. However, if greedy combine with other heuristic algorithm, we can get a better solution. Therefore, we propose integrate simulated annealing (e.g. simulated annealing + greedy) to solve result diversification in EBSNs. Simulated annleal-ing is a probabilistic technique for approximating the global optimum of a given function. Specifically, it is a meta-heuristic to approximate global optimization in a large search space [7].

We first make an initial arrangement of a set of users to a set of events that satisfy $|S_v| \leq \delta_v$, and the current total cost is f_0 ($old_f = f_0$). As the temperature decreases, we will random select a user u, suppose u has been matched to v_i, then random select an event v_j guarantee that $i \neq j$ and $|S_{v_j}| < \delta_{v_j}$, and assign u to v_j. The total cost of this arrangement is new_f. Let $\Delta f = new_f - old_f$, if $\Delta f \leq 0$, assign u to v_j with probability $p = 1$; if $\Delta f \geq 0$, assign u to v_j with probability $p = e^{-\frac{|\Delta f|}{RT}}$. When we found that there are continuous $|\frac{T}{2}|$ times $\Delta f \geq 0$, rise the temperature until it finds a new_f that satisfy $\Delta f \leq 0$. Finally, we set the temperature until it is decrease to zero, the process stops. This approach did not guaranteed optimal solution. When we found a optimization among all solutions space, we can not guarantee that no better solution exists. Therefore, in order to get better results, we take a total cost that it is the minimum potential gain of all solutions space. This process iterative repeatedly, when the temperature drops to zero, the process stops. More details are showed in Algorithm 2.

Details of each iteration are as follows. Let $M(u)$ denotes the arrangement of user u, (u, v_i) presents the pair of user u and event v_i in the current iteration. If v_j is another event and v_j is not full, i.e. $|S_{v_j}| < \delta_{v_j}$, and u is arranged to v_i in the current step, i.e. $u \in S_{v_i}$. We then try to change the arrangement of u, and u is arranged to v_j, i.e. $u \in S_{v_j}$. Otherwise, other users can not change his/her current arrangement. More specifically, let M_i denotes the arrangement of ith, and M_j presents the arrangement of jth, if $|S_{v_j}|$ can accommodate one more users, i.e. $M_i - (u, v_i) = M_j - (u, v_j)$, and we can compute the total cost of M_i and M_j. For each u in U, we have the cost of arrangement M_i is equal to the cost of arrangement M_j. Each iteration, change the arrangement of one user, this process can be repeatedly with temperature decrease.

We illustrate the procedure in Algorithm 2. In line 1, let S_v denotes the set of users in event v, and initialize constant R and temperature T. In lines 2–3, random assign a set of users to a set of events satisfy $|S_v| \leq \delta_v$, note the current value is f_0. In lines 6, random choose an user u and randomly change its assigned event to an available event v_j. The total cost of current arrangement is new_f. Let $\Delta f = new_f - old_f$. In lines 7–11, compare between the current solution and neighbour solution. If the arrangement cost of neighbour solution is smaller than the arrangement cost of current arrangement, then with $p = 1$ select the neighbour solution; otherwise with $p = e^{-\frac{|\Delta f|}{RT}}$ select the neighbour solution. In lines 12–17, if there are consecutive n times $\Delta f > 0$, rise the temperature until

---

Algorithm 2. Integrate Simulated Annealing (ISA)

1: $S_v \leftarrow \emptyset$ for all v, $f\_increasing\_count = 0$, initialize $R, T_0, \Delta T, n$;

2: Randomly assign all users to events satisfying $|S_v| \le \delta_v$ with total cost f_0.

3: $new_f = f_0$, $T = T_0$

4: **while** $T > 0$ **do**

5:    $old_f = new_f$

6:    Randomly choose an user u, randomly change its assigned event to an available event v_j, and get a new total cost new_f with $\Delta f = new_f - old_f$.

7:    **if** $\Delta f \le 0$ **then**

8:       u with $p = 1$ match to v_j.

9:    **else**

10:       u with $p = e^{-\frac{|\Delta f|}{RT}}$ match to v_j.

11:    **end if**

12:    **if** Δf is positive for consecutive n times **then**

13:       **while** $new_f - old_f > 0$ **do**

14:          $T = T + \Delta T$, $old_f = new_f$.

15:          Randomly choose an user u, randomly change its assigned event to an available event v_j, and get a new total cost new_f with $\Delta f = new_f - old_f$.

16:       **end while**

17:    **end if**

18:    $T = T - \Delta T$

19: **end while**

20: Finally select the minimum cost of all solutions, and record the arrangement of users and events.

---

it produces $\Delta f \le 0$. Finally, when the temperature dropped to zero, we select the minimum total cost of all solutions.

Example 3. Running our algorithm in Example 1. Random assign five users to two events, current arrangement is (u_1, v_2), (u_2, v_1), (u_3, v_2), (u_1, v_2), (u_5, v_1) that satisfy $|S_v| < \delta_v$, and current total cost is 1.55. Random select an user u_3, the arrangement of current step is v_2, and let the arrangement of u_3 from v_2 to v_1, other users can not change. And the total cost of this step is 1.53. Since $1.53 < 1.55$, u_3 with $p = 1$ match to v_1, and u_3 with $p = e^{-\frac{|\Delta f|}{RT}}$ stay v_2. Therefore, the current arrangement is (u_1, v_2), (u_2, v_1), (u_3, v_1), (u_1, v_2), (u_5, v_1). Moreover, random select an user u_2, and the arrangement of current step is v_1, and the arrangement of u_2 from v_1 to v_2, the current arrangement is also satisfy $|S_v| < \delta_v$. In this step, the arrangement is (u_1, v_2), (u_2, v_2), (u_3, v_1), (u_1, v_2), (u_5, v_1) and the total cost is 1.4. The process is repeatedly as the temperature decrease until it drops to zero. Finally, the arrangement is (u_1, v_2), (u_2, v_1), (u_3, v_2), (u_1, v_2), (u_5, v_1) and the total cost is 1.045.

Complexity Analysis. For the initialization step, random assign a set of users to a set of events and compute the spacial distance of users and events, this step spend $O(|U||V|)$ time to find pair of users and events. Thus, the time complexity of the initialization step is $O(|U||V|)$. Then random select each user and each event, the number of iterations are at least $\Omega(|U||V|)$. However, the time complexity of the next step is according to its running time and effect. In overall, the time complexity of Integrate Simulated Annealing algorithm is at least $\Omega(|U||V|)$.

5   Experimental Evaluation

5.1   Experiment Setup

In this subsection, we evaluate our proposed algorithms. We use both real and synthetic datasets for experiments.

Real Datasets. We use the Meetup dataset from [9] as real dataset. In the Meetup dataset, each user is associated with some tags and a location. The events are not explicitly associated with tags, but each event is associated with some tags. Thus, for each event, we use the tags of the group who creates it as the tags of the event itself. We use the after processed dataset from [13], similar to [13] we use three datasets from VA, Auckland and Singapore. They are consists of 225 activities and 2012 users, 37 activities and 569 users and 87 activities and 1500 users, respectively. Since capacity is not given in the dataset, we generate the capacity of events following Normal and Uniform distribution. Statistic and configuration of real datasets are illustrated in Table 1.

Synthetic Datasets. For synthetic datasets, we generate the number of users, the number of events, the balance parameters α and the capacity of events δ_v according to normal distribution. Statistic and configuration of synthetic datasets are illustrated in Table 2.

Furthermore, synthetic datasets are created by Python, all algorithms are implemented in C++, under Linux Ubuntu and the experiments were performed on a machine with Intel Xeon E5620 2.40 GHz with 16-core CPU and 12 GB memory.

5.2   Experiment Results

In this subsection, we mainly evaluate the total cost of all proposed algorithms. We computed the total cost according to $Cost(U, V, \alpha) = \alpha \cdot \sum_{i \in v_i} d(i, v_i) + (1 - \alpha) \cdot \sum_{e=(u_i, u_j) \in v, v_{u_i} = v_{u_j}} w(u_i, u_j)$, and tested our proposed algorithms via

Table 1. Real dataset

City	V	U	δ_v	α, β, γ
VA	225	2012	Normal: $\mu = 50, \sigma = 25$	0.1, 0.2, 0.3, 0.4, 0.5, 0.6, 0.7, 0.8, 0.9
Auckland	37	569	Normal: $\mu = 50, \sigma = 25$	0.1, 0.2, 0.3, 0.4, 0.5, 0.6, 0.7, 0.8, 0.9
Singapore	87	1500	Normal: $\mu = 50, \sigma = 25$	0.1, 0.2, 0.3, 0.4, 0.5, 0.6, 0.7, 0.8, 0.9

Table 2. Synthetic dataset

Factor	Setting
$|V|$	10, 20, 50, **100,** 200, 500
$|U|$	100, 200, 500, **1000,**, 2000, 5000
δ_v	**Normal:** $N(25, 25), N(50, 25), N(75, 25), N(100, 25), N(125, 25)$;
	Uniform: [1,20], [1,50], [1,100], [1,150], [1,200]
α	0.1, 0.2, 0.3, 0.4, 0.5, 0.6, 0.7, 0.8
$|d|$	0.1, 0.3, 0.5, 0.7, 0.9
R	0.005, 0.0005
T_0	1000, 10000
n	10, 20, 30
Δ_T	1, 10

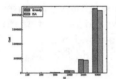 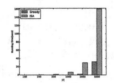 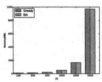

Fig. 2. Results on varying $|U|$

varying following parameters: the size of U, the size of V, the capacity of events δ_v and the balance parameter α.

Effect of $|U|$. We first study the effect of varying $|U|$, and set users are 100, 200, 500, 1000, 2000 and 5000, respectively, the number of events (20) are fixed. We then present the results of arrangement cost, running time and memory cost in Fig. 2. We first can observe that the total cost values generally increase with U increases. We then observe that *ISA* perform better in total cost than *Greedy*. Finally, the running time and memory cost increase as $|U|$ becomes larger.

Effect of $|V|$. We then study the effect of varying $|V|$, set events are 10, 20, 50, 100, 200 and 500, respectively, the number of users (1000) are fixed. We present the result of arrangement cost, running time and memory cost in Fig. 3. We can observe that the total cost decrease as $|V|$ increases. The reason is that the number of users is limited and thus when the number of events increases, more users are available to each event on average. We also observe that *greedy* perform better than *ISA* in running time. Finally, the memory cost of effect of V is not particularly obvious.

Effect of $|d|$. We then study the effect of varying $|d|$. We present the result of arrangement cost, running time and memory cost in Fig. 4. We can observe that the total cost decrease as $|d|$ increases. The reason is that users have more

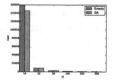
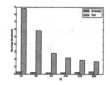
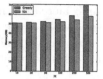

Fig. 3. Results on varying $|V|$

friends, they attend a same event with a larger probability. We also observe that memory cost increases as $|d|$ increases, and the time of effect of V is not particularly obvious change.

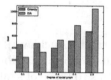
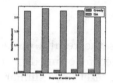
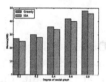

Fig. 4. Results on varying $|d|$

Effect of δ_v. We then study the effect of vary δ_v following Normal and Uniform distribution. We present the results of arrangement cost, running time and memory cost in Fig. 5. We can first observe that the arrangement did not particularly obvious change when δ_v generated following Normal and Uniform distribution. Second, in the running time, greedy preforms better than ISA, since ISA has more iterations. Third, varying δ_v has little effect on the memory cost of all algorithms.

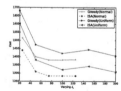
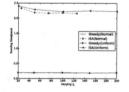
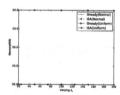

Fig. 5. Results on varying δ_v

Effect of α. We study the effect of vary α. Particularly, we vary the balance parameter α is 0.1, 0.3, 0.5, 0.7, 0.9. We present the results of arrangement cost, running time and memory cost in Fig. 6. we obverse that the total cost decrease when the balance parameter α increases. The reason is that the balance parameter α increases, the objective function places extra emphasis on the social cost. Also, the *integrate simulated annealing* (ISA) perform better in total cost than *Greedy*. Finally, the running time and memory cost has little effect of all algorithms when varying α.

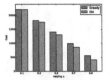
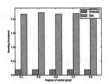
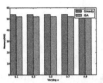

Fig. 6. Results on varying α

Real Dataset. Figure 7 shows the results on real dataset (Auckland) when the capacity values are generated following Normal and Uniform distribution. Notice that the results on real dataset have similar patterns to those of synthetic data. Similar patterns are observed on the other two real datasets and when the capacity values are generated following Normal and Uniform distribution, and we omit the results due to limited space.

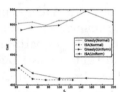
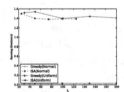
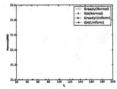

Fig. 7. Results of varying δ_v on real dataset

6   Related Work

In this section, we will review the related works in three categories, spatial matching, event-based social networks, and result diversification.

Spatial Matching. With the rapid development of mobile Internet, the issue of spatial matching has more and more applications in crowdsourcing [15,16] and social networks [8]. Recently, there have been a series of works about spatial matching, such as [17,18,20,21]. These works are mainly about spatial information and capacity constraints in the matching scenario. However, these works ignore the information of social networks.

Event-Based Social Networks. There are a lot of works about event-based social networks(EBSNs), [9] is the first work of the unique features in EBSNs. Recently, [8] introduces the social event organization (SEO) problem, which assigns users to activities such that maximizes the overall innate and social affinities. However, the solution of [8] considers two factors, the similarity of attributes and social friendship among users, they neglect the spatial influence between activities and users. Furthermore, there is a novel approach in EBSNs, [2] introduces multi-criteria social graph partitioning: a game theoretic approach in EBSNs, which consider two factors are as follows: the distance between users and event, the friendship of users. The model of [2] partitions a social network into a set of input events, so that users in the same event are socially connected, and at the same time they have high similarities to the same event. Clearly, [2] cannot handle the situation that the event has capacity.

Result Diversification. Diversification has been studied for Web search [1,3–6] and information retrieval [10]. These earlier work has mostly focused on the result of web search on assessing relevance and diversity of the result. The prior work often adopts specific objective functions according to their similarity between each others. However, these works did not consider the information of social network, they are significantly different with our work.

7 Conclusion

This paper studies result diversification in EBSNs, which assign a set of users to a set of events so that the distance between users and their events and the social friendship of users in a same event are minimized. To achieve efficiency, we devise a model of result diversification in EBSNs, and develop a baseline and propose a optimization algorithm to enhance its performance, *greedy* and *Integrate Simulated Annealing*. In addition, We implement our proposed methods in both real datasets and synthetic datasets, and we observe that the *integrate simulated annealing* perform better in total cost than *Greedy*.

References

1. Agrawal, R., Gollapudi, S., Halverson, A., Ieong, S.: Diversifying search results. In: WSDM 2009, pp. 5–14 (2009)
2. Armenatzoglou, N., Pham, H., Ntranos, V., Papadias, D., Shahabi, C.: Real-time multi-criteria social graph partitioning: a game theoretic approach. In: SIGMOD 2015, pp. 1617–1628 (2015)
3. Drosou, M., Pitoura, E.: Search result diversification. ACM SIGMOD Rec. **39**(1), 41–47 (2010)
4. Jiang, D., Leung, K.W.T., Vosecky, J., Ng, W.: Personalized query suggestion with diversity awareness. In: ICDE 2014, pp. 400–411 (2014)
5. Jiang, D., Leung, K.W.T., Yang, L., Ng, W.: Query suggestion with diversification and personalization. Knowl. Based Syst. **89**, 553–568 (2015)

6. Jiang, D., Ng, W.: Mining web search topics with diverse spatiotemporal patterns. In: SIGIR 2014, pp. 881–884 (2013)
7. Kirkpatrick, S., Gelatt Jr., C., Vecchi, M.: Optimization by simulated annealing (1983)
8. Li, K., Lu, W., Bhagat, S., Lakshmanan, L.V., Yu, C.: On social event organization. In: SIGKDD 2014, pp. 1206–1215 (2014)
9. Liu, X., He, Q., Tian, Y., Lee, W.C., McPherson, J., Han, J.: Event-based social networks: linking the online and offline social worlds. In: SIGKDD 2012, pp. 1032–1040 (2012)
10. Ounis, I., Macdonald, C., Santos, R.L.: Search result diversification. Found. Trends Inf. Retrieval **9**(1), 1–90 (2015)
11. Santos, R.L., Macdonald, C., Ounis, I.: Exploiting query reformulations for web search result diversification. In: WWW 2010, pp. 881–890 (2010)
12. She, J., Tong, Y., Chen, L.: Utility-aware social event-participant planning. In: SIGMOD 2015, pp. 1629–1643 (2015)
13. She, J., Tong, Y., Chen, L., Cao, C.C.: Conflict-aware event-participant arrangement. In: ICDE 2015, pp. 735–746 (2015)
14. She, J., Tong, Y., Chen, L., Cao, C.C.: Conflict-aware event-participant arrangement and its variant for online setting. IEEE Trans. Knowl. Data Eng. **28**(9), 2281–2295 (2016)
15. Tong, Y., Cao, C.C., Chen, L.: Tcs: Efficient topic discovery over crowd-oriented service data. In: SIGKDD 2014, pp. 861–870 (2014)
16. Tong, Y., Cao, C.C., Zhang, C.J., Li, Y., Chen, L.: Crowdcleaner: data cleaning for multi-version data on the web via crowdsourcing. ICDE **2014**, 1182–1185 (2014)
17. Tong, Y., She, J., Ding, B., Chen, L., Wo, T., Xu, K.: Online minimum matching in real-time spatial data: experiments and analysis. Proc. VLDB Endow. **9**(12), 1053–1064 (2016)
18. Tong, Y., She, J., Ding, B., Wang, L., Chen, L.: Online mobile micro-task allocation in spatial crowdsourcing. In: ICDE 2016, pp. 49–60 (2016)
19. Tong, Y., She, J., Meng, R.: Bottleneck-aware arrangement over event-based social networks: the max-min approach. World Wide Web J. **19**(6), 1151–1177 (2016)
20. U, L.H., Yiu, M.L., Mouratidis, K., Mamoulis, N.: Capacity constrained assignment in spatial databases. In: SIGMOD 2008, pp. 15–28 (2008)
21. Wong, R.C.W., Tao, Y., Fu, A.W.C., Xiao, X.: On efficient spatial matching. In: VLDB 2007, pp. 579–590 (2007)

Complicated-Skills-Based Task Assignment in Spatial Crowdsourcing

Jiaxu Liu[1]([✉]), Haogang Zhu[1], and Xiao Chen[2]

[1] State Key Laboratory of Software Development Environment,
Beihang University, Beijing, China
{liujiaxu,haogangzhu}@buaa.edu.cn
[2] School of Computer Science and Technology, Beijing University of Posts
and elecommunications, Beijing, China
xiaochen@bupt.edu.cn

Abstract. Spatial crowdsourcing is an activity consisting in outsourcing spatial tasks to a community of online, yet on-ground and mobile, workers. Presently an increasing number of spatial crowdsourcing applications emerges due to the related technologies tends to maturity. Distinct from traditional crowdsourcing dualistic entities, task and worker, a special kind of applications imports the third one of skill. Consequently, a novel assignment problem called *multiple skills assignment problem* (MSAP) is generated which extends the entity relationship from 2 to 3 dimensions. Inspired by group strategy we first propose a lightweight algorithm GMA that could achieve approximate optimal solution quickly. However, GMA exists a defect of ignoring that workers with multiple skills can decrease total travel distance significantly. Thus we propose a revised algorithm RGMA to cut down distance cost. With synthetic datasets, we empirically and comparatively evaluate the performance of the baseline and two proposed algorithms.

1 Introduction

With the rapid development of crowdsourcing techniques [1–4], an increasing number of spatial crowdsourcing applications, such as Uber and Baidu Delivery etc., have facilitated our daily life, of which a special kind of applications with extra condition that task has multiple skills requirement are emerging. As a burgeoning form of spatial crowdsourcing, we will introduce an application instance to interpret it. For example, middle and small TV stations are at a disadvantageous status competed with large stations for the reason that they lack money and high-qualified personnel. These two disadvantages can be solved by multiple skills spatial crowdsourcing platform effectively. If a small TV station plans to make a real-time news topic on Syrian refugees, it could publish questions to platform whose duty is finding optimal workers locates in German, Turkey etc. with skills to form teams and perform the task. These skills requirement comprises driving, camera shooting, interviewing and translating etc. which are hardly mastered by a single worker. Once receives primitive videos sent from abroad, the station can utilize platform again to recruit local workers with special skills such as captioning, reviewing etc. to finish the video post-production.

S. Song and Y. Tong (Eds.): WAIM 2016 Workshops, LNCS 9998, pp. 211–223, 2016.
DOI: 10.1007/978-3-319-47121-1_18

The whole procedure is a cooperation of idle labor resources distributed around the world with low cost. Other similar spatial crowdsourcing applications with skills requirement are all in the embryonic stage, such as volunteer recruitment in large-scale events and motorcade recruitment in wedding ceremony on spatial crowdsourcing platform.

This paper puts research focus on assignment problem. Different from traditional crowdsourcing dualistic entities, tasks and workers, this new form of spatial crowdsourcing imports the third entity, that is skills which increase the difficulty of assignment greatly. Multiple skills assignment problem (MSAP) can be depicted with a tasks-skills-workers tripartite graph which the intermediate layer, skills, play a constraint role in the assignment between tasks and workers [5].

The following prerequisites related to tasks have been taken into consideration in this paper. (1) Workers are expected to reach the location of the task before the arrival deadline (2) the required skills of task should be fully covered by skills of its assigned workers. (3) Any worker is assigned to only one spatial task, that is each worker can't be assigned to multiple tasks simultaneously. Similar to tasks, workers also propose assignment constraints as follows. (1) Each worker has a maximum moving distance. (2) Any worker have rights to refuse the assigned tasks that platform sends. The objective of multiple skills assignment problem is minimizing the total travel distance on the condition that all the prerequisites mentioned above are satisfied.

Note that, existing works on spatial crowdsourcing focused on assigning workers to tasks to maximize the total number of completed tasks [6], the number of performed tasks for a worker with an optimal schedule [8], or support assignments for dynamic real-time scenarios [10,11]. However, they did not take into account multi-skill covering of complex spatial tasks, time/distance constraints. Thus, we cannot directly apply prior solutions to solve our MSAP problem.

The contributions of this paper are follows. (1) We provide a formal definition of multiple skills assignment problem, where the goal is to minimize universe travel distance under spatial-temporal constraint. (2) We propose a lightweight algorithm GMA to achieve the optimal solution quickly. (3) Considering the fact that workers with multiple skills can decrease total travel distance of team significantly, we propose a revised algorithm RGMA that cut down distance cost by 10–15 % compared with GMA. (4) We conduct extensive experiments that verify the efficiency and effectiveness of the proposed approaches.

2   Related Work

Recently, with the popularity of GPS-equipped smart devices and wireless networks (e.g., Wi-Fi and 4 G), the spatial crowdsourcing that performs location-based tasks has emerged and become increasingly important in both academia and industry. Task assignment as the key problem of spatial crowdsourcing has been well studied. In this section, we review the related works on task assignment problem in spatial crowdsourcing and team formation problem in social networks.

Spatial Crowdsourcing. L. Kazemi et al. [6] first proposes the issue of task assignment in spatial crowdsourcing and introduces a taxonomy which divided spatial crowdsourcing into Worker Selected Tasks (WST) mode and Server Assigned Tasks (SAT) mode. D. Deng et al. [8] proposes two exact algorithms and three approximation algorithms to find a schedule for each worker that maximizes the number of performed tasks on the WST mode. Besides, participatory sensing [9,16], a particular type of WST based on spatial crowdsourcing has been studied to involve human workers to perform sensor-dependent tasks. Because the practical applications of task assignment is usually used in dynamic scenario, Tong et al. [10,11] studies the issue of online task assignment in spatial crowdsourcing. Furthermore, in [12–15], the spatio-temporal conflict constraint in task assignment in spatial crowdsourcing is also studied. However, these existing works on participatory sensing cannot be generalized to any type of spatial crowdsourcing. Contrary to WST mode that worker select task autonomously, SAT mode requires spatial crowdsourcing server conducting assignment to achieve universal optimal objective in budget, travel cost etc. L. Kazemi et al. [6] refer a spatial crowdsourcing platform requires workers to physically move to some specific locations of tasks, and perform the requested services with the objective of maximizing the overall number of assigned tasks.

Team Formation on Social Networks. MSAP problem is similar with team formation problem (TFP) in social networks except that MSAP needn't to consider communication costs among workers. T. Lappas et al. [17] first studied the team formation problem on social networks and propose two different communication-cost functions to measure the effectiveness of a team. Since then, many researchers extended this work. A. Anagnostonpoulos et al. [18] presented a general framework for the team formation problem on social networks. M. Kargar et al. [19] studied the online team formation problem on social networks and explored the problem of discovering the top-k teams of experts with/without a leader on social networks. S. Datta et al. [20] considered the capacities of experts and studied the problem of forming a capacitated team on social networks. L. Y. Li et al. [21] proposed fast algorithms for team member recommendation. All the works above tend to arrange members with intimidate relationships into the same group, while MSAP problem in spatial crowdsourcing emphasizes on skills coverage constraint rather than communication constraint.

3   Problem Statement

In this section, we first introduce notations and concepts used in this paper, and then present the formal definition of the multi-skill spatial crowdsourcing, in which we assign multi skills workers with spatial-temporal constrained complex spatial tasks. The notations of symbols are summarized in Table 1.

Table 1. Summary of symbol notations

Notation	Description
$W = \{w_1, ..., w_{|W|}\}$	Set of workers
$T = t_1, ..., t_{|T|}$	Set of tasks
$S = s_1, ..., s_{|S|}$	Set of skills
$Team_t$	Set of workers that answer task t
S_w	Set of skills that worker w owns
S_t	Set of skills that task t requires
l_t, l_w	Location of task t or worker w
e_t	Deadline of task t
$arrivaltime(t, w)$	Timestamp that worker w arrive at the location of task t
$dist(l_t, l_w)$	Distance from worker w to task t
d_w	Maximum moving distance of worker w
$Team_t^{opt}$	Optimal team to perform task t
w_t^s	A worker member of team satisfies the skill s Requirement of task t

Definition 1 *(Team). For any $t \in T$, we define a team $Team$ as a set of workers that could satisfy the skills requirement of t. That is,*

$$Team_t = \{w_1, ..., w_n\} \quad s.t. \quad \bigcup_{w \in Team_t} S_w \supseteq S_t$$

Definition 2 *(Team Travel Distance). For any $t \in T$, Team Travel Distance (TTD) is the sum of distance from each member location to task location. That is,*

$$TTD(Team_t) = \sum_{w_i \in team_t} dist(l_t, l_{w_i})$$

Definition 3 *(Optimal Team). We define a best team $Team_t^{opt}$ with the minimum team travel cost among all teams. That is,*

$$Team_t^{opt} = \min TTD(Team_t)$$

Problem 1 *(MSAP).* The problem of multi-skills assignment problem (MSAP) is to assign the available workers $w_i \in W$ to spatial tasks $t_j \in T$, and to obtain a task assignment instance set R with the following constraint.

1. Any worker $w_i \in W$ is assigned to only one spatial task $t_j \in T$ such that his/her arrival time at location l_{t_j} before the arrival deadline e_{t_j}.
2. The moving distance is less than the worker's maximum moving distance d_{w_i}.
3. All workers assigned to t_j have skill sets fully covering S_{t_j}.

The goal of MASP is minimize the total travel distance of the task assignment result set R with the three constraint above. Its formalization is as follows.

$$\min \sum_{(t_j, Team_{t_j}^{opt}) \in R} \sum_{w_i \in Team_{t_j}^{opt}} dist(l_{t_j}, l_{w_j})$$

$$s.t. \quad dist(l_{t_j}, l_{w_i}) \leq d_{w_i}$$

$$arrivaltime(t_j, w_i) \leq e_{t_j}$$

4   GMA Algorithm

Our first method is the Group Matching Algorithm (GMA). Algorithm 1 presents the pseudo-code of GMA. GMA accepts two sets as inputs. Task set T contains all the spatial tasks and worker set W contains those workers haven't assigned with any task yet. Set R is utilized to reserve the final assignment solution with a set of task-team pairs, e.g., an arbitrary entry $(t_j, Team_{t_j}^{opt})$ in R denote task t_j and its corresponding the best team.

The detailed algorithm is as follows. Initially, we set R empty and maintain a link list V_w whose initial value is $\emptyset$ for each worker w to keep the task trace that w has traversed. GMA algorithm assigns workers with spatial tasks for multiple rounds. In each round, each worker in set W is traversed and grouped to its nearest task. If only any worker satisfies skill requirement and has less travel cost, it will be the new optimal team member and recorded in the assignment result set R. Specifically, at the beginning of each round, we set variable terminal as the default value 1 which indicates while-loop would terminate after this round, provided that variable terminal doesn't change to 0. We create a list

---

Algorithm 1. Group Matching Algorithm (GMA)

Input: Set T, Set W
Output: A feasible assignment result set R
 1: $R \leftarrow \emptyset; V_w \leftarrow \emptyset, \forall w;$
 2: **while** $true$ **do**
 3:     $G_t \leftarrow \emptyset, \forall t; terminal = 1;$
 4:     **for all** $w \in W$ **do**
 5:         $t \leftarrow arg\min_{t \in T} dist(l_t, l_w);$
 6:         **if** $t \notin V_w$ **then**
 7:             $G_t \leftarrow w; V_w \leftarrow t;$
 8:             **for all** $s_i \in S_w \cap S_t$ **do**
 9:                 **if** $dist(l_t, l_w) < dist(l_t, l_{w_t^{s_i}})$ **then**
10:                     $W \leftarrow w_t^{s_i}; w_t^{s_i} = w;$
11:                     remove w from $W;$
12:                     $terminal = 0;$
13: **if** $teminal$ equals to 1 or W is empty **then**
14:     **return** $R;$

---

for each task to store all of workers in group with its initial value is empty, i.e. $G_t \leftarrow \emptyset, \forall t$ (Line 3). The statements in the forall loop (Line 4) obtain the nearest task t of worker w has never visited presently, and then assign w to group G_t dominated by t. If worker w owns skills t required and is nearer to t than corresponding members in $Team_t^{opt}$, these members are replaced by w and then reinserted into W (Lines 8–11). Alteration in R implies assignment solution hasn't converged yet, so terminal needs to be assigned to 0 to keep iteration procedure continuously (Line 12). When all of the workers in W have no abilities to alternate result set R, the procedure is terminated (Lines 13–14).

Although it is efficient in team formation for multiple tasks, GMA can be improved further if we utilize the fact that multiple skills worker could reduce total travel distance significantly, e.g., we have task $t(s_1, s_2, s_3)$ and 3 workers of $w_1(0.2, s_1), w_2(0.2, s_2)$ and $w_3(0.3, s_1, s_2)$. GMA will enroll w_1 and w_2 in optimal team of t because they both have shorter distance than w_3. Therefore, the team travel distance is 0.4. In fact, if we only choose w_3 to form team, w_3 is more competent with a shorter distance of 0.3. Thus w_3 is the most suitable candidate for task t, which unfortunately GMA is incapable of discovering. In the Sect. 5 we propose a revised algorithm to shorten the distance by considering the fact that multiple skills worker can reduced travel distance.

5   RGMA Algorithm

Since multi-skills workers have an advantage of team reduce travel distance effectively, we propose a revised algorithm which combines GMA with the idea of assigning tasks to multi-skilling workers as far as possible to realize the reduction of the global travel distance. However, the assignment problem become complicated for the reason that we have to take combinative distance into consideration. For example, task t requires skills range from s_1 to s_5, initially t form team with 5 members and each of them meets one skill requirement. Then a new worker w_6 with four skills required is grouped with t. If the condition of $d_1 + d_2 + d_3 + d_4 > d_6$ satisfied, worker w_6 will replace four workers from w_2 to w_5 and reform another better team together with w_1. Suppose w_7 is a subsequent upcoming worker in group t with shorter distance than w_6 and w_2, obviously w_7 couldn't replace w_6 alone because this would increase travel cost from d_6 to $d_6 + d_7$. Therefore, w_7 must unionize other vicinal workers to replace w_6, e.g., w_3, w_4 and w_5 accompany with d_7 compete against w_6. If w_6 is winner they are obliged to wait for other upcoming workers to form a more competitive team to beat optimal team. The more workers arrive, the more combinations burst, which has a serious influence on the running times of platform. In this section we propose a revised group matching algorithm (RGMA) to solve multiple skills assignment problem mentioned above. To achieve relatively accurate result we allocate an additional list for each task to save those workers who has potential to become optimal team member in future. Meanwhile, we import pruning method to constrain the scale of computation (Table 2).

Table 2. Distance comparison between optimal team and candidate team

Sequence	Optimal team					Candidate team					Distance comparison
	s_1	s_2	s_3	s_4	s_5	s_1	s_2	s_3	s_4	s_5	
1st	w_1	w_2	w_3	w_4	w_5		w_6	w_6	w_6	w_6	$d_1 + d_2 + d_3 + d_4 > d_6$
	(d_1)	(d_2)	(d_3)	(d_4)	(d_5)		(d_6)	(d_6)	(d_6)	(d_6)	
2nd	w_1	w_6	w_6	w_6	w_6		w_7				$(d_7 + d_3 + d_4 + d_5, d_6)$
	(d_1)	(d_6)	(d_6)	(d_6)	(d_6)		(d_7)				
3rd	w_1	w_6	w_6	w_6	w_6			w_8		w_8	$(d_7 + d_8 + d_4, d_6)$
	(d_1)	(d_6)	(d_6)	(d_6)	(d_6)			(d_8)		(d_8)	
4th	w_1	w_6	w_6	w_6	w_6		w_9		w_9		$(d_9 + d_8, d_6)$,
	(d_1)	(d_6)	(d_6)	(d_6)	(d_6)		(d_9)		(d_9)		$(d_9 + d_3 + d_5, d_6)$,
											...

5.1 Candidate Set and Candidate Team Computation

We introduce a set called *candidate set* which comprises a *compound entry* and several *worker entries*. Compound entry locates at the rear of candidate set and records the nearest worker for each skill that task requires. Worker entry records any multi-skills worker who has potential to join in optimal team. e.g., we assume $CS_t = \{(N, w_3, N, w_3, w_3), (w_1, N, w_2, w_1, N)\}$ is a giving candidate set with two entries, of which (N, w_3, N, w_3, w_3) is the worker entry while (w_1, N, w_2, w_1, N) is the compound entry. Through merging all of the entries in candidate set, we achieve a candidate team which maybe superior than optimal team. The pseudo-code of merging computation is depicted in Subroutine 1.

To achieve high-quality candidate team, we sort all of the worker entries in descending order (Line 1), e.g., we assume candidate set $CS_t = \{(N, N, w_4, N, w_4), (N, w_3, N, w_3, w_3), (w_1, N, w_2, w_1, N)\}$, and then we get $CS_t = \{(N, w_3, N, w_3, w_3), (N, N, w_4, N, w_4), (w_1, N, w_2, w_1, N)\}$ after sorting operation. Notice that compound entry always locates at the rear of set and candidate team is the result of merging computation with the direction from left to right.

Subroutine 1. CompCandiTeam

Input: Candidate set CS_t of task t
Output: Candidate team of CS_t
1: Sort worker entries in CS_t by the number of skills in descending order
2: $CandiTeam \leftarrow \emptyset$;
3: **for all** $e \in CS_t$ **do**
4:    preprocess $CandiTeam$ and e;
5:    **if** $TTD(CandiTeam) > TTD(e)$ **then**
6:       $CandiTeam = e$;
7: **return** $CandiTeam$;

Additional pretreatments are exerted on two operands of merging computation (Line 4): (1) Merging two operands each other to complement vacant team members as many as possible, e.g., we assume $CS_t = \{(N, w_3, N, w_3, w_3),$ $(w_1, N, w_2, w_1, N)\}$, after preprocessing two entries are $e_1 = (\boldsymbol{w_1}, w_3, \boldsymbol{w_2}, w_3, w_3)$ and $e_2 = (w_1, \boldsymbol{w_3}, w_2, w_1, \boldsymbol{w_3})$. The workers in bold are complementary members achieved by utilizing the information of opposite operand. (2) Finding all of the members with only one-skill workload in both operands and replacing those whose travel distances are farther than opposite members, e.g., we assume $e_1 = (w_1, \boldsymbol{w_5}, \boldsymbol{w_2}, w_1, \boldsymbol{w_3})$ and $e_2 = (w_1, \boldsymbol{w_5}, w_4, w_1, w_4)$ are operands of another merge computation. Travel distances of workers in e_1 are $(1, 1, 2, 1, 3)$, while in e_2 are $(1, 1, 2, 1, 2)$. Workers w_5, w_2, w_3 only complete one-skill workload of task and are marked in bold. Worker w_2 should be replaced by w_4 in e_2 since w_4 is multiple skills worker with highly combination ability in subsequent merging computations. Another worker w_3 should be replaced by w_4 of e_2 as well since w_4 has a shorter distance of 2 to task than w_3. Thus, we can achieve $e_1 = (w_1, w_5, \boldsymbol{w_4}, w_1, \boldsymbol{w_4})$ and $e_2 = (w_1, w_5, w_4, w_1, w_4)$ after preprocessing these one-skill workload worker.

RGMA compares travel distances of two teams and select the lower one as candidate team (Lines 5–6).

5.2 Prune Strategy

In RGMA algorithm those multi-skills workers, who temporarily have no capability to substitute current optimal team members, are kept in candidate set, which causes candidate set expands in proportion to the number of workers. In order to limit the scale of candidate set, we introduce pruning strategy to delete redundant workers from candidate set if only substitution happens between optimal team and candidate team.

We take $Team_t^{opt}$ as a ruler to prune worker entries in candidate set CS_t (Lines 1–9 in Subroutine 2). The variable flag indicates whether a worker entry e should be eliminated from CS_t and value 1 represents deletion operation. $w_t^{s_i}$ denotes the member in optimal team or candidate team corresponding to skill

---

Subroutine 2. Prunning

Input: Candidate set CS_t of task t; current optimal team $Team_t^{opt}$ for t
Output: Candidate set CS_t after pruning operation
1: **for all** worker entry e in CS_t **do**
2:     $flag = 1$;
3:     **for all** skill s_i that e owns **do**
4:         $w_1 = $ worker $w_t^{s_i}$ in $Team_t^{opt}$; $w_2 = $ worker $w_t^{s_i}$ in e;
5:         **if** $dist(l_t, l_{w_1})) > dist(l_t, l_{w_2})$ **then**
6:             $flag = 0$;break;
7:     **if** $flag$ equals to 1 **then**
8:         delete e from CS_t
9: **return** CS_t;

s_i (Line 4). The entry e will be deleted from CS_t when all members to task are farther than those in optimal team.

5.3   Algorithm Description

RGMA adopts group strategy to decompose an enormous and complex problem into some trival and feasible problems. After multiple rounds of group operation, RGMA algorithm will gradually promote result accuracy rates till it converges to approximate optimal solution. Each task and surrounding workers are divided into the same group, and then each task searches optimal team within its group scope. Each worker w in W is grouped to its nearest task (Line 5) and then is directly examined by task $t_{nearest}$ whether w should join the optimal team. Specifically, if w owns the skills that $t_{nearest}$ required, we insert w into candidate set of $t_{nearest}$ and then get a candidate team through calling Subroutine 1 (Line 9). Once candidate team is superior to current optimal team we execute following operations: (1) Decompose optimal team and reinsert the retired workers into W (2) Nominate candidate team as the best team for task. (3) Prune candidate set to prevent it grows to a large scale (Lines 11–14). The variable terminal is used to verify the result whether has converged, and value 1 represent that RGMA has achieve a convergent result and the procedure should be terminal rather than continuous group operations. The detail Pseudo-code of RGMA is shown in Algorithm 2.

Algorithm 2. Revised Group Matching Algorithm(RGMA)

Input: Task set T, worker set W
Output: A feasible assignment result set R
1: $R \leftarrow \emptyset$;
2: **while** *true* **do**
3:    $terminal = 1$;
4:    **for all** $w \in W$ **do**
5:      group w with the nearest task $t_{nearest}$;
6:      $c = S_w \bigcap S_{t_{nearest}}$;
7:      **if** c is non-empty **then**
8:        Insert w into set $CS_{t_{nearest}}$
9:        $CandiTeam = CompCandiTeam(CS_{t_{nearest}})$;
10:        **if** $TTD(Team_t^{opt}) > TTD(CandiTeam)$ **then**
11:          Remove $(t, Team_t^{opt})$ from R
12:          decompose $Team_t^{opt}$ and reinsert scattered workers into W
13:          $Team_t^{opt} = CandiTeam$; insert $(t, Team_t^{opt})$ into R
14:          $Prunning(CS_{t_{nearest}}, Team_t^{opt})$;
15:          $terminal = 0$;
16:    **if** $terminal$ equals to 1 or W is empty **then**
17:      **return** R;

---

6   Experiments

We conduct experiments on synthetic dataset and generate locations of workers and tasks in a 2D data space $[0, 1000]$ following random distribution. Then, for skills of each worker and required skills of each task, we associate them with skills whose number is within range $[1, 5]$ following Gaussian distribution. In order to apply algorithms to real scenario we attached two constraints, worker maximum moving distance and task deadline, to GMA and RGMA. Table 3 depicts our experimental settings, where the default values of factors are in bold font. In each set of experiments, we vary one factor, while setting other factors to their default values.

We choose random algorithm as baseline and evaluate RANDOM, GMA and RGMA algorithms in terms of running time and average travel distance. These algorithms are implemented in C++, and the experiments were performed on a Windows 10 machine with Intel Core(TM) i5-2400 CPU@3.10 GHz and 4 GB memory.

We vary the number of workers from 1 K to 10 K and plot the performance of algorithms GMA, RGMA and RANDOM. RGMA perform best in the travel distance among these three algorithms while RANDOM does worst. Note that RGMA reduces the average team travel distance at 10–15 % than GMA at the cost of running times increasing showed in Fig. 1a. Therefore, RGMA is suitable for small scale multi-skills spatial crowdsourcing applications for it causes less travel cost and ignorable increasing of algorithm running times compared with GMA.

Figure 2a shows the influence on running time of algorithms with the increasing number of tasks. All the algorithms running times increase significantly for massive tasks complicate assignment problem. Since the number of workers (10 K) is sufficient for tasks to form local optimal teams even if the number of tasks achieve the highest 3 K in our settings, the average travel distance of all teams of GMA and RGMA increases slightly In Fig. 2b.

The maximum moving distance of workers has an insignificant influence on running times in GMA and RGMA. We can observe from Fig. 3a that time cost of the two algorithms increase remarkably with the expansion of workers' moving scope. Figure 3b demonstrate another measurement, team average travel

Table 3. Experiments settings

Factor	Setting
The number of tasks m	0.5 K, 1 K, 1.5 K, 2 K, 2.5 K, **3 K**
The number of workers n	1 K, 2 K, 3 K, 4 K, 5 K, 6 K, 7 K, 8 K, 9 K, **10 K**
The maximum number of skills worker owns	1, 2, 3, 4, **5**
The maximum number of skills task requires	**5**
Worker maximum moving distance	**100**, 200, 300, 400, 500, 600, 700, 800, 900, 1 K
Task deadline	[0, 0.5], [0.5, 1], [1, 2], [2, 5], **[5, 10]**

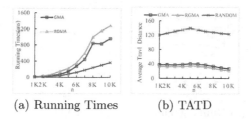

(a) Running Times   (b) TATD

Fig. 1. Effect of the number of workers n.

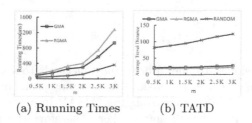

(a) Running Times   (b) TATD

Fig. 2. Effect of the number of tasks m.

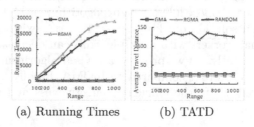

(a) Running Times   (b) TATD

Fig. 3. Effect of moving distance.

distance, almost has no relation to the variation of workers' maximum moving scope on the premise that plenty of workers wait to be assigned.

Figure 4 reports the effect of deadline range over the synthesis data. Since it is impossible for spatial crowdsourcing platforms to obtain real velocities of all workers, we uniformly set worker travel velocity a fixed value of 10 per unit time for simplicity. As a result, constraint on task deadline can be converted to distance constraint, e.g., if the deadline of task t is 0.5, which indicate the task only can be answered by workers within circular space centered at t with radius 5 to form team. In other words, those workers outside the scope are ignored by the task because they couldn't arrive task location on time. Thus deadline constraint could decrease the running times of algorithms as shown in Fig. 4. Note that although GMA and RGMA run fast with deadline setting [0, 0.5], we should recognize that successful assignment numbers decrease dramatically and simultaneously.

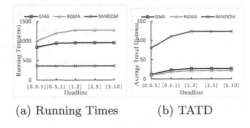

(a) Running Times   (b) TATD

Fig. 4. Effect of task deadline.

7   Conclusion

In this paper, we dedicate to find approximate methods to solve a novel spatial crowdsourcing problem called multi-skills assignment problem (MSAP). Unlike other assignment problem in crowdsourcing, MSAP imports a new role called skill besides the traditional task and worker. Therefore the assignment problem get complicated for the reason that we must consider how multiple workers form universal optimal teams to meet skills requirement of tasks. Inspired by group strategy we first propose a lightweight algorithm GMA that could achieve approximate optimal solution quickly. However, GMA exists a shortcoming of ignoring the fact that workers with multiple skills can decrease total travel distance of team significantly. So we propose a revised algorithm RGMA that cut down distance cost by 10–15 % compared with GMA. Finally, we conduct extensive experiments that verify the efficiency, effectiveness and scalability of the proposed approaches.

References

1. Cao, C., She, J., Tong, Y., Chen, L.: Whom to ask? Jury selection for decision making tasks on micro-blog services. Proc. VLDB Endowment **5**(11), 1495–1506 (2012)
2. Cao, C., Tong, Y., Chen, L., Jagadish, H.V., WiseMarket: a new paradigm for managing wisdom of online social users. In: SIGKDD, pp. 455–463 (2013)
3. Tong, Y., Cao, C., Zhang, C., Li, Y., Chen, L., CrowdCleaner: data cleaning for multi-version data on the web via crowdsourcing. In: ICDE, pp. 1182–1185 (2014)
4. Tong, Y., Cao, C., Chen, L.: TCS: efficient topic discovery over crowd-oriented service data. In: SIGKDD, pp. 861–870 (2014)
5. Gao, D., Tong, Y., She, J., Song, T., Chen, L., Xu, K.: Top-k team recommendation in spatial crowdsourcing. In: Cui, B., Zhang, N., Xu, J., Lian, X., Liu, D. (eds.) WAIM 2016. LNCS, vol. 9658, pp. 191–204. Springer, Heidelberg (2016). doi:10. 1007/978-3-319-39937-9_15
6. Kazemi, L., Shahabi, C.: Geocrowd: enabling query answering with spatial crowdsourcing. In: SIGSPATIAL GIS, pp. 189–198 (2012)
7. Alt, F., Shirazi, A.S., Schmidt, A., Kramer, U., Nawaz, Z., Location-based crowdsourcing: extending crowdsourcing to the real world. In: NordiCHI, pp. 13–22 (2010)

8. Deng, D., Shahabi, C., Demiryurek, U.: Maximizing the number of worker's self-selected tasks in spatial crowdsourcing. In: SIGSPATIAL GIS, pp. 314–323 (2013)
9. Hull, B., Bychkovsky, V., Zhang, Y., Chen, K., Goraczko, M., Miu, A., Shih, E., Balakrishnan, H., Madden, S.: Cartel: a distributed mobile sensor computing system. SenSys, pp. 125–138 (2006)
10. Tong, Y., She, J., Ding, B., Chen, L., Wo, T., Xu, K.: Online minimum matching in real-time spatial data: experiments and analysis. Proc. VLDB Endowment. $9(12)$, 1053–1064 (2016)
11. Tong, Y., She, J., Ding, B., Wang, L., Chen, L.: Online mobile micro-task allocation in spatial crowdsourcing. In: ICDE, pp. 49–60 (2016)
12. She, J., Tong, Y., Chen, L., Cao, C.: Conflict-aware event-participant arrangement. In: ICDE, pp. 735–746 (2015)
13. She, J., Tong, Y., Chen, L.: Utility-aware social event-participant planning. In: SIGMOD, pp. 1629–1643 (2015)
14. She, J., Tong, Y., Chen, L., Cao, C.C.: Conflict-aware event-participant arrangement and its variant for online setting. IEEE Trans. Knowl. Data Eng. $28(9)$, 2281–2295 (2016)
15. Tong, Y., She, J., Meng, R.: Bottleneck-aware arrangement over event-based social networks: the max-min approach. World Wide Web J. $19(6)$, 1151–1177 (2016)
16. Bulut, M., Yilmaz, Y., Demirbas, M.: Crowdsourcing location-based queries. In: PERCOM Workshops, pp. 513–518 (2011)
17. Lappas, T., Liu, K., Terzi, E.: Finding a team of experts in social networks. In: SIGKDD, pp. 467–476 (2009)
18. Anagnostonpoulos, A., Becchetti, L., Castillo, C., Gionis, A., Leonard, S.: Power in unity: forming teams in large-scale community systems. In: CIKM, pp. 599–608 (2010)
19. Kargar, M., An, A.: Discovering top-k teams of experts with/without a leader in social networks. In: CIKM, pp. 985–994 (2011)
20. Datta, S., Majumder, A., Naidu, K.: Capacitated team formation problem on social networks. In: SIGKDD, pp. 1005–1013 (2012)
21. Li, L., Tong, H., Cao, N., Ehrlich, K., Lin, Y.: Replacing the irreplaceable: fast algorithms for team member recommendation. In: WWW, pp. 1206–1215 (2015)

Market-Driven Optimal Task Assignment in Spatial Crowdsouring

Kaitian Tan[✉] and Qian Tao

State Key Laboratory of Software Development Environment,
Beihang University, Beijing, China
tankaitian@gmail.com, taoqian1993@buaa.edu.cn

Abstract. With the popularity of mobile devices and Online To Offline
(O2O) marketing model, various spatial crowdsourcing platforms, such
as Gigwalk, WeGoLook, TaskRabbit and gMission, are getting popular.
An important task of spatial crowdsourcing platforms is to allocate spatial tasks to suitable workers. Existing approaches only simply focus on
maximizing the number of completed spatial tasks but neglect the influence of supplies and demands from real crowdsourcing market, which
leads to different optimal objectives for crowdsourcing task assignments.
In this paper, to address the shortcomings of the existing approaches, we
first propose a more general spatial crowdsourcing task assignment problem, called *Market-driven Optimal Task Assignment (MOTA)* problem,
consisting of two real scenarios, Excess Demand of Crowd Workers and
Insufficient Supply of Spatial Tasks, in daily life. Unfortunately, we prove
that the two variants of this problem are NP-Hard. Thus, we design two
approximation algorithm to solve this problem. Finally, we verify the
effectiveness and efficiency of the proposed methods through extensive
experiments on synthetic datasets.

1 Introduction

With the rapid development of mobile devices, crowdsourcing techniques are
becoming popular recently [3,4,17,18]. Particularly, spatial crowdsourcing has
attracted much attention from both academia (e.g., the database community)
and industry (e.g., Uber). Specifically, a spatial crowdsourcing platform is in
charge of assigning of a number of workers to nearby spatial tasks, such that
workers need to physically move towards to some specified locations to finish
tasks (e.g. taking photos). Several instances of such mobile crowdsourcing mar-
kets have emerged commercially including Gigwalk, FiledAgent, and TaskRabbit
(see Table 1), typical tasks pay users a few dollars for capturing photos of build-
ings or sites, price checks, product placement checks in stores, traffic checks,
location-aware surveys, and so on.

Spatial Crowdsourcing has recently been attracting attention in both the
research communities (e.g., [6,8,10,13,19,20]) and industry (e.g., TaskRabbit,
Gigwalk). A recent survey in this area distinguishes spatial crowdsourcing from
related fields, including crowdsourcing, participatory sensing, volunteered geographic information. In [10], a spatial crowdsourcing framework in static scenario

© Springer International Publishing AG 2016
S. Song and Y. Tong (Eds.): WAIM 2016 Workshops, LNCS 9998, pp. 224–235, 2016.
DOI: 10.1007/978-3-319-47121-1_19

whose goal is to maximize the number of assigned tasks is proposed. In [19,20], two variants of online task assignment problem in dynamic real-time spatial crowdsourcing scenarios are proposed and provide the theoretical guarantee for the proposed solutions. In contrast to our study, in [2,5,11,13,22], the workers need to travel to the tasks locations, which may take a long time in rush hour. Consequently, the workers may reject their assigned tasks. Recently, the problem of assigning and scheduling tasks for multiple workers that maximizes the utility of assigned tasks is proposed in [14]. Although different work has done about spatial crowdsourcing, no one takes the labor dynamics in market into consideration. One of the challenges with crowdsourcing is to aggregate the workers responses with varying degree of accuracy. One of the well-known mechanisms is majority voting, which accepts the result supported by the majority of workers. Another work aims to tackle the issue of trust by having tasks performed redundantly by multiple workers [5]. In the scope of this study, we assume that selected workers provide trustworthy data. If there are multiple reports for one task the crowdsourcing system will send all the related workers to the task requester.

Crowdsourcing hyper-local information has been presented in both research (e.g., mPING [1]) and industry (e.g., mPing4, WeatherSignal5). mPing and WeatherSignal are two popular crowdsourcing system. mPing is the state-of-the-art system for crowdsourcing weather reports; anyone with a smart phone can report precipitation observations. However, mPing neglects to consider the cost of selecting workers to perform tasks. Also, the task assignment in mPing is often suboptimal as workers do not have a global system view. Workers typically choose nearby tasks to them, which may cause multiple workers to cover the same task unnecessarily while many other tasks remain uncovered. WeatherSignal uses phone sensors, such as temperature and humidity sensors, to measure local atmospheric conditions. Particularly, an algorithm was employed to translate phone battery temperature into the ambient temperature. However, the algorithm accuracy is the main concern. In addition, WeatherSignal is restricted to the availability of phone sensors, which currently cannot detect various kinds of hyper-local information, such as air pollution, noise level, light intensity and precipitation level such as rain, snow, drought.

The paper is organized as follows. In Sect. 2, we presents some formal definition about spatial crowdsourcing and introduce our problem focus. In Sect. 3, we discuss the MNST problem and give an algorithm. In Sect. 4, we discuss the MNST problem and give an algorithm as well. In Sect. 5, we show some extensive experiments on synthetic datasets with results on covered tasks, assigned workers, execute time and memory. The related works are presented in Sect. 6. We finally conclude this paper in Sect. 7.

2   Problem Statement

We first introduce two basic concepts, spatial task and crowd worker, then formally define two different kinds of problem statements.

Table 1. Some active mobile crowdsourcing platforms

Platform	Tasks
Filed agent	Price checks, Service Assessment, Photography, Verification, Property Evaluation
Gigwalk	Data collection, Photography, Focus Groups, Store Audits, IT, Final Services
NeighborFavor	Deliveries, Rides, Groceries
TaskRabit	Odd Errands, Groceries, Moving, Deliveries, Cleaning
WeGoLook	Verify Properties, Automobiles, Datas, Boats, Heavy Equipment

Definition 1 (Spatial Task *[10]*). *A spatial task t, denoted by $< l_t, e_t >$, is a crowdsourcing task t to be finished at location l_t in the 2D space before its expired time e_t.*

Definition 2 (Crowd Worker *[10]*). *A crowd worker w is denoted by w_i, with time or distance budget b_i. A worker can finish more than more tasks, but the time or distance he spends has to be less than the budget.*

Definition 3 (Feasible Solution). *Given m spatial tasks $t_1, t_2 ..., t_m$, and k crowd workers w_1, $w_2 ...$, w_k, and $m >> k$, this problem is assign tasks to work-ers such that (1) each task is finished before its expired time; (2) the total time of performed tasks for each worker is less than his/her time or moving distance budget. For each worker w_i, we can calculate a service Set $S_i = \{s_{i1}, s_{i2} ..., s_{im}\}$, and s_{ij} represents the set of tasks that w_i can performe. Thus, the task assign-ment problem is reduced to pick sets from the Service Sets but with the additional restriction that at most one set picked form each Service Set.*

Problem 1 (Light Load). Maximizing The Number of Completed Spatial Tasks (MNST). The objective of the problem is to find a feasible solution such that the number of tasks that can be serviced is maximized. More precisely, MNST is to find a subset $H \subset \{s_1, s_2 ..., s_k\}$, such that $|H| \leq k$, and $|H \bigcap s_i| \leq 1$, for $1 \leq i \leq k$. And the number of covered tasks is maximized.

Problem 2 (Heavy Load). Minimizing The Number of Selected Crowd Workers (MNCW). The objective of the problem is to find a feasible solution such that the number of workers is minimized while all tasks can be serviced. More precisely, MNCW is to find a subset $H \subset \{s_1, s_2 ..., s_k\}$, such that $|H| \leq k$, and $|H \bigcap s_i| \leq 1$, for $1 \leq i \leq k$. And the number of assigned workers is minimized.

3   Greedy Algorithm for MNST

In this section we show that simple greedy algorithms give constant factor approximation ratios for MNST.

In spatial crowdsourcing system, the number of subsets of X, is exponential in the number of workers and the sets are defined implicitly. In such cases we require a polynomial time algorithm with some desirable properties. Here, we give such an algorithm. We assume there exists an algorithm Φ that takes as input, a subset

---

Algorithm 1. FDS

1: Input X, S_i
2: Initialize $H = \bigcup s_i$
3: **for all** $s \in S_i$ **do**
4:     **if** $|X \bigcap s| \leq |X \bigcap H|$ **then**
5:         $H = s$
6:     **end if**
7: **end for**
8: **return** H.

X' of X, and an index i. $\Phi(X', i)$ outputs a set $s_{ij} \in G_i$ such that $|s_{ij} \bigcap X'|$ is maximized over all sets in S_i. And Φ is an α-approximate algorithm if $\Phi(X', i)$ outputs a set $s_{ij} \in S_i$ such that $|s_{ij} \bigcap X'| \geq \frac{1}{\alpha} max_D|D \bigcap X|$. In this paper, we show that a simple greedy algorithm gives a constant factor approximation for MNST. First, we assume we have an α-approximate algorithm described above, the greedy algorithm we consider is a natural generalization of the greedy algorithm of the maximum coverage problem. It iteratively picks sets that cover the maximum number of uncovered elements, however it considers sets only from those groups that have not already had a set picked. The algorithm is described below.

Without loss of generality, we assume that Greedy picks a set from Service Set S_i in the jth iterator. Let OPT denote some fixed optimal solution and let $i_1 < i_2 < i_k$ be the indices of the groups that OPT picks sets from. We set up a bijection ϱ form $\{1, 2, ..., k\}$ to $\{i_1, i_2, ..., i_k\}$ as follows. For $1 \leq h \leq k$, if $h \in i_1, i_2, ..., i_k$, then we set $\varrho(h) = h$. We choose ϱ to be any bijection that respects this constrain.

Algorithm 2. Greedy

1: Initialize heap $H, \leftarrow \emptyset, X' \leftarrow X$.
2: **for** $j = 1, 2, ..., k$ **do**
3:     **for** $i = 1, ..., k$ **do**
4:         **if** a set from S_i has not been added to H **then**
5:             $A_i \leftarrow FDS(S_i, X')$
6:         **else**
7:             $A_i \leftarrow \emptyset$
8:         **end if**
9:     **end for**
10:     $r \leftarrow argmax_i|A_i|$
11:     $H \leftarrow H \bigcup \{A_r\}, X' \leftarrow X' - A_r$
12: **end for**
13: **return** H.

We assume C_j be the set that Greedy picks from S_j, and let O_j be the set that OPT picks from $G_{\varrho(j)}$. We let

$$P_j = P_j - \bigcup P_h$$

denote the set of new elements that greedy adds in the jth iteration. Therefor,

$$C = \bigcup A_j, O = \bigcup O_j$$

denote the number of elements that Greedy and OPT pick.

Lemma 1. For $1 \leq j \leq k, |P_j^i| \geq \frac{1}{\alpha}|O_j - C|$

Proof. If $|O_j - C| = 0$, then the lemma holds, otherwise, when *Greedy* picked A_j, the set O_j was available to be picked. Greedy did not pick O_j because $|A_j'|$was at least $\frac{1}{\alpha}|O_j \bigcup A_h|$. Since $\bigcup A_h \subset C$, the lemma holds.   □

Theorem 1. Greedy is an $(\alpha + 1)$-approximation algorithm for MNST with an α-approximation pick-up algorithm.

Proof. Concluded from Lemma 1, we have that

$$|C| = \sum_j |A_j'| \geq \sum_j \frac{1}{\alpha}|O_j - C| \geq \frac{1}{\alpha}(|\bigcup O_j| - |C|) \geq \frac{1}{\alpha}(|O| - |C|)$$

Hence

$$|C| \geq \frac{1}{\alpha + 1}|O|$$

It is easy to prove that 2 is the upper bound for MST, and the ratio of 2 is tight. And when we consider the whole set system, the problem is hard to approximate within a factor of $\frac{e}{e-1}$. [1] [18] has proved that a matching ratio can obtained.

4   Greedy Algorithm for MNCW

Theorem 2. Greedy is an $(\log n + 1)$-approximation algorithm for MNCW.

Proof. We can assume that the optimal value of MNCW is OPT, and there exists such an optimal set cover O^*, subject to

$$max_i |O^* \bigcap S_i| \leq OPT.$$

Then we can create a solution of MNST, we assume the budget of each worker is OPT. The tasks picked up by greedy algorithm covers more than half of the tasks. In each iteration the number of set added from any worker is less than OPT. Let be G_i be the number of sets added from each worker by greedy, thus,

$$G_i \leq (\log n + 1)OPT$$

5 Experiment Evaluation

5.1 Experiment Setup

In this section, we evaluate our proposed algorithm. We use synthetic datasets for experiments. We generate the number of workers and the number of tasks. All workers and tasks are random distributed. Statistic and configuration of synthetic datasets are illustrated in Table 2.

Furthermore, synthetic datasets are created by Python, all algorithms are implemented in C++, under Linux Ubuntu and the experiments were performed on a machine with Intel Xeon E5620 2.40 GHz with 16-core CPU and 16 GB memory.

5.2 Experiment Results of MNST

In this subsection, we mainly evaluate the greedy algorithm compared to the OPT. We compute the number of covered tasks for MNST and the number of worker assigned for MNCW, and we tested our proposed algorithm via varying the number of workers and the number of tasks.

MNST. For MNST, we evaluate the performance of greedy algorithm. We study the effects of varying $|T|$, First, the number of workers (20) are fixed, and the number of tasks are 200, 300, 400, 500,600 respectively. The result of the experiment is showed in Fig. 1. Then we add the number of workers up to 50, 80, 100 respectively, and the result of the experiment is showed in Figs. 2, 3 and 4. We can observe that the number of covered tasks of OPT and greedy algorithm is in accordance with our theorem above. What' more, we evaluated our problem by the execution time and memory in four settings describe above, and the results are showed in Figs. 5, 6, 7, 8, 9, 10, 11 and 12. As is showed in the figure, *greedy* algorithm can solve the problem more quickly and use less memory.

Table 2. Synthetic dataset

Factor	Setting
$|W|$	20, 50, 80, 100, 200, 300, 400, 500
$|T|$	30, 50, 80, 100, 200, 300, 400, 500, 600

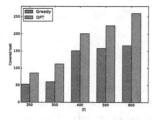

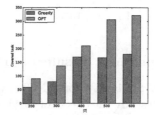

Fig. 1. $|W| = 20$ **Fig. 2.** $|W| = 50$

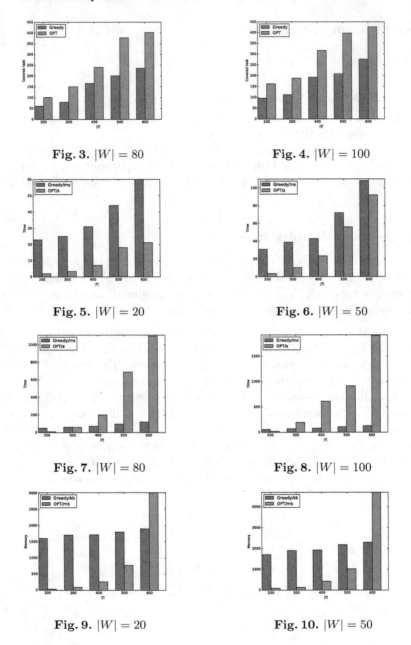

Fig. 3. $|W| = 80$

Fig. 4. $|W| = 100$

Fig. 5. $|W| = 20$

Fig. 6. $|W| = 50$

Fig. 7. $|W| = 80$

Fig. 8. $|W| = 100$

Fig. 9. $|W| = 20$

Fig. 10. $|W| = 50$

5.3   Experiment Results of MNST

MNCW. Similarity, for MNCW, we evaluate the performance of greedy algorithm. We study the effects of varying $|T|$. First. The number of workers (200) are fixed, and the number of tasks are 30, 50, 80, 100 respectively. The result of the experiment is showed in Fig. 13. Then we add the number of tasks up to 300, 400,

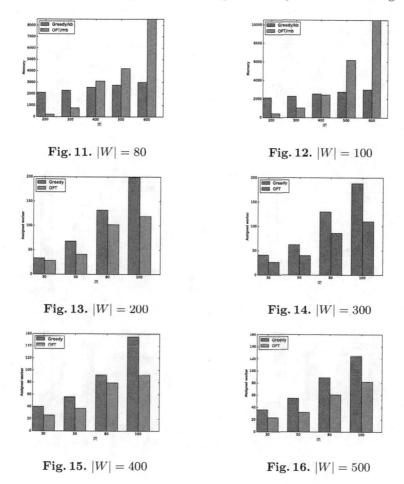

Fig. 11. $|W| = 80$ **Fig. 12.** $|W| = 100$

Fig. 13. $|W| = 200$ **Fig. 14.** $|W| = 300$

Fig. 15. $|W| = 400$ **Fig. 16.** $|W| = 500$

500, 600 and the result of the experiment is showed in Figs. 14, 15 and 16. Also, We can observe that the number of assigned worker of OPT and greedy algorithm is in accordance with our theorem. Similar to MNST, we evaluated our problem by the execution time and memory in four settings describe above, and the results are showed in Figs. 17, 18, 19, 20, 21, 22, 23 and 24. As is showed in the figure, *greedy* algorithm can solve the problem more quickly and use less memory.

6   Related Work

Set Cover and Maximum Coverage Problems. Set Cover and maximum coverage problem are fundamental algorithmic problems that arise frequently in a variety of settings. Their importance is partly due to the fact that many *covering* problems can be reduced to those problems. The *greedy* algorithm iteratively picks the set that covers the maximum number of uncovered elements is a ($\log n$ +1) approximation from the set cover problem and $\frac{1}{e} + 1$ approximation for the

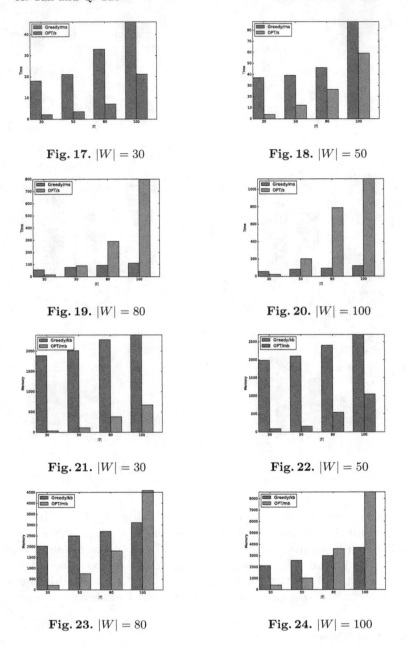

Fig. 17. $|W| = 30$

Fig. 18. $|W| = 50$

Fig. 19. $|W| = 80$

Fig. 20. $|W| = 100$

Fig. 21. $|W| = 30$

Fig. 22. $|W| = 50$

Fig. 23. $|W| = 80$

Fig. 24. $|W| = 100$

maximum coverage problem [11]. Feige [7] showed that these ratios are optimal unless NP is constrained in quasi-polynomial time. In a number of applications the set system is implicitly defined and the number of sets is exponential in the number of elements. However, the greedy algorithm can still be applied if a polynomial time algorithm that returns a set with good properties is available.

Spatial Matching Problem. In recent years, there have been a series of works about spatial matching. Such as [12,23,24]. These works are mainly about spatial information and capacity constrains in the matching scenario. The solution of [12] is aim at the min-max matching distance, and [23] introduces the assignment of capacity constrained. Furthermore, the spatial matching problems are extended and integrated to the social networks, the extended issues are called event-participate arrangement [14–16,21].

Vehicle Routing Problem. Considering the budget, we should get the shortest path visiting the location of multiple tasks. Thus, the task assignment problem is quite similar to the Traveling Salesman Problem (TSP) and Vehicle Routing Problem (VRP). The goal of VRP is to minimize the cost of delivering goods located at a central depot to customers who have placed orders for such goods with a fleet of vehicles. However, different from our problem, there is only one worker in TSP. And in our problem, each worker has a different source location, while in VRP, all workers start from the same depot.

Pickup and Delivery Problem. Different from Vehicle Problem, Pickup and Delivery Problem has two locations, source and destination. Worker has to move from origins to destinations without transshipment at intermediate locations. The Pickup and Delivery Problem is introduced to deal with vehicle transportation, including tasks with a set of origins and a single destination, tasks with a single origin and a set of destination and tasks with different start and end locations. Many applications on practical pickup and delivery situations are demand sensitive. That means when new task comes up, the system has to take it into consideration. Thus we have to re-optimize the problem as the same time as the new request emerge.

Online Task Assignment. The aforementioned issues of task assignment are assumed in offline scenarios, where all information is known in advance. In the online scenario where workers and tasks arrive dynamically, it becomes more challenging to achieve the global optimal solution. Since the server dose not have prior knowledge about future workers and tasks, it tries to optimize task assignment locally at every time period. However, the optimization within every time periods, similar to MNST or MNCW, is also NP-hard. A randomized algorithm [9] was proposed to achieve a competitive ratio of $1 - \frac{1}{e}$. Meanwhile, the result in [9] shows that the greedy algorithm has the competitive ratio of $\frac{1}{2}$, which is lower than that of the proposed randomized algorithm. Recently, Tong et al. [20] extended the issue of online task assignment to the two-sided dynamic scenario and devised a deterministic algorithm framework with $\frac{1}{4}$-competitive ratio under the online random order model, which means the average-case analysis.

7 Conclusion

In this paper, we discuss about task assignment problem in spatial crowdsourcing problem in real market. Existing approaches only simply focus on maximizing the number of completed spatial tasks but dose not take the labor dynamics into consideration, which leads to different optimal objectives for crowdsourcing task assignments. In this paper, we first propose a more general spatial crowdsourcing task assignment problem, consisting of two real scenarios, Excess Demand of Crowd Workers and Insufficient Supply of Spatial Tasks, in daily life. Thus, we design two approximation algorithm to solve this problem. What's more, we prove our algorithm has theoretical garuntee. Finally, we verify the effectiveness and efficiency of the proposed methods through extensive experiments on synthetic datasets. The results show that our algorithm can get remarkable effect in terms of the number of covered tasks and assigned workers and in less time and less memory.

References

1. Alt, F., Shirazi, A.S., Schmidt, A., Kramer, U., Nawaz, Z.: Location-based crowdsourcing: extending crowdsourcing to the real world. In: NordiCHI 2010, pp. 13–22 (2010)
2. Arora, S., Karakostas, G.: A 2+epsilon approximation algorithm for the k-mst problem. In: SODA 2000, pp. 754–759 (2000)
3. Cao, C.C., She, J., Tong, Y., Chen, L.: Whom to ask?: jury selection for decision making tasks on micro-blog services. Proc. VLDB Endow. 5(11), 1495–1506 (2012)
4. Cao, C.C., Tong, Y., Chen, L., Jagadish, H.V.: Wisemarket: a new paradigm for managing wisdom of online social users. In: SIGKDD 2013, pp. 455–463 (2013)
5. Chekuri, C., Kumar, A.: Maximum coverage problem with group budget constraints and applications. In: Jansen, K., Khanna, S., Rolim, J.D.P., Ron, D. (eds.) RANDOM 2004 and APPROX 2004. LNCS, vol. 3122, pp. 72–83. Springer, Heidelberg (2004)
6. Deng, D., Shahabi, C., Demiryurek, U.: Maximizing the number of worker's self-selected tasks in spatial crowdsourcing. In: GIS 2013, pp. 324–333 (2013)
7. Feige, U.: A threshold of $\ln n$ for approximating set cover. J. ACM 45(4), 634–652 (1998)
8. Gao, D., Tong, Y., She, J., Song, T., Chen, L., Xu, K.: Top-k team recommendation in spatial crowdsourcing. In: Cui, B., Zhang, N., Xu, J., Lian, X., Liu, D. (eds.) WAIM 2016. LNCS, vol. 9658, pp. 191–204. Springer, Heidelberg (2016). doi:10.1007/978-3-319-39937-9_15
9. Karp, R.M., Vazirani, U.V., Vazirani, V.V.: An optimal algorithm for on-line bipartite matching. In: STOC 1990 (1990)
10. Kazemi, L., Shahabi, C.: Geocrowd: enabling query answering with spatial crowdsourcing. In: GIS 2012, pp. 189–198 (2012)
11. Khuller, S., Moss, A., Naor, J.: The budgeted maximum coverage problem. Inf. Process. Lett. 70(1), 39–45 (1999)
12. Long, C., Wong, R.C.W., Yu, P.S., Jiang, M.: On optimal worst-case matching. In: SIGMOD 2013, pp. 845–856 (2013)
13. Pournajaf, L., Xiong, L., Sunderam, V.S., Goryczka, S.: Spatial task assignment for crowd sensing with cloaked locations. In: MDM 2014, pp. 189–198 (2014)
14. She, J., Tong, Y., Chen, L.: Utility-aware social event-participant planning. In: SIGMOD 2015, pp. 1629–1643 (2015)
15. She, J., Tong, Y., Chen, L., Cao, C.C.: Conflict-aware event-participant arrangement. In: ICDE 2015, pp. 735–746 (2015)

16. She, J., Tong, Y., Chen, L., Cao, C.C.: Conflict-aware event-participant arrangement and its variant for online setting. IEEE Trans. Knowl. Data Eng. **28**(9), 2281–2295 (2016)
17. Tong, Y., Cao, C.C., Chen, L.: TCS: efficient topic discovery over crowd-oriented service data. In: SIGKDD 2014, pp. 861–870 (2014)
18. Tong, Y., Cao, C.C., Zhang, C.J., Li, Y., Chen, L.: Crowdcleaner: data cleaning for multi-version data on the web via crowdsourcing. In: ICDE 2014, pp. 1182–1185 (2014)
19. Tong, Y., She, J., Ding, B., Chen, L., Wo, T., Xu, K.: Online minimum matching in real-time spatial data: experiments and analysis. Proc. VLDB Endow. **9**, 1053–1064 (2016)
20. Tong, Y., She, J., Ding, B., Wang, L., Chen, L.: Online mobile micro-task allocation in spatial crowdsourcing. In: ICDE 2016, pp. 49–60 (2016)
21. Tong, Y., She, J., Meng, R.: Bottleneck-aware arrangement over event-based social networks: the max-min approach. World Wide Web J. **19**(6), 1151–1177 (2016)
22. Tsitsiklis, J.N.: Special cases of traveling salesman and repairman problems with time windows. Networks **22**(3), 263–282 (1992)
23. Yiu, M.L., Mouratidis, K., Mamoulis, N.: Capacity constrained assignment in spatial databases. In: SIGMOD 2008, pp. 15–28 (2008)
24. Wong, R.C.W., Tao, Y., Fu, A.W.C., Xiao, X.: On efficient spatial matching. In: VLDB 2007, pp. 579–590 (2007)

SemiBDMA 2016

A Shortest Path Query Method Based on Tree Decomposition and Label Coverage

Xiaohuan Shan[1], Xin Wang[1], Jun Pang[2], Liyan Jiang[1],
and Baoyan Song[1(✉)]

[1] School of Information, Liaoning University, Shenyang, China
bysong@lnu.edu.cn
[2] School of Information Science and Engineering,
Northeastern University, Shenyang, China
pangjun@research.neu.edu.cn

Abstract. The shortest path query is one of core contents in graph theory study, various problems in the real world can be transformed into it to solve. With the increase of network scale, classic shortest path query algorithms cannot meet the query demand on large-scale graphs by reason of query efficiency, storage costs, etc. In order to solve above problems, we lucubrate on previous works, and propose a novel method based on tree decomposition and label coverage (TDLC-SP) which consists of two phases: offline pretreatment phase and online query phase. In the pretreatment phase, we propose a novel acceleration index method TDLC, it maps the graph into a tree, allocates minimum label coverage for each vertex to reduce redundant data storage and vertices traversal range; In the query phase, utilizing the TDLC index, query is completed by traversing the tree structure only once, it further improves the query efficiency. Experimental results on several real-world networks and synthetic datasets demonstrate the efficiency and effectiveness of the proposed methods.

Keywords: Shortest path query · Large graph · Pretreatment · Tree decomposition · Label coverage

1 Introduction

The shortest path problem is one of the core contents on network optimization and graph theory, it is widely applied to transportation networks, social networks [1–3], computer science and many other fields. In the real world, various problems can be converted to the shortest path problem. For example, how to quickly find the fastest route to drive from Shenyang to Tibet and how to get in touch with a stranger via the least friends utilizing Facebook, etc. The series of problems all can be abstracted to the shortest path query.

However, with the fast development of network and computer technology, the scale of graph data is increasing rapidly, and processing requirements on large-scale graph are more and more widely. Take example for the social network, for the global largest social network Facebook, active user number has exceeded 1 billion per month; WeChat which is designed by Tencent in early 2011 has covered nearly more than 90%

S. Song and Y. Tong (Eds.): WAIM 2016 Workshops, LNCS 9998, pp. 239–248, 2016.
DOI: 10.1007/978-3-319-47121-1_20

smartphones all over the country, and the active user number has also achieved 549 million per month. For the real applications, if you want to get interpersonal connections of some users on Facebook or WeChat, namely, it is necessary to process large graphs, calculate the distance between any two vertices and related information. Calculating the shortest path between any two vertices on the graph is a fundamental problem in computer science, but the shortest path problem on large-scale graph is facing great challenges in terms of storage overheads, query response time and so on. Classical shortest path algorithms such as Dijkstra [4], Floyd [5] have problems on query efficiency increasingly which leads to that they cannot meet query demand query on large-scale graphs within acceptable time and storage costs [6]. In recent years, a series of heuristic acceleration strategies combines with the shortest path query which has become the main research direction.

In this paper, aiming at large graphs, we consider storage and operational capability of current computers adequately, in order to improve the shortest path query efficiency this paper proposes a novel shortest path query method based on tree decomposition [7] and label coverage TDLC-SP. It consists of an offline pretreatment phase and an online query phase. In the pretreatment phase, we propose a novel acceleration index method TDLC. It divides the graph into finite vertex sets and maps the graph into a tree structure according to the relationship among vertex sets; constructs minimum label coverage for tree decomposition, optimizes them to reduce redundant data and improve the efficiency of pretreatment. In the process of index construction, it reduces the pretreatment cost by narrowing the scope of BFS. In the query processing phase, utilizing the TDLC index, the query only need traverse the tree structure once, so that it further improves the query efficiency.

2 Related Work

Large-scale graphs have been widely applied in many emerging areas, such as XML database, complex network analysis, geographic navigation, etc. However, in these areas, the shortest path query has become one of optimal technologies to solve related problems. With graph growth, classic shortest path query algorithms have already been unable to meet the query demand on large-scale graphs. Data preprocessing is the novel solution, which can accelerate the shortest path query. This is a typical practice of trading space for time, and its main idea is to extract part of useful information, which can accelerate the later shortest path query.

Among the preprocessing of the shortest path query, the simplest method is to pre-calculate the shortest path between any two vertices, then all paths and distances about all vertices was stored in memory or hard disk with table structure, after that, we can get the shortest path between any two vertices from data tables. However, with data scale enlarging gradually, data tables which are used to store paths and distances expand extremely, the storage capacity of existing computers cannot complete storage tasks of such large-scale data, resulting in the decrease of query efficiency.

At present, during the data preprocessing, various speedup methods are combined to improve the efficiency of preprocessing and query. The A* algorithm [8] selects landmark points, and preprocess shortest paths from all vertices to all landmark points which is leading to oversize storage overhead. In allusion to the problem, literature [9] reduces the memory consumption by only storing shortest paths from part of vertices to landmark points, which solves the problem of oversize storage overhead, but causes several serious problems, such as the decrease of query accuracy, the increase of query complexity and so on. PCD algorithm [10] is a pruning method, which limits the search process by preprocessing distances between clusters, and its basic idea is to limit search areas by eliminating vertices and edges on the optimal path which have less possibility or little effect on the result. The efficiency of PCD is related to the number of clusters k, and during the preprocessing phase, single source shortest paths need to be calculated k times, but the efficiency of abandoning vertices and edges is quite low with the increase of graphs, and it causes oversize time complexity, so it cannot be suitable for query on large graphs. TDEI [11] maps the graph into a tree structure and creates an index for any vertex in each node package on the tree based on the technology of limiting searching areas, which stores the shortest path between any two vertices, the bottom-up approach is used to query the shortest path on the tree decomposition. The problem of this method is that it causes redundant vertices on tree decomposition, and it is possible that the number of redundant vertices exceeds vertex number on the original graph, for large graphs, the storage overhead of the index is extremely large and the query need traverse tree decomposition repeatedly, so it is unable to query on large-scale graphs. LBFS method [12] is based on optimal coverage and landmark breadth traversals, although it can keep high query efficiency on large graphs, but it ignores the dispersion of landmarks during optimal coverage selection and shorter distance between vertices during the query, therefore, it can't satisfy the need of query accuracy.

Above all, previous works have some problems with query efficiency, storage cost and query accuracy when they are applied on large-scale graphs.

3 TDLC Index Construction

On the offline preprocessing phase, we propose a novel index acceleration method TDLC. Its primary mission is to extract and save the "useful" information which can support online query work by utilizing the theory of trading space for time.

3.1 Tree Decomposition

Given an initial graph $G = (V, E)$, V is the set of vertices and E is the set of edges. Before introducing tree decomposition, we first present two important definitions.

Definition 1 MapTree: $G (V, E)$ is an undirected and unweighted graph, map G into tree structure MT which is called MapTree. Nodes on MT consist of subgraphs of G, the node on MT is called node package.

Definition 2 MapTree Breadth: A maximum of node package cardinality on each MT is MapTree Breadth, namely MT-B.

In this paper, we utilize the heuristic elimination method to construct *MT*. Elimination operations decompose the graph by deleting vertices which start from the minimum degree vertex. Threshold k is set to limit executions of the elimination. The reason to set k is that with the process of elimination, more and more edges need to be added during deleting vertices, which reduces the efficiency of pretreatment. The specific process of elimination is as follows:

1. Find vertex v which meets current degree demand and all its neighbors $v_1, v_2 \ldots v_k$;
2. Judge whether deleting v will cause information loss, if it will lose, add edges between v's neighbors to save connected information about them, else just delete v directly;
3. Push v and all its neighbors into stack S;
4. If there are still vertices that meet current degree demand, return to 1; otherwise degree plus one, and return to 1 until the degree reaches k.

After elimination, we construct *MT* according to the order of elimination in stack S. The specific process is as follows:

1. Make the remaining vertices on the subgraph after elimination as the root node package of *MT*, and its id is 0;
2. Pop top element, and make it as a node package of *MT*, its id is deleted vertex's id;
3. Find the node package which involves deleted vertex and all its neighbors as the father node;
4. Return to 2 until the stack is empty, *MT* construction is finished.

Figure 1 shows the specific implementation process of tree decomposition.

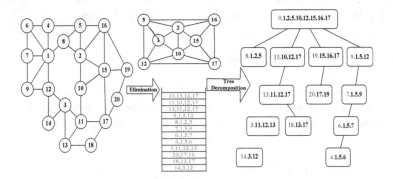

Fig. 1. The sample of tree decomposition

In this paper, we choose deleted vertex id as the node package id on *MT* construction. That is because when the shortest path query is coming, it only need traverse tree decomposition once that can find source and destination node packages instead of traversing tree decomposition repeatedly to find node packages including the same vertex and then confirm the root node package. So it not only improves the efficiency of *MT* construction, but also improves pretreatment and query.

3.2   Label Coverage Construction

After mapping the graph into *MT*, we construct minimum label coverage for each node package, and optimize label structure to reduce the amount of redundant data, thus improve the efficiency of pretreatment and query.

Definition 3 Distance Label: For any vertex u in a node package, the distance label structure is $L(u) = \{(v_1, \text{sdist}(u, v_1)), (v_2, \text{sdist}(u, v_2)), \ldots, (v_k, \text{sdist}(u, v_k))\}$, and the $\text{sdist}(u, v_k)$ shows the shortest distance from any vertex u to vertex k.

Definition 4 Node Package Label Coverage: Distance label set of all vertices in the node package that supports the path query of any two vertices in the node package.

The specific procedure of constructing minimum label coverage for each node package on *MT* is as follows:

1. BFS starts from deleted vertex in the node package until all vertices in it have been accessed, after that it can get distance labels from deleted vertex to other vertices;
2. According to distance labels got from 1, BFS starts from non-deleted vertex. Judge whether the shortest path got from current BFS(BFS-SP) less than that is combined by distance labels in the existing label set(Label-SP), if BFS-SP is less than Label-SP, classify BFS-SP into the corresponding label set in label structure;
3. Return to 2 until all vertices in the node package is finished performing.

We take *MT* in Fig. 1 for an example. Firstly, we construct label coverage for leaf node packages. According to the characteristics of tree decomposition, any leaf node package is the source or destination point of deleted vertex in it, there is only one possibility to query in leaf node package, that is from deleted vertex in leaf node package to other vertices, so label coverage of any leaf node package only need store key values related to delete vertex. Such as the label coverage of node package 8 are $L_8(1) = \{(8,1)\}$, $L_8(2) = \{(8,1)\}$, $L_8(5) = \{(8,1)\}$. Due to non-leaf node packages may be as query source or destination points or transitional node package, so it cannot just save the records related to deleted vertices. For example in our method the label coverage of non-leaf node package 3 are $L_3(11) = \{(3,1)\}$, $L_3(12) = \{(3,1)\}$, $L_3(13)\{(3,1), (11,1)\}$, thus TEDI [8] need save larger with $L_3(3) = \{11,1), (12,1), (13,1)\}$, $L_3(11) = \{(3,1), (12,2), (13,1)\}$, $L_3(12) = \{(3,1), (11,2), (13,2)\}$, $L_3(13) = \{(3,1), (11,1), (12,2)\}$.

The proposed method adds judging conditions in label coverage construction, at the same time optimizes the label coverage of leaf node packages according to the characteristics of query path on *MT*, which can reduce the storage of redundant data.

4   TDLC-SP

In this section, we will introduce TDLC-SP method in detail, the method consists of offline pretreatment and online query. On the pretreatment phase introduced in the previous section we proposed a TDLC method to manage graph data and extract the "useful" information, and then to speed up the online shortest path query. Figure 2 shows the frame of TDLC-SP method. Thereinto, P1 and P2 constitute the process of tree decomposition, P3 and P4 is the process of minimum label coverage construction, above processing steps are the offline pretreatment. P5 is online query.

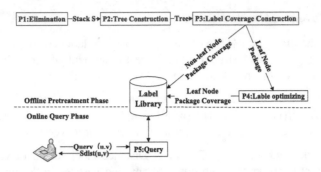

Fig. 2. TDLC-SP frame

The shortest path query is conducted on *MT*, thereinto the path on *MT* is consisted of the shortest path in the nodes package (internal edges) and the path among node packages (middle edges). The query on *MT* adopts a bottom-up method to find the simple path among nodes packages and the precomputed shortest path in the node package. The specific implementation shows in Algorithm 1.

The following also takes Fig. 1 as an example to illustrate the specific process of TDLC-SP. Suppose you want to query the shortest path between vertex 14 and 4. Firstly judge whether two vertices are in the same node package, we can find it doesn't meet the condition; Secondly find each vertex which corresponding query initial position is node package 14 or 4 by traversing *MT* and the youngest parent of two nodes package as the root node package; Finally query from node package 14 to 11 and from 4 to 9 respectively by bottom-up, the result must in the intersection set of root node and package 11 or root node and package 9. Figure 4 shows the path mapping of two intersections, The shortest paths in Path (*u*) and Path (*v*) come from label coverage of node packages 14, 3, 13, 11 and 4, 6, 7, 9 respectively, shortest paths in X-Y come from label coverage of root node. In all paths mapping combination, we output the minimum path, which is 14->12->1->4 (Fig. 3).

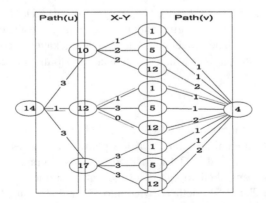

Fig. 3. Sample of query path mapping

Algorithm 1.TDLC-SP

```
Input: vertices u,v
Output: SP(u,v)
If (Ru ==Rv )   return path(u,v)
root= youngest common ancestor of Ru and Rv
while(Ru!=root){
  XP=Ru.Parent
  for(all t∈XP\X1){
    for(all s∈XP∩X1){
       if du(t))>du(s)+sdist(s, t)
       then  du(t)=du(s)+sdist(s, t)   p(t)=s; } }
  X1=XP
  end while
  X1=X1∩root;
  Rv equals Ru
  X2=X2∩root
  mdist={min du(x)+dv(y)+sdist(x, y)}
  output:sdist(u,v)
```

5 Experiments

5.1 Experimental Environment and Datasets

This section experimentally evaluates our TDLS-SP method against TEDI method [8] on the pretreatment time and pretreatment storage overhead. We implement our algorithm in JAVA, all of our experiments are conducted on a machine with an Inter®Pentium®2.90 GHz CPU, 4 GB RAM, running Windows 7 Ultimate (32 bit/SP1). We perform experiments on several real-world networks such as biological information networks, social networks, computing network, information networks and synthetic datasets. Synthetic datasets are generated by BA model, the number of vertices starts from 1000 to 10000. Due to the tree width definition during tree decomposition is a NP problem, in order to show the optimal contrast effects, we apply heuristics to choose the optimal tree width. Tables 1 and 2 show the real-world networks and synthetic datasets after tree decomposition. Among them, N and TreeN represent the number of vertices in the graph and node packages on MT, R is the size of root node package.

5.2 Analysis of Experimental Results

In this section, we compare the proposed method TDLC-SP with TEDI method [8] on pretreatment time and storage overhead.

Table 1. Real-world networks

Graph	N	TreeN	K	R
G_1	1000	808	3	194
G_2	2000	1730	5	272
G_3	3000	2641	6	361
G_4	4000	3559	7	443
G_5	5000	4460	8	542
G_6	6000	5355	9	612
G_7	7000	6292	9	710
G_8	8000	7201	9	801
G_9	9000	8089	9	913
G_{10}	10000	8983	9	1091

Table 2. Synthetic datasets

Graph	N	TreeN	K	R
Gemo	3621	3000	10	623
Epa	4253	3637	7	618
Dutsc	3621	3442	9	258
Eva	4475	4475	9	75
Cal	5925	5095	14	832
Erdos	6927	6690	9	405
PPI	1458	1359	11	101
Yeast	2284	1770	6	516
Homo	7020	5778	10	1244
Inter	22442	21757	10	687

The comparison of pretreatment time with TDLC-SP and TEDI on real-world networks and synthetic datasets shows in Fig. 4(a) and (b). We can find that the pretreatment time with two methods both increases gradually following the increase of dataset scale, but the pretreatment time of TDLC-SP is superior to TEDI obviously. This is because TEDI need create the shortest path index for all vertices in the node package, and then the proposed method only need create minimum label coverage for each node package. Although we need judge whether the current shortest path is less than it in distance label library during label coverage construction, and it need some time to query label library, but with the current distance label set growing stronger and stronger, the number of accessing vertices will be less and less that may decrease pretreatment time greatly. Especially when it creates minimum label coverage for the root node package which includes more vertices, the advantage of our method emerges obviously.

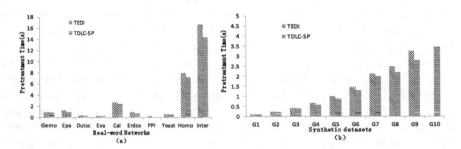

Fig. 4. Pretreatment time comparison on real-world networks and synthetic datasets

The comparison of pretreatment storage overhead with TDLC-SP and TEDI on real-world networks and synthetic datasets shows in Fig. 5 (a) and (b) respectively. Because of adding the judging conditions during the process of creating minimum label coverage for tree decomposition, so it guarantees that redundant distance labels in the label coverage of each node package don't exist. Relative to save the shortest path of

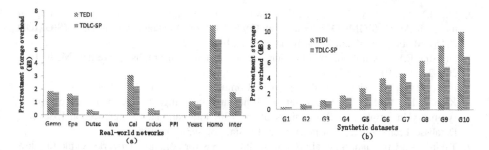

Fig. 5. Comparison of storage overhead on real-word networks and synthetic datasets

any two vertices with TEDI, our method can decrease storage overhead effectively, and the gap will increase with the graph scale growth.

Experimental results on real-world networks and synthetic datasets indicate that the proposed method is superior to TEDI on pretreatment time and storage overhead, and then the gap will increase linearly with the growing of dataset scale.

6 Conclusion

In this paper, we study the problem of the shortest path query on large-scale graphs. The problem has many practical applications. We proposed a solution (TDLC-SP) consisting of an offline pretreatment phase and an online query phase. The low cost acceleration index (TDLC) is built in the offline phase, it is based on tree decomposition and label coverage to reduce the storage of redundant data and improve the efficiency of pretreatment. On the online query phase, with optimizing the tree decomposition and label structure, TDLC-SP only need traverse tree structure once to finish the query. It further improves the efficiency of query so that it can be applied to the shortest path query on large-scale graphs efficiently.

Acknowledgements. This work was supported by National Natural Science Foundation of China under Grant (Nos. 61472169, 61502215); Science Research Normal Fund of Liaoning Province Education Department (No. L2015193); Doctoral Scientic Research Start Foundation of Liaoning Province (No. 201501127); Young Research Foundation of Liaoning University under Grant (No. LDQN201438).

References

1. Tong, Y.X., She, J.Y., Meng, R.: Bottleneck-aware arrangement over event-based social networks: the max-min approach. World Wide Web-internet Web Inf. Syst. **19**, 1–27 (2015)
2. Tong, Y.X., She, J.Y., Chen, L.: Towards better understanding of app functions. J. Comput. Sci. Technol. **30**(5), 1130–1140 (2015)

3. She, J.Y, Tong, Y.X., Chen, L.: Utility-aware event-participant planning. In: Proceedings of the 34th ACM SIGMOD International Conference on Management of Data (SIGMOD 2015), Melbourne, Victoria, Australia, pp. 1629–1643 (2015)
4. Dijkstra, E.W.: A note on two problems in connexion with graphs. J. Numer. Math. $1(1)$, 269–271 (1959)
5. Floyd, R.W.: Algorithm 97: shortest path. Commun. ACM $5(6)$, 345–348 (1962)
6. Xiao, Y.H., Wu, W.T., Pei, J., et al.: Efficiently indexing shortest paths by exploiting symmetry in graphs. In: Proceedings of the 12th International Conference on Extending Database Technology, Saint Petersburg, Russia, pp. 493–504 (2009)
7. Takuya, A., Christian, S., et al.: Shortest-path queries for complex networks: exploiting low tree-width outside the core. In: Proceedings of the 15th International Conference on Extending Database Technology, New York, USA, pp. 144–155 (2012)
8. Goldberg, A.V., Werneck, R.: Computing point-to-point shortest paths from external memory. In: Proceedings of the 7th Workshop on Algorithm Engineering and Experiments, London, pp. 26–40 (2005)
9. Schultes, D.: Route planning in road networks, Ph.D. thesis, Universitat Karlsruhe (2008)
10. Maue, J., Sanders, P., et al.: Goal directed shortest path queries using precomputed cluster distances. J. Exp. Algorithms $14(32)$, 1–27 (2009). ACM
11. Fang, W.: TEDI: efficient shortest path query answering on graphs. In: Proceedings of the 2010 ACM SIGMOD International Conference on Management of data, pp. 99–110 (2010)
12. Konstantin, T., Abel, A.-C., et al.: Fast fully dynamic landmark-based estimation of shortest path distances in very large graphs. In: Proceedings of the 20th ACM International Conference on Information and Knowledge Management, vol. 278(13), pp. 1785–1794 (2011)

An Efficient Two-Table Join Query Processing Based on Extended Bloom Filter in MapReduce

Junlu Wang[1], Jun Pang[2], Xiaoyan Li[1], Baishuo Han[1], Lei Huang[1], and Linlin Ding[1(✉)]

[1] School of Information, Liaoning University, Shenyang 110036, China
dinglinlin@lnu.edu.cn
[2] School of Information Science and Engineering, Northeastern University,
Shenyang 110819, China
pangjun@research.neu.edu.cn

Abstract. With the development of Cloud Computing, the Internet of things and some similar technologies, a large amount of data has been produced. MapReduce as a processing architecture for Cloud Computing has been widely used. It can achieve large-scale data processing. However, when connecting two tables on the data processing model of MapReduce, there will be a great deal of data that do not meet the conditions of the connection. These data will also be transferred from the map side to the reduce side. It will bring more time overhead and I/O cost at shuffle stage, which will result in low efficiency. Therefore, how to improve the join query processing algorithm based on the MapReduce has been an urgent problem. In this paper, we put forward two-table join query processing and optimization strategies for the above problems. The optimized method can achieve the expansion of the Bloom Filter. Meanwhile it can reduce the time of shuffle phase, and improve the efficiency of the system.

Keywords: Mapreduce · Bloom Filter · Join query processing and optimization

1 Introduction

With the arrival of the era of big data and the wide application of connection and query operations, it has become a trend for large-scale data connection and query on the MapReduce programming model. However, when connecting two tables on the MapReduce, there will be a great deal of data that are not satisfied with the join condition from Map to Reduce side, and it will bring greater network communication cost and I/O cost. How to improve the join query processing algorithm based on MapReduce has been an urgent problem.

In this paper, we propose the strategy about two tables joining query processing and optimization strategy aimed at the shortcomings of connection based on MapReduce. Based on the data structure of Bloom Filter, we create a M bit array and initialize all bits to 0. Then let each join k counted k times according to hash function, and calculate the results with "XOR" operation. In an array of new settlement, the bit position corresponding results of "XOR" will be set 1. By compressing the value of connection attribute used the improved Bloom Filter data structure, we can effectively filter out a

© Springer International Publishing AG 2016
S. Song and Y. Tong (Eds.): WAIM 2016 Workshops, LNCS 9998, pp. 249–258, 2016.
DOI: 10.1007/978-3-319-47121-1_21

lot of tuples that are not satisfied with the join condition. The connection task for two tables, we can respectively pick up the property of connection in both tables, and compressed by extended bloom Filter, then it will form two compressed files to mutually filter the tuples that not satisfied with connected condition in the two tables. The main contribution of this paper:

(1) The misjudgment rate of Bloom Filter data structure is reduced.
(2) Joining two tables based on MapReduce model, it reduces the cost of shuffle stage and improves the efficiency of the system.

2   Related Work

In view of the wide application of connection query operations, database researchers conduct a lot of groundbreaking research in this area and put forward many ways to optimize the connection, such as Hash Join [1], Sort-merge Join [2], Nested-loop Join [3], and so on. On this basis, a series of improvements and extensions have been done to optimize join queries algorithm [4–7]. Currently, there already exist some optimization schemes about connection query operations for massive data at home and abroad. The following are the main types:

About KNN connection scheme, the literature [8] proposes the effectively parallel KNN connection algorithm. Based on the KNN algorithm, the connection of large-scale data on MapReduce performs KNN connection efficiently by constructing the partition function for H-zkNN algorithm. Basing on k Nearest Neighbor Joins, literature [9] proposes the effective mapping mechanism and uses pruning rules in the distance filter to reduce the cost of shuffle stage and calculation. For the cost of shuffle stage, the paper proposes two approximation algorithms to minimize the number of copies. Experiments have proved that the optimization strategy is efficient, robust and elasticity.

In the connection scheme of the parallel Top-k, the literature [10] proposes the approximation connection algorithms based on the data processing model of MapReduce. This paper uses the control algorithm and limits branch algorithm to reduce the data transmission by dividing the key/value pairs both special and all.

About Theta-Joins scenario, the literature [11] proposes optimized Theta-Joins algorithm. It designs a random algorithm SEJ to optimize the multi-channel Theta-Joins and a dynamic algorithm, which can further optimize the multi-channel Theta-Joins. The experiments have proved the feasibility and effectiveness of this method. Literature [12] analyzes and improves the efficiency of the binary Theta-Joins based on MapReduce. Literature [13] proposes a new model and a random algorithm to optimize the Theta-Joins, namely 1-Bucket-Theta.

In the Map-based side connection, literature [14] conducts a further study. This optimized method directly does connection function in the Map side, and do not need to send key/value pairs to Reduce side. Theoretically, the way can improve the efficiency of the system and reduce the time of inquiring. But this method only adapts to the data connection under certain circumstances. Literature [15] proposes optimization strategy of broadcast connection. The idea of this method is as follows. When connecting a small data set with a large data set, if the small data set table is small enough that can be placed in memory, the mechanism of Hadoop Distributed Cache will transfer small data sets to every Mapper node. In the Map phase, the records of large data sets that do not satisfied with the conditions of connection will be filtered through a small data set table, and then make connection directly and improve the efficiency of the system. However, this method is only suitable for the connection of the big data table and small data table. When two large tables are connected, the efficiency of this approach will be low. So it does not have universality.

In the Reduce-based side connection, the literature [16] proposes an optimized algorithm for large data sets connection based on Hadoop. This paper compresses the connection attribute values by the method of Bit-map compression. But the collision rate of this approach is too high, and the article doesn't analyze the misjudgment rate. Literature [17] proposes a pre-sorting connection algorithm. The tuples with the same property of the two tables are assigned to the same Map node to sub-connecting, and the final connection is done in Reduce side. This practice will bring a lot of network overhead during pre-treatment.

3   Two-Table Join Query Processing Based on Bloom Filter

3.1   Extended Bloom Filter Algorithm

There is an example that to explain the extended Bloom Filter algorithm (Fig. 1).

Customer

Id	Name	Telephone
5	Lengu	555-555-556
7	David	123-456-789
9	Jose	245-562-478
14	Edwark	526-856-234
15	Eark	412-856-234

Orders

Id	Cus-id	Price	Data
A	5	95	Jun-2013
B	6	85	May-2013
C	9	74	Jan-2013
D	14	96	Nov-2013
E	59	75	Nov-2012

Fig. 1. Two-table join

Taking the connection attribute value of 5 and 7 in the Customer table as an example (Cus-id). There are two hash functions: $h_1(x), h_2(x)$, which are the array of 13 bits in the memory. Let two hash functions be: $h_1(x) = x, h_2(x) = 2x + 3$ (3.1) and $H_1(x) = h_1(x)$ mod13, $H_2(x) = h_2(x)$mod13 (3.2). The hash address is 5 corresponded to the index are 5 and 0. The address is 7, and the index is 7 and 4. The consequence is calculated by data elements. The position corresponding to the hash address of each join key should be set 1. As shown in Fig. 2, if we want to check the key-value of join key 59 of other table whether it meets the join condition, we will get the address 7 and 5 by taking hash calculation to the join attribute value 59. And retrieving the compressed file BF (C), all the bit of the hash address is 1. This will produce false positive.

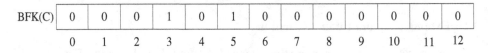

BF(C)	1	0	0	0	1	1	0	1	0	0	0	0	0
	0	1	2	3	4	5	6	7	8	9	10	11	12

Fig. 2. Filter file

Extended Bloom Filter algorithm: Basing on the data structure of Bloom Filter, we create a bit array of M bit and initialize all bits to 0. Then operate the consequence of the k hash values by "exclusive" or operation. In the array of new location, the bit position corresponding to the results will be set 1. Using the above example, we create another bit array of 13bit in the memory. Under twice hash calculation, do exclusive OR respectively with the hash address, namely: H1 (X) $\oplus$ H2 (X) (3.4). Set 1 in the corresponding exclusive OR address in the bit array. The result of the exclusive OR address is: 5:5 $\oplus$ 0 = 5;7:7 $\oplus$ 4 = 3. As Fig. 3, set 1 to the corresponding position in the new bit array. Checking the tuple59 in another table, calculate it by exclusive OR with hash address and get the result. While the position of new bit array (BFK) is 0, we check the bit array in Fig. 3. So the element of 59 is inconformity and be filtered.

BFK(C)	0	0	0	1	0	1	0	0	0	0	0	0	0
	0	1	2	3	4	5	6	7	8	9	10	11	12

Fig. 3. Extended Filter file

3.2 Two-Table Join Query Processing Algorithm

The following is the integral structure diagram of connecting two tables (Fig. 4).

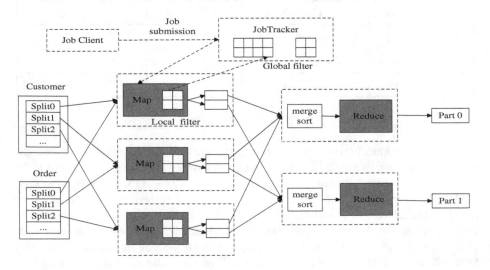

Fig. 4. Two-table join

Extended Bloom Filter's process:

(1) The first map stage: Create two bit arrays of M bits for each split in the memory. Read each piece of input information, parse it into the key/value pairs and label reasonable signs on them. Each split is divided into two compressed files, the split o1 of table order creates BF(o1) and BFK(o1)), namely the join key of each split forms its own local filter files by compression processing.

(2) The generate of global filter file stage: Create two bit arrays for each table which need connection. When receiving all local filter files, we do "OR" operation with these of the two tables respectively. Finally do the "AND" operation to get the final filter file.

(3) The second stage of Map: After receiving the global filter file (BF(CO), BFK(CO)) when the Map task carrying out. Filtering can be divided into two stages. First, use the normal Bloom filter algorithm to calculate the result. If one of the bits not equal to one, filter out it. If all of the bits equal to one, enter to the second stage, then calculate these strategies with "XOR" operation and search in the BFK table to insure whether it equals to one (Fig. 5).

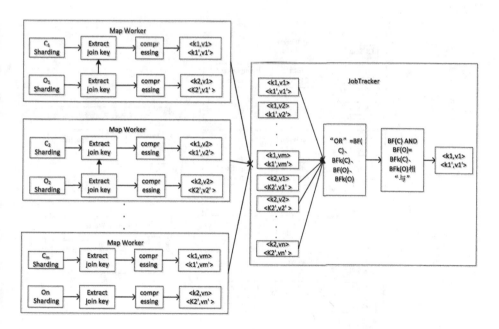

Fig. 5. Map process

The pseudo-code of this algorithm shows in this picture:

```
the filter algorithm in the stage of map
```

```
1.obtain the join key from the values that parse out from the
input split;
2.get the hash address of specific value :h1,h2,h3,..., hk;
3.If  ( BF(CO)[ h1]= BF(CO)[ h2]=.... BF(CO)[ hk]=1 )
4.{ H1=h1⊕h2⊕..... ⊕hk;
5. If ( BFk(CO)[H1]=1 )
6.Text join-key = generateGroupkey (avalue) ;
7.output.collect (newTextPair(key.toString(),tag), newTextPair (
values.
toString( ),tag) );

}
```

---

(4) The stage of reduce: Change the output of the map stage into the input of the reduce stage. Join the same key values from the different labels, and output the final values (Fig. 6).

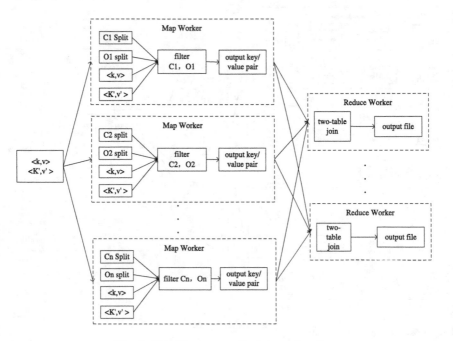

Fig. 6. The filter process in the map function

The pseudo-code of side-connection based on reduce showed:

```
the process in the reduce function
```

```
1.ArrayList<Text> T1 = new ArrayList<Text>();
2.Text tag = key . getSecong();
3.TextPair value = null;
4.While(values . hasNext())
5.{  value = value. next() ;
6.   if(value . getSecond().compareTo(tag)==0)
7.   {  T1. Add(value. getFirst());

}
```
---

4   Experiment and Analysis

4.1   Environmental and Experimental Data

The research of this paper is based on Hadoop platform. The Hadoop cluster environment of experiment in this article is established by ten ordinary PC and a client machine. Let one of the 10 ordinary PC be the Job Tracker, and the rest of the 9 work as Task Trackers. Each PC has 4G memory size, four thread 2.5 GHZ dual-core CPU, and the 2 TB hard disk. The operating system used in this experiment is 32-bit Ubuntu10.10, Java JDK version 1.7.0_07, and the version of Hadoop software is 0.20.2.

All data adopted in this paper are from TPC-H [18] during the experiment. The obtained data in the TPC - H contain eight associated data sheets.

4.2   Performance Test and Analysis

Using the table of CUSTOMER and ORDERS verifies the feasibility of optimization. The experiment selects six different size of the data set, as showed in Table 1. First compare the time of shuffle stage respectively with BRSJ and BF-BRSJ two methods.

Table 1. Size selected six groups of data sets

Data set	1	2	3	4	5	6
Customer	165 M	225 M	525 M	750 M	900 G	1.1 G
Orders	170 M	275 M	400 M	770 M	950 G	1.2 G

Through the comparison of the experimental results in Fig. 7, we can see that the time of the improved method is much reduced in the shuffle stage. So using this method, the cost of time at shuffle stage will reduce a lot, it can be even reduced by more than half of the time of traditional method.

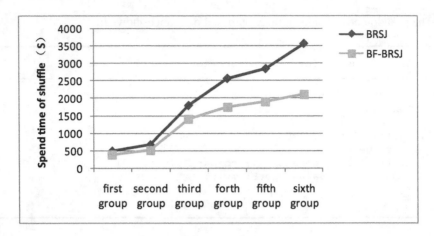

Fig. 7. Comparison of shuffle phase time

Next, do connection to six different size data sets. As shown in Fig. 8, by comparing the results of the experiment, we can find that the efficiency of the improved strategy is much better than the traditional method based on Reduce side. Such as the results of the first experiment group are shown in Fig. 8, therefore it does not reflect the advantages of the improved method. But with the increasing of the data, the data that are not satisfied with the connection condition will be further increased in the two tables. By improving mutual filtering policy, most of the tuples in two tables that are not satisfied with the connection condition can be filtered out. Thus, the more data, the more superiority it can reflect of the optimization strategy of two-table join.

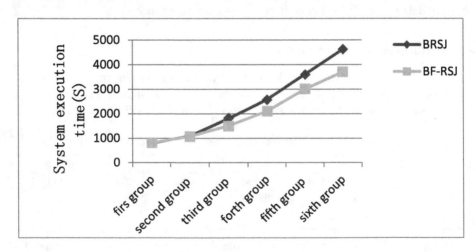

Fig. 8. Compare with the results of two-table join

5 Conclusion

In two-table join strategy based on Map Reduce, the paper proposes the mutual filtering policy based on the extended Bloom Filter. It respectively extracts the join key from two tables, and forms a compressed file of connection properties by the compressing of the extended Bloom Filter. Then it filters out the tuples in the two tables by compressing files which don't meet the join condition. This method reduces a lot of time of the shuffle and I/O overhead, which optimizes the efficiency of connecting two tables through twice of the Map stage and once of the Reduce stage. The experiment proves that this optimized method can efficiently perform the tasks of two-table join.

Acknowledgements. This work is supported by National Natural Science Foundation of China under Grant (Nos. 61472169, 61502215); Science Research Normal Fund of Liaoning Province Education Department (L2015193); Doctoral Scientific Research Start Foundation of Liaoning Province (201501127); the Young Research Foundation of Liaoning University under Grant No. LDQN201438.

References

1. Mishra, P., Erich, M.H.: Join processing in relational databases. ACM Comput. Surv. **24**, 63–113 (1992)
2. Ramakrishnan, R.: Database Management Systems. McGraw -Hill Inc, New York (1997)
3. Garcia-Molina, H., Widow, J., Ullman, J.D.: Database System Implementation. Prentice-Hall, Inc., Upper Saddle River (1999)
4. Kwan, S.C., Baer, J.-L.: The I/O performance of multiway merge sort and tag sort. IEEE Trans. Comput. **34**, 383–387 (1985)
5. Fushimi, S., Kitsureqawa, M., Tanaka, H.: An overview of the system software of a parallel relational database machine GRACE. In: Proceedings of the Very Large DataBases Conference, pp. 209–219 (1986)
6. Dewitt, D.J., Katz, R.H., Olken, F., et al.: Implementation techniques for main memory database systems. In: Proceedings of the ACM SIGMOD International Conference, pp. 1–8 (1984)
7. Stamos, J.W., Young, H.C.: A symmetric fragment and replicate algorithm for distributed joins. IEEE Trans. Parallel Distrib. Syst. **4**(12), 1345–1354 (1993)
8. Zhang, C., Li, F., Jestes, J.: Efficient parallel kNN joins for large data in mapreduce
9. Lu, W., Shen, Y., Chen, S., Ooi, B.C.: Efficient processing of k nearest neighbor joins using mapreduce
10. Zhang, C., Li, J., Wu, L.: Optimizing Theta-Joins in a mapreduce environment. Int. J. Database Theory Appl. **6**(4), 91–108 (2013)
11. Koumarelas, I.K., Naskos, A., Gounaris, A.: Binary Theta-Joins using mapreduce: efficiency analysis and improvements
12. Okcan, A., Riedewald, M.: Processing Theta-Joins using mapreduce
13. White, T.: Hadoop: The Definitive Guide, 2nd edn. O'Reilly Media, Inc., California (2011). pp. 247–249
14. Blanas, S., Patel, J.M., Ercegovac, V., Rao, J., Shekita, E.J., Tian, Y.: A comparison of join algorithms for log processing. In: Proceedings of the 2010 ACM SIGMOD International Conference on Management of Data (SIGMOD 2010), pp. 975–986 (2010)

15. Hui, S.: Large data set connection optimization algorithm based on Hadoop framework. Nanjing University of Posts and Telecommunications (2013)
16. Lin, Y., Agrawal, D, Chun, C., et al.: Llama: leveraging columnar storage for scalable join. In: Proceedings of SIGMOD 2011. ACM, New York (2011)
17. Yang, H.-C., Dasdan, A., Hsiao, R.-L., Parker, D.S.: Map-reduce-merge: simplified relational data processing on large clusters. In: Proceedings of the 2007 ACM SIGMOD International Conference on Management of Data (SIGMOD 2007), pp. 1029–1040 (2007)
18. http://www.tpc.org/tpch/

An Improved Community Detection Method in Bipartite Networks

Fan Chunlong, Song Yan, Song Huimin, and Ding Guohui[✉]

Liaoning Provincial Key Lab of Large-scale Distributed System,
The University of Shenyang Aerospace, Shenyan 110136, China
songyanpaper@sina.com, dinggh.sau@gmail.com

Abstract. Bipartite networks is one of the important research object in complex networks. At present, the bipartite networks community partition is mainly aimed at how to carry out accurate community structure, while the study in the community merge strategy is relative rare. In this paper, we study community partition merger principle in single nodes of bipartite networks to propose an improved bipartite networks community detection method, which is based on Page Rank algorithm, information spreading probability model and combined with the modularity. The information of single nodes is calculated by information diffusion matrix. The value of information diffusion matrix larger than the threshold are merged every time, which quickly reduces the dimensions of the information diffusion matrix to speed up the merger of the community significantly. By comparing and analyzing experimental result of this method with other typical bipartite networks community partition algorithm on South women data set, We demonstrate the effectiveness of the proposed method.

Keywords: Page Rank · Bipartite networks · Community division · Information diffusion

1 Related Research

Since the 1990s, complex networks research rapidly infiltrate into many different disciplines which are mathematical disciplines, life science and engineering disciplines etc. Bipartite networks is an important form of complex networks and exist universally in the real world, such as P2P Internet formed by computer terminal and data [1]; company shares net formed between the investors and their holdings [2, 3]; actor cooperation networks by an actor with their films [4]; A club members participate in the activities to form the activity network [5] and so on.

1.1 Research Status

Bipartite networks community partition is roughly divided into two types: mapping method and non-mapping method. Mapping method firstly map the bipartite networks into a single network and use unipartite networks community discovery algorithm to divide community. Mapping method is also called projection method. The way map

© Springer International Publishing AG 2016
S. Song and Y. Tong (Eds.): WAIM 2016 Workshops, LNCS 9998, pp. 259–268, 2016.
DOI: 10.1007/978-3-319-47121-1_22

bipartite networks to unipartite networks by unweighted projection and weighted projection, but the experiment proved that no matter what methods of projection will lead to lack of information and the inaccurate community detection. The non-mapping method directly divide on the bipartite networks community. So we should do further study to bipartite networks' nature and divide bipartite networks by imitating community detection thought of unipartite networks.

Current research on bipartite networks community detection algorithm, Zhang Peng and others proposed bipartite networks' edge clustering coefficient, meanwhile, giving the corresponding hierarchical clustering community detection algorithm based on edge clustering coefficient [6], they put the binary modularity as a condition for stopping the algorithm; Lehmann et al. extend bipartite networks based on k-group algorithm and propose Ka, b group algorithm to detect the bipartite networks [7]; After Guimera proposed bipartite networks modularity definition, he divide the actual network using unweighted projection (UWP), weighted projection (WP), binary algorithm (B) [8, 9]; Barber used modularization matrix's feature to propose BRIM algorithm [10]; literature [11] the thought based on fast Newman algorithm improved the BRIM algorithm and proposed the modularity agglomeration algorithm (MAB); Literature [10] combining adaptive BRIM maximize bipartite modularity to obtain community partition; Recent Raghavan et al. [12] proposed label propagation method (LPA) for community detection. Later Murata and others in the literature [13] at the expense of the accuracy to make LPA algorithm suit for large bipartite networks and large bipartite networks' parallel real-time community analysis; NDu et al. [14] proposed BiTector algorithm based on maximum two molecules filter, Wang Yang proposed clustering method based on the comparative definitions and community force to divide the bipartite networks.

1.2   Page Rank Algorithm

Page Rank algorithm [15] is proposed and published by Larry Page and Sergey Brin in 1998, it is used to rank pages, the centre algorithm is to calculate scores and rank. Page Rank algorithm use mathematical statistical model - "random walk" by markov chain (or markov process) to form. Markov property refers to when a random process is under the condition of given current state and the state of all the past and the conditional probability distribution of the future state only depends on the current state. Page Rank algorithm's iterative process is represented by formula (1):

$$R^{n+1} = R^n P \tag{1}$$

P is the transition probability matrix (if the node i to node j has the edge, then $P_{ij} = 1$, otherwise is 0). R^n is to the N-th iterative score vector. Page Rank algorithm use the markov property independence: network users are random walk without considering each user's every click link behavior to get each web page residing on the number of users as each web page scores (i.e. the PR value). Page Rank algorithm is widely used, such as rumors network, the paper reference network, citation index, trust network [16], influence analysis [17], index analysis of social network [18–21] etc.

The formation of social networks is based on the information communication between individual behavior. Individual in social system spread information and receive information all the time, i.e. the information diffusion. The essence of information diffusion is a process that using communication media spread information from the sender to the recipient. When information diffuse many times, information can be accepted by more and more individuals. Information diffusion method in social networks and PR value by multiple iterative page rank algorithm have the same meaning. Page's PR value is directly proportional to the importance of the page. Similarly after a lot of information diffusion, important individuals have more information in the network. Community structure is detected by the information of individual.

1.3 Information Spreading Probability Model

A bipartite networks partition algorithm based on information diffusion probability is proposed. First of all, the model transforms bipartite networks to a transition probability matrix and regards each node of one side bipartite networks as a community. Initialize and distribute information on both sides of the nodes. Then using one side bipartite networks nodes' relation begin to transfer probability, which can obtain the probability transition relation of single nodes. The information spread both sides of nodes to form information diffusion matrix. The value of information diffusion matrix is regarded a judgment for community merger. Combine Barber for the definition of bipartite modularity as a community partition standard. Stopping community merger when Q value is maximum can get the optimal community partition result.

1.4 Modularity

For unipartite networks, Newman put forward with the modularity [22] to evaluate the quality of community partition. The physical meaning of modularity: the proportion of the edge by connecting two same type nodes (the edges in the same community) minus the expectation of the proportion of random connecting two nodes' edge in the same community structure. Modularity is the larger, the community structure is the more obvious. Generally in the actual network, modularity is between $0.3 \sim 0.7$. Its definition is shown below by formula (2):

$$Q = \sum_{i,j}^{n} (e_{ii} - a_i^2) \tag{2}$$

Because the algorithm in this paper also depends on the relation of the single nodes to divide community, it still use the modularity to evaluate the quality of community partition. Assume that the set U and set V are two types nodes of bipartite networks G, the set U are divided into l communities, set V is divided into m communities. As for the adjacency matrix A_{m*n} of two parts graph A, Murata modularity is defined as:

$$Q = \sum_{l} (e_{lm} - a_l a_m) \tag{3}$$

e_{lm} represent the collection of the communities of vertex U connecting the communities of vertex V. Define a K × K matrix and a_l is the element of the matrix.

$$e_{lm} = \frac{1}{2m} \sum_{i \in U_l} \sum_{j \in V_m} A_{ij} \tag{4}$$

$$a_l = \frac{1}{2m} \sum_{i \in U_l} \sum_V A_{ij} \tag{5}$$

2   Description of the Methods

2.1   The First Stage

The key of first phase is the calculation of information diffusion matrix Sn after n-th iteration. Bipartite networks can be represented as a two parts graph G = (U, V, E), U and V are respectively bipartite networks G's two kinds of nodes, E is the edges in G. Through the definition of bipartite networks, we know a collection of U's (or V) nodes is not connected in interior. To arbitrary edge of set $E(u_i, v_j)$, there will be $u_i \in U, v_j \in V$. The number of nodes in set U is m, The number of nodes in set V is n, bipartite networks adjacency matrix can be expressed as:

$$A = \begin{bmatrix} 0_{m*m} & A_{m*n} \\ A^T_{n*m} & 0_{n*n} \end{bmatrix} \tag{6}$$

The matrix A is the sub matrix of bipartite networks G's adjacency matrix, A is the relation matrix of set U to V. A^T is the relation matrix of set V to U. Assume that information spread from the set U to V, the initial unit information matrix I of set U is expressed as:

$$I^U_{ij} = \begin{cases} 1 & i = j \\ 0 & i \neq j \end{cases} \tag{7}$$

Through A and A^T, information left matrix R_{m*n} from set U to V and information right matrix T_{n*m} can be obtained. Information spread from set U to V and then passed back to U, which complete an information transfer. The information diffusion matrix of information passed n times is expression:

$$S^n_{m*m} = (I^n_{m*m} \cdot R_{m*n} \cdot T_{n*m})^n \tag{8}$$

S^n_{m*m} is the information diffusion matrix after information spread n times, S^n_{ij} indicates that the node u_i receives information from the node u_j. According to the value S^n_{ij} of in the information diffusion matrix S^n_{m*m}, we can divide one side networks.

2.2   The Second Stage

The key of this stage is to take a fast nodes merger according to the value of the information diffusion matrix S_n to improve the efficiency of the combination. Traditional algorithms merge two or more nodes into a community each time, which lead to community merger speed too slow. As for this situation, this paper proposes a new merger strategy, by determining a z value and found several values that are larger than z in the information diffusion matrix to merge several communities simultaneously to improve the efficiency of the merger. In this paper, we determine $z = \mu + \sigma$ which is used to choose nodes to merge the by calculation of the average and variance, average reflect the average level of the data, the variance reflect the volatility of the data. We just find out the value in that is lager than z each time and choose them. In this way, We can get many nodes to merge community. In initial moment, regard each node of set U as a community and traverse information diffusion matrix S_n except on the diagonal element, find out the value that is larger than z. We deal with these values from big to small and select n value at different rows and column to merge, This can guarantee to merge two nodes to a community in every community merger, at the same time there are multiple nodes to take community merger.

After finding the combined community every iteration, we must to deal with information diffusion matrix. The reason is that the merger of the node will be as a whole with the other nodes to remerge in the next calculation. In this algorithm, when a group of nodes are merged, the matrix merge the corresponding the i-th row and the j-th row, the i-th column and the j-th column. As for the update strategy for the maximum, in the row direction, the maximum value of the same column is used as the outflow information of the i community. In the column direction, the average value of the same row is used as the information of the other nodes flowing into the i community. Meanwhile, delete the j-th row and the j-th column of the information diffusion matrix. N group communities are merged at the same time and delete n rows, n column, which make the dimension of the information diffusion matrix decrease rapidly. Continue to traverse the updated information diffusion matrix, recalculate the mean and variance of the updated information diffusion matrix to find the value that is larger than z, you can find many groups nodes can be combined at a time to improve the speed of community merge greatly. According to the above principle of merger, until all communities are merged into one community, the result of the merger is a tree. Combined with the Barber definition for bipartite modularity as the standard of the community division, when the Q value is maximum to stop community merger, the community partition is the best result. The following Table 1 is the diagonal element set 0 information diffusion matrix of 10 nodes after six times information diffusion. The mean of, = 0.1026, std = 0.0861, z == 0.1887. Bold values in the Table 1 are the multiple nodes found by > z value. But these values are not all that we should merge the node pairs. Our strategy is merging two nodes to a community, so we first find the maximum value in these value, $S_{89} = 0.2986$, delete the values that it is in the same

Table 1. Information diffusion matrix of 10 nodes

	1	2	3	4	5	6	7	8	9	10
1	0.0000	**0.1870**	0.0979	**0.1850**	0.1367	0.0591	0.0149	0.0064	0.0185	0.0114
2	**0.2805**	0.0000	0.0979	0.1826	0.1360	0.0601	0.0163	0.0077	0.0207	0.0122
3	**0.2938**	**0.1958**	0.0000	**0.1888**	0.1340	0.0512	0.0095	0.0029	0.0112	0.0079
4	**0.2776**	0.1826	0.0944	0.0000	0.1381	0.0630	0.0176	0.0082	0.0222	0.0131
5	**0.2051**	0.1360	0.0670	0.1381	0.0000	0.0975	0.0596	0.0459	0.0841	0.0360
6	0.0886	0.0601	0.0256	0.0630	0.0975	0.0000	0.1343	0.1274	**0.1999**	0.0703
7	0.0223	**0.2806**	0.0047	0.0176	0.0596	0.1343	0.0000	**0.1961**	**0.2778**	0.0863
8	0.0097	0.0077	0.0014	0.0082	0.0459	0.1274	**0.1961**	0.0000	**0.2986**	0.0877
9	0.0185	0.0138	0.0037	0.0148	0.0561	0.1332	**0.1852**	**0.1991**	0.0000	0.0872
10	0.0341	0.0244	0.0079	0.0262	0.0719	0.1405	0.1727	0.1755	**0.2617**	0.0000

Table 2. The merged information diffusion matrix

	(3, 1)	4	5	6	(7, 2)	(8, 9)	10
(3, 1)	0.0000	0.1888	0.1340	0.0512	0.0047	0.0014	0.0079
4	0.0472	0.0000	0.1381	0.0630	0.0088	0.0041	0.0131
5	0.0335	0.1381	0.0000	0.0975	0.0298	0.0230	0.0360
6	0.0128	0.0630	0.0975	0.0000	0.0672	0.0637	0.0703
(7, 2)	0.0024	0.0176	0.0596	0.1343	0.0000	0.0981	0.0863
(8, 9)	0.0007	0.0082	0.0459	0.1274	0.0981	0.0000	0.0877
10	0.0039	0.0262	0.0719	0.1405	0.0863	0.0877	0.0000

row and column, then find the value of the second largest $S_{31} = 0.2938$. In this way, you can find merger nodes at the same time for the first time: (8, 9) (3, 1) (7, 2).

The following Table 2 is the information diffusion matrix's result by merging (8,9) (3,1) (7,2).

From Table 2, we can see that 3 groups of communities are merged at the first time, the above example in this paper is in order to illustrate the combined process, although only 10 nodes, three groups communities have been combined in one time, so in large data sets can be combined with more community. In large data set, this method can merge more communities at the same time to make the dimension of the information diffusion matrix decrease rapidly.

2.3   Algorithm Process

```
Algorithm
Input:
I,R,T,m,iteration;
Output:
List, CutPoint;
```

1:        Initialize(I);
2:        for each $n \in [0, \text{iteration}]$ do
3:            $S^n_{m*m} = (I^n_{m*m} \cdot R_{m*n} \cdot T_{n*m})^n$;
4:        end for
5:        while(dimensionality(S)>0) do
6:            mean(S(:));
7:            std(S);
8:            z=mean(S(:))+std(S);
9:            for each $i \subseteq [0, \text{dimensionality}(S)]$ do
10:           $S^n_{ij}=0$;
11:        end for
12:        if value(S^n_{ij})>z
13:            List+=(i,j);
14:        end
15:        S^n_{ik} =max(S^n_{ik}, S^n_{jk});
16:        S^n_{ki}=mean(S^n_{ki} , S^n_{kj});
17:        delete(S^n_{j*} , S^n_{*j});
18:        end while
19:        {List1, List2,$\cdots$, Listn-1}$\leftarrow$(List);
20:        for each $r \in [1, N-1]$ do
21:            Q(List$_r$);
22:        end for
23:        CutPoint=max(Q(List$_x$));

3 Experimental Results

3.1 The Test of Southern Women Data Set

South African women's network data set [23] by Stephen is an actual network which consist of southern United States women's activities. The bipartite networks are composed of 18 women and 14 activities, the number 1–18 stands for women nodes, 19–32 stands for active nodes. If women take part in an activity, then the corresponding node in the network will be connected, as shown in Fig. 1:

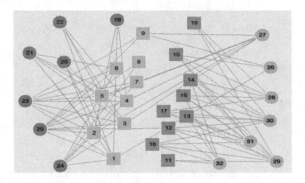

Fig. 1. South women - activity network

The experimental results show that the network is divided into four communities, in which women 1–9 and 10–18 each are into a community, 19–26 and 27–32 are divided into another communities. That is same as Davis's original observations. The results of several other algorithms is compared with this method and use modularity as a standard of evaluation. The experimental results as shown in Table 3.

Table 3. Women - events on both sides of network node partition results compared with Q value

Algorithm	Community partition results	Q value
IPSF	{1–9}{10–18}{19–26}{27–32}	0.5857
BRIM	{1–6}{7, 9, 10}{19–26}{27–32} {11–15}{8, 16–18}	0.4749
LPA	{1–7, 9}{810–18} {19–24}{25–26}{27, 29}{28, 30–32}	0.5666
MP	{1–6}{7–10}{11–18}{19–24}{25–2}{29–32}	0.4364
ACODC	{1–9}{10–18}{19–26}{27–32}	05856

As is shown in Table 3, the method in both nodes respectively partition get the maximum module value by comparing the results of the various partition algorithms. The partition result is better than ACODC algorithm and is same as Davis's original observations. Therefore, the method can effectively identify the bipartite networks community structure and can obtain the accurate community partition and the high quality Modularity. The community structure of this method on Southern Women - activity network is shown in Fig. 2 below:

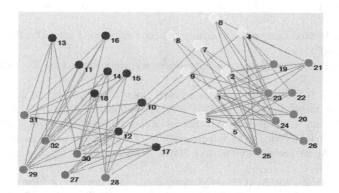

Fig. 2. Southern Women community detection structure

4 Conclusion

For bipartite networks community partition, an method for improving the efficiency of community division in bipartite networks is proposed in this paper. Different from the previous method, our method is based on Page Rank algorithm and the information diffusion principle. In the community merge strategy, we proposed that determining a threshold z enables multiple nodes to merge simultaneously to improve the efficiency of community division. The experimental results show that the algorithm not only accurately detect the bipartite networks community structure, but also obtain Ideal community partition results. How to determine the value of z to further improve the efficiency of community partition is left for our future work.

References

1. Guillaume, J.-L., Latapy, M., Le-Blond, S.: Statistical analysis of a P2P query graph based on degrees and their time-evolution. In: Sen, A., Das, N., Das, S.K., Sinha, B.P. (eds.) IWDC 2004. LNCS, vol. 3326, pp. 126–137. Springer, Heidelberg (2004)
2. Robins, G., Alexander, M.: Small worlds among interlocking directors: network structure and distance in bipartite graphs. Comput. Math. Organ. Theor. **10**, 69–94 (2004)
3. Battiston, S., Catanzaro, M.: Statistical properties of corporate board and director networks. Eur. Phys. J. B **38**, 345–352 (2005)
4. Watts, D.J., Strogatz, S.H.: Collective dynamics of small world networks. Nature **393**, 440–442 (1998)
5. Ergun, G.: Human sexual contact network as a bipartite graph. Phys. A **308**, 483–488 (2002)

6. Zhang, P., Li, M., Mendes J.F.F., et al.: Empirical analysis and evolving model of bipartite networks. http://arxiv.org/abs/0804.3854v1

7. Lehmann, S., Schwartz, M., Hansen, L.K.: Biclique communities. Phys. Rev. E **69**(2), 026113 (2004)

8. Guimera, R., Sales-Pardo, M., Amaral, L.A.: Module identification in bipartite and directed networks. Phys. Rev. E **76**, 036102 (2007)

9. Borgattia, S.P., Everettb, M.G.: Network analysis of 2-mod data. Soc. Netw. **19**, 243–269 (1997)

10. Barber, M.J.: Modularity and community detection in bipartite networks. Phys. Rev. E **76**, 066102 (2007)

11. Gao, M.: Binary network community discovery algorithm study. D. Hefei Industrial university, pp. 31–37 (2012)

12. Raghavan, U.N., Albert, R., Kumara, S.: Near linear time algorithm to detect community structures in large-scale networks. Phys. Rev. E **76**, 036106 (2007)

13. Liu, X., Murata, T.: Community detection in large-scale bipartite networks. In: Proceedings of the 2009 IEEE/WIC/ACM International Joint Conference on Web Intelligence and Intelligent Agent Technology (WI-IAT 2009), Washington, DC, pp. 5–8. IEEE Computer Society (2009)

14. Du, N., Wang, B., Wu, B., et al.: Overlapping community detection in bipartite networks. In: International Conference on IEEE Web Intelligence and Intelligent Agent Technology, 2008, WI-IAT 2008, vol. 1, pp. 176–179. IEEE/WIC/ACM (2008)

15. Page, L., Brin, S., et al.: The pagerank citation ranking: bringing order to the web. R. Stanford Digital

16. Richters, O., Peixoto, T.P.: Trust transitivity in social networks. PLoS ONE **6**(4), 1–14 (2011)

17. Yue, Z., Hongli, Z., Weizhe, Z., Junjia, L.: Identifying the influential users in network forum. J. Comput. Res. Dev. **50**(10), 2195–2205 (2013)

18. Yuanzhe, C., Miao, Z., Dijun, L., Chris, D., Sharma, C.: Low-order tensor decompositions for social tagging recommendation. In: Proceedings of the 4th ACM International Conference on Web Search and Data Mining, New York, pp. 695–704 (2011)

19. Tong, Y., She, J., Chen, L.: Towards better understanding of app functions. J. Comput. Sci. Technol. **30**(5), 1130–1140 (2015)

20. She, J., Tong, Y., Chen, L.: Utility-aware event-participant planning. In: Proceedings of the 34th ACM SIGMOD International Conference on Management of Data (SIGMOD 2015), pp. 1629–1643 (2015)

21. She, J., Tong, Y., Chen, L., Cao, C.C.: Conflict-aware event-participant arrangement. In: Proceedings of the 31st International Conference on Data Engineering (ICDE 2015), pp. 735–746 (2015)

22. Newman, M.E.J.: Modularity and community structure in networks. Proc. Nat. Acad. Sci. U. S.A. **103**(23), 8577–8582 (2006)

23. Davis, A., Gardner, B.B., Gardner, M.R.: Deep South. University of Chicago Press, Chicago (1941)

Community Detection Algorithm
of the Large-Scale Complex Networks Based
on Random Walk

Ding Guohui[(⊠)], Song Huimin, Fan Chunlong[(⊠)], and Song Yan

Liaoning Provincial Key Laboratory of Large-Scale Distributed System,
Shenyang Aerospace University, Shenyang 11036, China
sumeng235@163.com, fanchl@sau.edu.cn

Abstract. Community detection on large-scale complex networks has become a popularly discussed topic with the development of the social network. In this paper, we proposed a community detection algorithm based on the random walk theory. We assume each node has the energy value and the random walk process is considered as energy transfer. According to the transition probability matrix, nodes transfer energy in the network. We divide two nodes which transfer the most energy to each other into one community. The algorithm can obtain accurate division results on small data sets. However, when we applied it to the large-scale network, we find a problem that the sparse degree of matrix is reduced during the energy transfer process. We set the threshold to keep the energy matrix is still sparse in the process of transfer to solve this problem. We conduct extensive experiments on real-word large network provided by Stanford University and the results demonstrate the efficiency and effectiveness of our proposed algorithm.

Keywords: Complex networks · Random walk · Power-law distribution · Sparse matrix

1 Introduction

With the development of web computing, the scale of the social networks continuously increasing is a very challenging problem to carry out community division for large complex networks. The lack of reliable and real community that has been well divided has become a block for community division on complex network [1, 2]. Experimental data in this paper comes from the Stanford University and the data is effective and real community division. So we can compare the community division results which use the algorithm proposed in this paper to the real community structure. Recently, scholars have devoted a lot of efforts in the analysis of the social network and network analysis platforms have been developed such as Stanford University SNAP; Carnegie Mellon University Pegasus, AutoMap IBM, Microsoft Research and so on. We utilize Pajek which is an open source visualization of social network analysis software to analyze the network structure.

PageRank algorithm is designed for Web system, the basic idea is to increase the vale when a page link to another page. Inspired by the algorithm, we consider the nodes

© Springer International Publishing AG 2016
S. Song and Y. Tong (Eds.): WAIM 2016 Workshops, LNCS 9998, pp. 269–282, 2016.
DOI: 10.1007/978-3-319-47121-1_23

in the social network as web pages and consider the edge as web link. We assume that each node has the energy value of 1 (theoretical basis) and energy matrix is obtained by the node connecting with other nodes. According to the random walk theory, the stable energy matrix is obtained after initial energy matrix iteration for X times. We iterate for 6 times because the whole community meets the small world model which is in line with six degrees of separation theory [12, 13]. We find two nodes that send maximum energy to each other in energy matrix and divide them into a community. Merging two nodes can obtain the corresponding different community structures then calculate the corresponding values of community evaluation index Q until the entire network is divided a community. When we find the maximum Q value, the corresponding community division is the final result.

Analyzing cluster structure of complex networks by the random walk is suitable for small and high precision network clustering analysis. For fine-grained community network, it will not be able to give a better division and its operating efficiency is not high [5, 14–16]. The energy matrix increases with the increasing scale of network. Then the entire calculation time increases with the size of the network which is a problem that a lot of classical algorithms face. We know that nodes of community in the network are to obey power-law distribution [4]. That is, the degree of a few nodes is high but the degree of most nodes is small in community. That is, the energy transfer matrix is very sparse. However, during transmission of energy the sparse degree of matrix is reduced, which affects the efficiency of community division. This paper sets threshold to solve this problem so as to improve the computation efficiency of the whole community partition algorithm. We conduct extensive experiments on classic data sets such as the American University karate club, dolphin network, an American football team and the results demonstrate the efficiency and effectiveness of proposed algorithm.

2 Related Research

Community division has been studied unremittingly all these years and a particular large number of effective approaches have been proposed. In this study we just focus on the fundamental problem of non-overlapping and unknown the number of nodes in community.

2.1 Community Division Methods

The method unknown the number of node in community is also called hierarchical clustering method, which divides the network into several communities according to the measurement values (or similarity, affinity, distance, etc.) of nodes. The existing division methods: according to the way of community formation (add or remove edges), the method can be divided into splitting method and cluster method. Newman et al. proposed GN [6] algorithm which was a splitting algorithm. Its basic idea is to continuously remove the edge with the maximum betweenness. Cluster method is to calculate similarity between nodes of the social network then utilize the method of partial and greedy to obtain division results. Its basic idea is that: we assume that the node in the network is an independent community and to find two communities to divide continually until the entire network is divided into a community. Newman et al. proposed a fast GN algorithm based on GN algorithm (FastGN) [7] which was the cluster algorithm. Xiaowei Xu et al. Proposed SCAN which was based on structure similarity [22, 25] and utilized the neighbor node as the standard to divide the community.

2.2 Random Walk for Community Division

Dongen proposed a Markov clustering algorithm [20] which utilized the two matrices to iterate to find the cluster of the node but the drawback is that the number of nodes is less. Haijun utilized the random walk to define the distance between the nodes [21] and indicated that the distance closer to the node was likely to belong to the same community. Pascal et al. proposed [10] walktrap algorithm which utilized the iterative matrix to calculate the distance between the nodes then iteratively merging the closer nodes after that algorithm utilized community module function to obtain the final community. Xianghua et al. proposed [23] algorithm CD-TR and walk which utilized the threshold random walk method to discover the core community. Yueping et al. proposed the algorithm [24] to detect local community based on random walk which can obtain the community structure without the use of global information by adopting a new selection strategy.

3 Community Division Based on Energy Transfer

This chapter introduces the algorithm of PageRank and our community division algorithm which is based on energy transfer and the evaluation indicators of community division. We also apply our algorithm to the large-scale network for community division.

3.1 PageRank Algorithm

PageRank [19] algorithm was proposed and published by Larry Page and Sergey Brin in 1998. The core of the algorithm is to calculate the score of the page and to rank. It is based on the basic ideas that the more the users visit, the higher the quality of the web page is. Therefore we calculate the frequency that the page has been accessed by analyzing the topology consisting of hyperlinks. For each page Pi, its PageRank value is defined as Formula (1).

$$Pagerank(P_i) = \frac{1-q}{N} + q \sum_{P_i} \frac{pagerank(P_i)}{L(P_i)} \qquad (1)$$

In Formula (1), q is the damping coefficient and generally defined as 0.85; $p_1, p_2, \ldots,$ p_n are being studied pages; $L(P_i)$ is the number of links to a page; N is the number of all pages. The scores of each page are obtained by this formula.

3.2   The Basic Algorithm

In this paper, the basic idea of our algorithm is described below. The background of large-scale network is undirected and unweighted graph. $N = (V, E)$ denotes a complex network where V is a collection of nodes and n is the number of the nodes, E is a collection of edges and m is the number of the edges. The degree of the node in the graph is denoted as $(V_i) = \sum_j a_{ij}$, $i = 1, 2, \ldots n$.

Assume that each node has the initial energy and the nodes random walk along the edge of the network N. Before every step the nodes are based on the transition probability to choose the next step. If the nodes are located at the position i and the next step to reach its neighbor node j is defined as Formula (2). That is the transition probability.

$$P_{ij} = \frac{a_{ij}}{\sum_k a_{ik}} \tag{2}$$

After the initial matrix $D (d_1, \ldots d_n)$ transfer energy the energy transfer matrix is denoted as Formula (3)

$$P = D^{-1}A \tag{3}$$

Definition: (Homogeneous Markov chain). Tt is homogeneous Markov chain when transition probability of Markov chain $P\{X\_(t + 1) = j \mid X_t = i\}$ has nothing to do with the time and we denoted it as $P_{ij} = P\{X_{t+1} = j | X_t = i\}$.

Theorem: (Chapman-Kolmogotov equation). If $\forall s, t \geq 0$ there are $P_{ij}^{(s+t)} = \sum_{k \in s} P_{ik}^{(s)} P_{kj}^{(t)}$ and its matrix form $P^t = P \cdot P^{(t-1)} = P \cdot P \cdot P^{(t-2)} = \ldots = P^t$. Based on the theory of random walk we define iterative process as the Formula (4).

$$A_\tau = A_{\tau-1}P, \tau = 1, 2, 3 \ldots, 6 \tag{4}$$

In Formula (4), A_1 is the background of the adjacency matrix; P is the background of the energy matrix (transition probability matrix); expand the Formula (3) as follows.

$$P = \begin{vmatrix} 0 & \frac{1}{d(v_1)} & \cdots & \frac{1}{d(v_1)} \\ \frac{1}{d(v_2)} & 0 & \cdots & \frac{1}{d(v_2)} \\ \vdots & \vdots & \ddots & \vdots \\ \frac{1}{d(v_n)} & \frac{1}{d(v_n)} & \cdots & 0 \end{vmatrix}$$

If $P_{ij} \neq 0$, there is a link between node i and node j and Formula (2) can be expressed as Formula (5).

$$A_{\tau=}A_1 P^\tau, \tau = 1, 2, 3 \ldots, 6 \tag{5}$$

After the node has the energy through the energy matrix to transfer energy X times, A_τ is obtained and it is a stable and convergence energy transfer matrix. After obtaining the energy transfer matrix, we extract the max value for each row to find the node j respectively. That is, node i has passed its own max energy to the node j. Then continue to extract maximum value in the max value of each row, namely find from node j to separate the most energy of node i. In this way, we found a pair of node (i, j) which transfer the most energy to each other. That is, node i and j have a strong tendency to divide into a community compared with other nodes. We remove j from the matrix directly after i and j are divided into a community. That is, the value of the i th row the j th column is set to 0. But such a simple deletion can lead energy to loss, which will have an impact on follow-up division, leading to results into serious error. We utilize the thought of "line to take the most value, the column take the average" to extract the pair of node (i,j) to merger. In this way energy efficiently pass through node i to node j and produce accurate division result. We know the proposed algorithm is cluster of ideas. Initial network has N nodes and each node is considered as a community. We divide these nodes according to the algorithm proposed by this paper. Each division we conducts a Q value calculation until N nodes are divided into a community or several community. Extracting the largest Q value and the corresponding community division results is the final community structure.

3.3   The Evaluation Indicators of Community Division

In 2004, Newman et al. [6] proposed a network module evaluation function for the network clustering (Q function). Network is divided into k communities by some form of division. Define a $K*K$ symmetric matrix array $E = (e_{ij})$, where the elements e_{ij} represents the proportion that edges connection nodes coming from two different communities of all sides in the network and the two nodes are located in the i th and j th community. Set the sum of each element on the diagonal matrix $Tre = \sum_i e_{ii}$ and it gives edges connecting to a community inside nodes in the proportion of all edges in a network. Defining the sum of the elements of each row (or a column) is $a_{ii=}\sum_j e_{ij}$ and it shows that the edges that connected to nodes of i th community accounting for all the edges in the network. So define the module of measure standard [9] as Formula (6).

$$Q = \sum_i (e_{ii-}a_i^2) = Tre - \|e^2\| \tag{6}$$

In Formula (6), $| x |$ denotes the sum of all elements in matrix X. That is the proportion of linking two nodes inside community minus any link of these two nodes. Formula (7) is another kind of expression way of Formula (6) where n_c is the number of communities; l_c is the number of edges contained within the community c; d_c is the sum of degree of all the nodes in the community c; m is the total number of edges in the network. Measure Q value satisfies the practical significance of the community division, namely compared with the external nodes, linking internal nodes in the community is more closely. Corresponding division of community is the final result with the maximum value Q.

$$Q = \sum_{C=1}^{n_c} \left[\frac{l_c}{m} - \left(\frac{d_c}{2m} \right)^2 \right] \tag{7}$$

3.4 Apply to Large-Scale Complex Networks

With the development of society, the real-life social networks tend to be large. For such a huge social network, proposing a quick and efficient community division algorithm is a hot topic [26]. We know that in real life networks are a power-law distribution, namely the entire network is sparse. Utilizing this character, we can improve the computational efficiency of the algorithm. Energy matrix $A_{\tau} = A_{\tau-1} P, \tau = 1, 2, 3, \ldots 6$ sparse degree will reduce gradually during the energy transfer. Some small energy value in energy matrix after τ iterations is meaningless in the community division. In order to keep the sparse degree of matrix, we set the threshold θ in the process of energy transfer. If the value after matrix multiplication is less than this threshold, we will set the value to 0. We utilize sparse matrix storage mode CRS [18] to calculate, which is not only saves storage space, but also improving the efficiency of matrix calculations.

We conduct block multiplication calculation after the matrix block storage. Firstly, $P[i][]$ denotes the i th row of matrix P and $P[][j]$ denotes the j th column of matrix P and $P[i][j]$ denotes the value where the i th row and j th column of matrix $P.nzP()$ denotes the number of non-zero elements in the matrix P and $nz(P[i][])$ denotes the number of non-zero elements of i th row in the matrix P accordingly, and $nz(P[][j])$ denotes the number of non-zero elements of j th column in the matrix P. If matrix P meets $nzP() = O(n)$, matrix is sparse. The matrix P^2 is obtained by P multiplying P. Utilizing sparse matrix storage and matrix multiplication as Formula (8) improved the efficiency of large-scale iterative calculation of the energy matrix. Meanwhile, the iterative process is calculated by the Formula (8), which can also reduce unnecessary computation element and the elements will also be stored in accordance with the CSR format. At the same time, the energy matrix P iterates τ time to obtain the final energy matrix A_{τ}. We utilize matrix multiplication associative law to reduce the number of iterations, which is not only improving the efficiency of the algorithm, but also it makes the algorithm to work in a large social network division.

$$P[i][j] = \sum_{l=0}^{n-1} (A[i][l] \times B[l][j]) \tag{8}$$

In the existing large community of network dataset, there is no specific practical community division result. We experimented on the classic small sample set. Corresponding algorithm is described as follows:

Algorithm 1. Community detection C
1   Require: A A_1
2   Ensure: C
3   For i=1;size(A) do
4   P←A(i,:)/sum(A(I,:))
5   end for
6   for i=1:1 do
7   M←A×P
8   M(M<θ)=0
9   end for
10   while sum(sum(M))≠0 do
11   qi←0
12   List←max(i,j)
13   M(i,:) ←max(M(i,:),M(j,:))
14   M(:,i) ←(M(i,:),M(j,:))/2
15   M(j,:) ←0
16   M(:,j)←0
17   Q(qi) ←lc/(2×m)-(dc/(2×m))2
18   qi←qi+1
19   end while
20   max(θ)
21   get community C.

4   Experiments

In order to quantitatively analyze the performance of the algorithm, we test it on a real-world network (network of small-scale and large-scale) and give parameter analysis

4.1   Setting the Threshold Value

Since the real network and the computer-generated networks with different topological properties we choose to explore the threshold issue through three communities known divided structure on real network, thereby quickly and efficiently perform complex network of community division. Complex network node is a power law distribution, and then the whole complex network of energy matrix is sparse. But after energy transfer, sparse of matrix is gradually reduced. As shown in Fig. 1 where nz is the number of non-zero elements of the matrix.

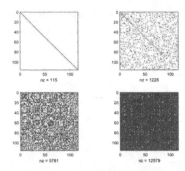

Fig. 1 Sparse degree of matrix

In order to take full advantage of matrix sparsity to improve the operational efficiency of the algorithm, we set value below the threshold to 0 to the node which contributes a smaller energy value in the entire energy matrix transfer process. The number of non-zero elements in matrix before setting the threshold shown in Fig. 2. The number of non-zero elements in matrix after setting the threshold shown in Fig. 3. Figures 2 and 3 denote the energy changes of energy matrix during the energy transfer process. In this paper, energy transfer matrix iterates for six times (left to right, top to bottom). In Figs. 2 and 3, abscissa denotes the energy transfer matrix energy value P_{ij} and ordinate denotes the number of P_{ij} at the entire energy matrix. By comparing Figs. 2 and 3, it can be seen that setting a threshold effectively maintains the characteristics of matrix.

In Fig. 4, abscissa denotes iteration times of matrix and ordinate denotes the number of non-zero after iterating. Star-shaped polyline indicates the number of non-zero elements during energy matrix in an iterative process before setting the threshold and it has

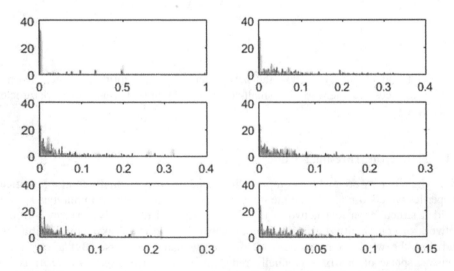

Fig. 2. Energy changes during the energy transfer process before setting the threshold

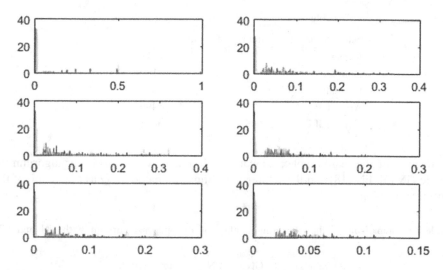

Fig. 3. Energy changes during the energy transfer process after setting the threshold

reached the dense at 3th time; Dot-dash line indicates the number of non-zero elements during energy matrix in an iterative process after setting the threshold. By setting the threshold, the sparse matrix maintained in a constant state, improving efficiency for large-scale network of community division algorithm. After several experiments, we fthat the threshold is inversely proportional to the maximum degree and diameter of the largest in the network and network.

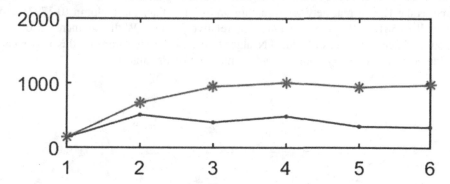

Fig. 4. The number of nonzero elements before and after setting the threshold value

4.2   Test Results of Sample Collection

In order to verify the performance of the algorithm we use currently five real networks which have been widely used as test data sets. The following chart shows the specific information that describes the data.

Table 1. Experiment used in the real network

Name of network	Number of nodes	Number of edges
Karate	34	78
Dolphin	62	160
Football	115	613
Arxiv GR_QC	5242	4496
Com-DBLP	36692	183831

In view of the real complex networks in Table 1, we compare our algorithm to classic GN [6], FN [18] and these three algorithms are based on Q as a function of the objective function.

Table 2. Comparison of the clustering quality of this paper algorithm with FN and GN algorithm

Q-values	GN	FN	Our algorithm
Karate	0.4013	0.2528	0.3715
Dolphin	0.4706	0.3715	0.3867
Football	0.5996	0.4549	0.5819
Arxiv GR_QC	_	0.3902	0.4356
Com-DBLP	_	0.5791	0.6102

From Table 2, "–" indicates that memory overflow or the results is not running out in 48 h. You can see that Q function value of GN algorithm in Karate and Dolphin networks are better than FN and the algorithm of this paper. This paper Q value is the same as the Karate real network structure and higher than the Dolphin (0.3722) and Football (0.5816). But the Q values are relatively close. With the increase of the amount of network data only the FN algorithm and the algorithm of this paper can divide community and Q value is higher than that of FN algorithm.

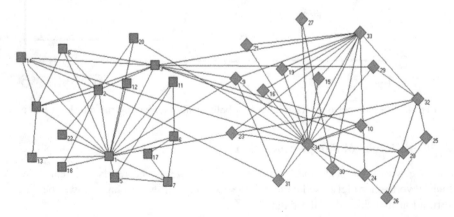

Fig. 5. Community division results on Zachary's karate club

In the above test the data sets such as karate, Dolphins, Football Network are known community structure. The following is comparative analysis of the algorithm of this paper and other algorithms in the community network in reference to the real social network structure of the data set. The first is the classic test data set Zachary's karate club. Nodes in the network indicate club members and edges are established by members of the relationship. In the past two years the club was eventually split two new clubs and formed two distinct networks with community structure. Utilizing the software of Pajek, we draw community division results by proposed algorithm shown in Fig. 5. We can see Zachary's karate club will be divided into node 3 on the left side of the community by our algorithm.

The second test data set is Dolphin Network. The network consists of 62 nodes and 159 edges, which are divided into two communities by the algorithm of this paper and the results are shown as Fig. 6.

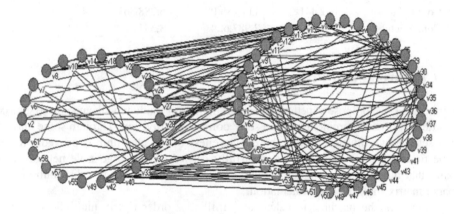

Fig. 6. Community division results on Dolphin Network

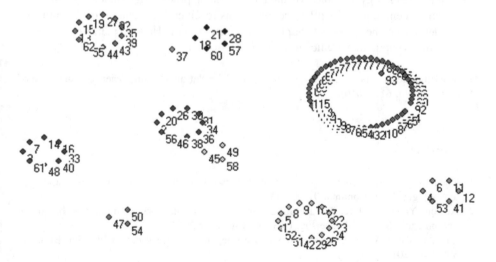

Fig. 7. Community division results on the American football team

The third test data set is the American football team and the results are shown as Fig. 7.

Through the above experimental results, we can see that the algorithm proposed in this paper can accurately carry out community division. Although FN algorithm can divide millions of nodes, with the development of information technology, with hundreds of millions of nodes of the complex network can be found everywhere (Table 3). FN algorithm can't deal with community division but the algorithm of this paper can still be effective in community division.

Table 3. Real network community data of Stanford University

Name	Number of node	Number of edges	Number of actual community
Com-Live Journal	3,997,962	34,681,189	287,512
Com-Friendster	65,608,366	1,806,067,135	957,154
Com-Orkut	3,072,441	117,185,083	6.288.363
Com-Youtube	1,134,890	2,987,624	8,385

5 Conclusion and Prospect

In this paper, we propose a method based on the idea of random walk and sparse matrix in order to solve the problem of large scale community division. Through some classic data sets known division results, the effectiveness of the algorithm is proved. At the same time, the algorithm can be successfully carried out on large-scale network utilizing that the nodes in complex networks is to obey the power law distribution and sparse matrix multiplication calculation.

Simple on the machine to calculate a million - order matrix multiplication, the efficiency is very low and even can't be calculated. MapReduce is a data parallel processing technology for massive data for parallel processing. Hadoop is an open source implementation of MapReduce and it has received the attention of the industry and academia. The next task is to apply the algorithm to the Hadoop platform to further improve the computational efficiency of the algorithm.

Acknowledgement. This research was supported by the National Natural Science Foundation of China (Grant No. 61303016).

References

1. Chengqi, Y., Bao Yuanyuan, X.: Social network data analysis framework and key technologies. ZTE Commun **01**, 5–10 (2014)
2. Aewon Yang, J., Jure, L.: Defining and evaluating network communities based on ground-truth. Stanford University, cs.SI 6, November 2012
3. Tang, J., Chen, W.G.: Deep analytics and mining for big social data. Chin. Sci. Bull. **60**, 509–519 (2015)

4. Siganos, G., Faloutsos, M., Faloutsos, P., Christos, F.: Power laws and the AS-level internet topology. IEEE/ACM Trans. Netw. **11**(4), 514–524 (2003)
5. Di, J., Bo, Y., Jie, L., Liu Dayou, H.: Ant colony optimization based on random walk for community detection in complex networks. J. Softw. **03**, 451–464 (2012)
6. Newman, M.E.J., Girvan, M.: Finding and evaluating community structure in networks. Phys. Rev. E **69**(2), 026113 (2004)
7. Newman, M.E.J.: Fast algorithm for detecting community structure in networks. Phys. Rev. **69**, 066133 (2004)
8. Fortunato, S.: Community detection in graphs complex networks and systems lagrange laboratory. ISI Foundation, Viale S. Severo 65 (2007)
9. Yang, J., Leskovec, J.: Defining and evaluating network communities based on ground-truth. In: ICDM (2012)
10. Pons, P., Latapy, M.: Computing communities in large networks using random walks. In: Yolum, P., Güngör, T., Gürgen, F., Özturan, C. (eds.) ISCIS 2005. LNCS, vol. 3733, pp. 284–293. Springer, Heidelberg (2005)
11. Kardes, H., Sevincer, A., Gunes, M.H., Yuksel, M.: Six degrees of separation among US researchers. In: IEEE/ACM International Conference on Advances in Social Networks Analysis and Mining, pp. 654–659 (2012)
12. Ke, X.: A social networking services system based on the "Six Degrees of Separation" theory and damping factors. In: Second International Conference on Future Networks, pp. 438–441 (2010)
13. Jiajia, L., Xiuxia, Z., Guangming, T., Mingyu, C.: Study of choosing the optimal storage format of sparse matrix vector multiplication. J. Comput. Res. Dev. (2014)
14. Ferrara, E.: A Large-Scale Community Structure Analysis in Facebook. Springer, Heidelberg (2012)
15. Liu, C., Ye, J., Ma, Y.: Storage and solving of large sparse matrix linear equations. In: ICCIS, pp. 673–677 (2012)
16. Kernighan, B.W., Lin, S.: A efficient heuristic procedure for partitioning graphs. Bell Syst. Tech. J. **49**, 291–307 (1970)
17. Girvan, M., Newman, M.E.J.: Community structure in social and biological networks. In: Proceedings of National Academy of Sciences of the United States of America, pp. 7821–7826 (2002)
18. Yan, G., Chen, C., Lv, J., Fu, Z.Q.: Synchronization performance of complex oscillator networks. Phys. Rev. E, **80**(5) (2009)
19. Lawrence, P., Sergey, B., Rajeev, M., Terry, W.: The PageRank Citation Ranking Bringing Order to the Web. Stanford Infolab, Stamford (1998)
20. Dongen, S.: Graph clustering by flow Simulation. Ph.D. thesis, University of Utrecht, Utrecht (2000)
21. Haijun, Z.: Network landscape from a Brownian particle's perspective. Phys. Rev. E. **67**(4), 041908 (2003)
22. Xu, X., Yuruk, N., Feng, Z., Schweiger, T.A.J.: SCAN: a structural clustering algorithm for networks. In: Berkhin, P. (ed.) Proceedings of the 13th ACM SIGKDD International Conference on Knowledge Discovery and Data Mining, pp 824–833. ACM, New York (2007)
23. Xianghua, F., Chao, W., Zhiqiang, W., Zhong, M.: Scalable community discovery based on threshold random walk. Comput. Inf. Syst. **8**(21), 8953–8960 (2012)
24. Yueping, L., Weikun, Z.: New algorithm for detecting local community based on random walk. Telkomnika. **12**(4), 1005–1016 (2015)

25. She, J., Tong, Y., Chen, L.: Utility-aware event-participant planning. In: Proceedings of the 34th ACM SIGMOD International Conference on Management of Data (SIGMOD 2015), pp. 1629–1643 (2015)
26. She, J., Tong, Y., Chen, L., Cao, C.C.: Conflict-aware event-participant arrangement. In: Proceedings of the 31st International Conference on Data Engineering (ICDE 2015), pp. 735–746 (2015)

Efficient Interval Indexing
and Searching on Cloud

Xin Zhou$^{(\boxtimes)}$, Jun Zhang, and GuanYu Li

School of Information Science and Technology,
Dalian Maritime University, Dalian, China
{zhouxin314159,zhgjun,liguanyu}@dlmu.edu.cn

Abstract. Interval queries are widely used in social networks, information retrieval and database domains. As an important query type, interval query has been explored in depth by researchers long ago. However,
the works to study interval indexing and querying on cloud platform are
few. The paper analyzes the shortcomings of existing work of interval
indexing and searching on key-value store. To reduce the space overhead
and respond time, we propose a new index structure and corresponding searching algorithms. The index structure takes full advantage of the
features of key-value store to improve the query performance. The extensive experiments based on real and simulated data sets show that our
approach is effective and efficient.

Keywords: Interval query · Indexing · Searching · Cloud

1 Introduction

As an important query type, interval queries are widely used in social networks,
information retrieval and database domains. We present a sample scenario to
introduce interval query.

Scenario 1: The crawler records the multiple versions of web pages. Each
version marks the crawled timestamp and the URL of web page. Each version
has a lifespan from the current crawled timestamp to the successor crawled
timestamp. There are 6 web pages which only has one version in Fig. 1. Now
consider a query of form "find all the versions of all the web pages that existed
during a given time interval q".

Interval query is similar with range query at a glance but different from range
query. The result set of interval query q is $\{v, w, x, y\}$ from Fig. 1. The result set
of range query q is $\{v, x, y\}$. The result set of range query is a subset of the results
for interval query. The indexing and searching approaches for range query are not
appropriate for interval query. Besides, the intervals of the objects are implicit.
So proposing the effective indexing and searching methods for interval queries
is challenges. Specifically, the volume of interval data is exploding today. The
interval objects are stored and managed on the cloud platform. It is a intractable
task to respond interval query real-time from huge interval data.

© Springer International Publishing AG 2016
S. Song and Y. Tong (Eds.): WAIM 2016 Workshops, LNCS 9998, pp. 283–291, 2016.
DOI: 10.1007/978-3-319-47121-1_24

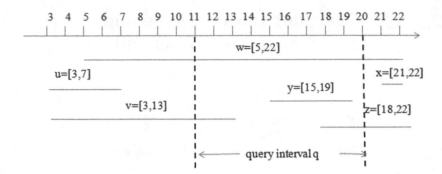

Fig. 1. The sample of interval query

In this paper, we study the related works of interval indexing and querying on key-value store and propose an solution to improve the performance of interval queries. The main contributions of this paper are 3 folders:

(1) We propose a novel index named EIIQHBase which takes full advantages of the characteristic of the cloud platform to support interval query efficiently.
(2) We present two new efficient algorithms to acquire the results for interval query separately based on EIIQHBase.
(3) The experimental evaluations based on real and simulated data set are shown that our approach exceeded the compared work both in space and time complexity.

2   Background

2.1   Interval Query

Interval query is an important query type applied in many applications. The goal of interval query is to find all the items from a data collection appeared during a given time interval.

Define 1: Interval query. Each entity o in data collection D is represented by binary group $o = <id, I>$, where id represents the identification of entity o, I is an interval denoted by $I = [begin, end]$ $(begin \leq end)$. Given a query $q = [a, b]$, the result of interval query is a data set O, for each o in O, $o.I$ make one of the tree query predicts as following is true. When $a = b$ in query q, interval query is transformed to **Stabbing query**.

$$begin \leq a \bigcap end \geq b \qquad (1)$$

$$begin \geq a \bigcap begin \leq b \qquad (2)$$

$$end \geq a \bigcap end \leq b \qquad (3)$$

From the definition 1, we can see that the entity set $O = \{v, w, y, z\}$ as the result for interval query q is returned.

Define 2: Interval relationship. There are 5 relationships between the interval data as following.

- Cover(CO): If the interval of the entity *o* satisfies the query predicate 1, we review that *o* covers *q*, i.e. the entity *w*.
- LeftIntersection(LI): If the interval of the entity *o* satisfies the query predicate 2, we review that *o* intersect *q* at the left part of *o*, i.e. the entity *z*.
- RightIntersection(RI): If the interval of the entity *o* satisfies the query predicate 3, we review that *o* intersect *q* at the right part of *o*, i.e. the entity *v*.
- Contain(CT): If the interval of the entity *o* satisfies both the query predicate 2 and 3, we review that *o* is contained by *q*, i.e. the entity *y*.
- NotIntersection(NI): If the interval of the entity *o* doesn't satisfy the tree query predicates, i.e. the entity *u* and *x*.

2.2 Indexing for Interval Query

As an important and popular query type, interval query is studied systemically in Temporal DB [1]. Temporal DB provides the storage, updating and accessing methods for temporal data. The work in paper [2] surveys the accessing methods for temporal data. The represented indexes for interval queries are divided to 3 class: (a) B-tree variant, such as Time Index [3], Time Index+[4], Interval B-tree [5] and TD-tree [6]; (b)R-tree variant, such as Segment R-tree [7] and 4R-tree [8]; (c)The indexes extend Segment tree such as Segment R-tree [7].

The above access methods and techniques for interval query based on disk are mature. However, few researchers study interval query on cloud. The paper [9] is the first work about interval indexing and querying on key-value store. Distributed Segment Tree(DST) [10] is a similar work to study range query and cover query based on peer-to-peer distributed system. As the popular cloud platform is based on master-slave architecture(such as Bigtable [11], HBase [12], PNUTS [13], Cassandra [14], Dynamo [15]), and key-value cloud store has a great majority among all the cloud platform. We focus our attentions on the interval indexing and querying on key-value store and the main related work is EPI+MRST [9]. Although EPI+MRST supports the effective indexing and querying for interval query, the drawbacks of EPI+MRST are as following:

- The building costs of MRST is huge. We got the source code from the authors of paper [9] and found that it is too difficult to build MRST for huge interval data on key-value store. Besides, the space overhead of MRST is also enormous.
- The maintenance of EPI+MRST is intractable as the different structure of EPI and MRST.
- Many duplicate results from MRST and EPI increase the searching cost and communicating cost.

To overcome the drawbacks of EPI+MRST, we present an index named EIIQH-Base to improve the performance of EPI+MRST. We introduce the indexing and storage structure of EIIQHBase in detail in Sect. 3.1, the algorithm for interval querying on EIIQHBase are presented in Sect. 3.2.

3   The New Solution for Interval Indexing and Querying

3.1   Index Structure of EIIQHBase

In EPI+MRST, the interval query $q = [a, b]$ is decomposed into two sub-queries:
(1) one is range query $q_R = [a, b]$ to find all the item which has a relationship
with q is LI, or RI, or CT; (2) the other is covered query $q_C = [a, b]$ to find all
the item which covers q. Two sub-queries can be efficiently processed in parallel.

EIIQHBase is motivated from EPI+MRST but different with EPI+MRST.
For a given interval query q, there are 4 interval relationships (LI, RI, CT, and
CO) between the result element and q. We find that the result element which
has a LI or CT relationship with q can be acquired using a common structure
named LICT and the result element which has a LI or CT relationship with q
can be acquired using RICO structure. The storage structure of LICT and RICO
in HBase shown in Fig. 2. EIIQHBase is consisting of two tiers: the LICT, an
inverted index on the interval left endpoints, and the RICO, an inverted index
on the elementary interval.

(1) **LICT:** We store the LICT in one HBase table with one column family.
For each row in LICT, the rowkey is the interval left endpoints and the notation
$id = end$ denotes each column with a column name of interval identifierid and a
column value of interval right endpointend.

(2) **RICO:** RICO is stored in another HBase table to implement the parallel
searching. To find the interval object which has a RI or CO relationship with
the query interval, the elementary intervals of all the intervals are used to form
the keywords of inverted index. To take full advantage of the characteristics of
HBase, the right endpoint of elementary interval as the rowkey of HBase table.
The inverted list of inverted index is consist of all the object whose interval covers
the elementary interval. For example, the second row of RICO with rowkey "7"
in Fig. 2 denoted the elementary interval $[5, 7]$, the interval object $u = [3, 7]$ and
$w = [5, 22]$ cover the elementary interval $[5, 7]$.

rowKey	LICT
3	u=7; v=13
5	w=22
15	y=19
18	z=22
21	x=22

(a) LICT

rowKey	RICO
5	u=[3,7]; v=[3,13]
7	u=[3,7]; w=[5,22]
13	w=[5,22]; v=[3,13]
15	w=[5,22]
18	w=[5,22]; y=[15,19]
19	w=[5,22]; y=[15,19]; z=[18,22]
21	w=[5,22]; z=[18,22]
22	w=[5,22]; x=[21,22]; z=[18,22]

(b) RICO

Fig. 2. The storage structure of EIIQHBase

EIIQHBase is simple but helpful to utilize the sorted key of HBase to improve the performance of interval queries. We custom the HBase scan operation to avoid huge false positives for interval queries. The detail process of handling the interval query on EIIQHBase is proposed in Sect. 3.2.

3.2   Interval Queries on EIIQHBase

Given a interval query $q = [a, b]$, EIIQHBase customs two scan operation to find all the items occur during q. Algorithms 1 and 2 propose the detail scan operation to handle $q = [a, b]$ in parallel.

Algorithm 1 customs and executes one scan operation on table LICT with a and b as the start rowkey and the stop rowkey (line 4–5). The interval data returned by the scan operation owns a LI or CT relationship with q. Another scan operation on table RICO in Algorithm 2 also set the a and b as the start rowkey and the stop rowkey (line 4–5). However, we only acquire the first row of the scan operation to get interval data owned a RI or CO relationship with q.

Algorithm 1. Interval Query On LICT

```
Input: the interval query q=[a,b]; the index table LICT
Output: The result set R
    1: List R = new ArrayList()
    2: HTable LICT =(HTable) getHTablePool().getTable("LICT");
    3: Scan scanLICT= new Scan();
    4: scanLICT.setStartRow(Bytes.toBytes(a));
    5: scanLICT.setStopRow(Bytes.toBytes(b));
    6: ResultScanner rs = LICT.getScanner(scanLICT);
    7: FOR (Result r: RS){
    8:     FOR (KeyValue kv: r.raw()){
    9:         R.add(kv.getQualifier());
   10:     }
   11: }
   12: rs.close();
   13: RETURN R;
```

Algorithm 2. Interval Query On RICO

```
    1: List R = new ArrayList()
    2: HTable RICO =(HTable) getHTablePool().getTable("RICO");
    3: Scan scanRICO= new Scan();
    4: scanRICO.setStartRow(Bytes.toBytes(a));
    5: scanRICO.setStopRow(Bytes.toBytes(b));
    6: scanRICO.setCaching(1);
    7: scanRICO.setMaxResultSize(1);
    8: ResultScanner rs = RICO.getScanner(scanRICO);
    9: FOR (Result r: RS){
   10:     FOR (KeyValue kv: r.raw()){
```

```
11:        R.add(kv.getQualifier());
12:    }
13: }
14: rs.close();
15: RETURN R;
```

Known the storage structure and searching algorithms of EIIQHBase, we review EIIQHBase and compare it with EPI+MRST [9]. EPI+MRST divide an interval query into one range query and one cover query. The range query executed on EPI gets all the item whose relationship with q is LI or RI or CT and the cover query executed on MRST gets all the item whose relationship with q is LI/RI or CO. There are duplicate items between range query and cover query. However, all the item returned by EIIQHBase is desired and not duplicative. Above all, EIIQHBase is superior to EPI+MRST in 3 folders:

(1) EIIQHBase is more efficient than EPI+MRST. EIIQHBase accesses less items than EPI+MRST and saves the communicate cost and compute time. The query performance of EIIQHBase exceeds EPI+MRST.
(2) The space overhead of EIIQHBase is less than EPI. The space overhead of LICT is half of EPI as it only records one endpoint rather than two endpoints. Although RICO stores one interval data several times, MRST stores the total tree structure also space consumed.
(3) EIIQHBase is more flexible than EPI+MRST. The basic index of EIIQH-Base is inverted index so that the mature techniques of inverted list and the sorted key of HBase are used to improve the performance of interval query on cloud. EIIQHBase takes full advantage of the characteristic of the cloud platform.
(4) EIIQHBase reduces the maintenance cost compared to the combined index EPI+MRST, especially MRST is difficult to build and update.

4   Experiment

We implement EIIQHBase using Java 1.6 version. All the experiments in this paper are finished in a distributed cluster consisting of 8 virtual machines. Each virtual machine has 8GB memory, 500G disk space. The distribute systems of EIIQHBase is HBase-0.94.2 based on Hadoop-1.0.4. There are real and simulated data sets to evaluate the performance of the index programs under different influence factors.

Two kinds data sets coined "DMOZ" and "SIMU" are employed to study the effect of interval lifespan and query selectivity on our approach. The former is a real category information in 7 years from site "dmoz.org", consisting of ≈ 2 million web pages, resulting ≈ 12 million intervals. The dataset coined "SIMU" is consist of total ≈ 10 million web pages which is simulated to describe the version change during 10 years, resulting ≈ 30 million intervals. We evaluate both the space and time overhead of EIIQHBase compared with EPI+MRST.

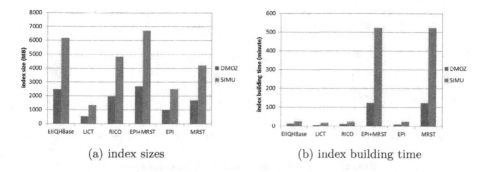

(a) index sizes   (b) index building time

Fig. 3. Index size and index building time.

Index Size. Figure 3(a) presents the index size of EIIQHBase and EPI+MRST. The index size of EIIQHBase is circa 2.4 GB and 6.0 GB for the DMOZ and SIMU datasets respectively. The index size of EPIMRST is slightly larger than EIIQHBase. RICO has an important contribution to the index size respect to LIRT in EIIQHBase.

Index Building Time. Figure 3(b) depicts the index building time result for DMOZ and SIMU. As is evident, the building/population of the MRST in EPI+MRST consumes the most time. It is nearly more than EPI one order of magnitude in SIMU dataset. The building time of LICT and RICO in EIIQH-Base is similar with EPI. Building EPI+MTST is more time-consuming than EIIQHBase. Above all, EIIQHBase is flexibly implemented on key-value store.

Respond Time. For query instances in our paper, we obey the rule of EPI+MRST and use three sample of 100 queries selected randomly from SIMU dataset to evaluate the respond time varying different interval span and three sample of 100 queries over the DMOZ dataset with different selectivity.

Figure 4(a) depicts the respond time of EIIQHBase and EPI+MRST for stabbing queries over SIMU dataset. We can see that both indexes can respond stabbing queries in millisecond scale. Besides, the performance of EIIQHBase is 4 times of EPI+MRST. We extend the span of interval queries shown in Fig. 4(b) and (c), we find that EIIQHBase exceeds EPI+MRST more than one order of magnitude, especially while the interval span is large as the duplicate data is huge.

We further survey the respond time of EIIQHBase and EPI+MRST, and Fig. 5 illustrates the experimental results with different selectivity over DMOZ dataset. We can see that EIIQHBase can respond most interval queries below 1 s but EPI+MRST require more than 10 s. Besides, the respond time of both indexes increase linearly with the selectivity increase, but EPI+MRST has a larger increase rate.

Above all, EIIQHBase reduces the space and time overhead compared with EPI+MRST, and more efficient and effective than EPI+MRST.

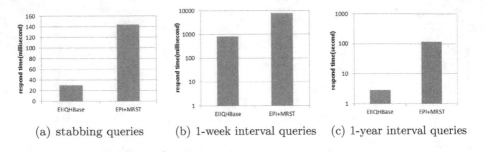

(a) stabbing queries   (b) 1-week interval queries   (c) 1-year interval queries

Fig. 4. respond time varying with interval span.

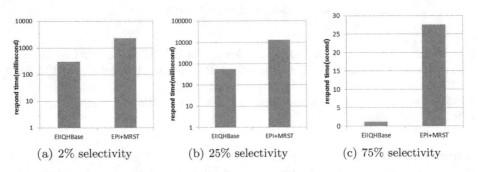

(a) 2% selectivity   (b) 25% selectivity   (c) 75% selectivity

Fig. 5. Respond time varying with selectivity.

5   Conclusion

Clouds and key-value stores are increasingly attracting attention for processing complex, intensive and huge queries, such as the interval queries. We survey the related work of interval queries on key-value store and few works are proposed to respond interval queries efficiently on cloud. EPI+MRST is the first work to research interval indexing and querying on key-value store. However, it is difficult to build MRST in distributed environment. Besides, EPI and MRST have many duplicate results which waste the cloud resource. So, we present a novel index structure named EIIQHBase to improve EPI+MRST. EIIQHBase divides the results for an interval query to two parts according to the interval relationship and presents two sub-indexes to handle each part. EIIQHBase owns the advantages of EPI+MRST but overcomes the drawbacks of EPI+MRST. Last, we present extensive experimental results, with real-world and simulated datasets, showing that our indexing and searching algorithms far outperform EPI+MRST, both the space and time overhead have a significant reduce. As is evident, our approach is more effective and efficient than the state of art.

Acknowledgments. Thank the author of paper [9] for sharing his source code. Our work is supported by "the Fundamental Research Funds for the Central Universities, No. 3132016031", and "National Natural Science Foundation of China, No. 61371090 and No. 61073057".

References

1. Kumar, A., Tsotras, V.J., Faloutsos, C.: Designing access methods for bitemporal databases. IEEE Trans. Knowl. Data Eng. **10**(1), 1–20 (1998)
2. Salzberg, B., Tsotras, V.J.: Comparison of access methods for time evolving data. ACM Comput. Surv. (CSUR) **31**(2), 158–221 (1999)
3. Elmasri, R., Wuu, G.T.J., Kim, Y.-J.: The time index: an access structure for temporal data. In: Proceedings of the 16th International Conference on Very Large Data Bases, pp. 1–12. Morgan Kaufmann, Brisbane (1990)
4. Kouramajian, V., Kamel, I., Elmasri, R., The, W.R.: Time index+: an incremental access structure for temporal databases. In: Proceeding of the Third International Conference on Information and Knowledge Management (CIKM), pp. 296–303. ACM, Gaithersburg (1994)
5. Ang, C., Tan, K.: The interval B-tree. Inf. Process. Lett. **53**(2), 85–89 (1994)
6. Stantic, B., Topor, R., Terry, J., Sattar, A.: Advanced indexing technique for temporal data. Comput. Sci. Inf. Syst. (COMSIS) **7**(4), 679–703 (2010)
7. Kolovson, C., Stonebraker, M.: Segment indexes: dynamic indexing techniques for multi-dimensional interval data. SIGMOD Rec. **20**(2), 138–147 (1991)
8. Bliujute, R., Jensen, C.S., Saltenis, S., Slivinskas, G.: Light-weight indexing of general bitemporal data. In: Proceedings of the 12th International Conference on Scientific and Statistical Database Management (SSDBM). IEEE Computer Society, Berlin, pp. 125–138 (2000)
9. Sfakianakis, G., Patlakas, I., Ntarmos, N., Triantafillou, P.: Interval indexing, querying on key-value cloud stores. In: Proceedings of 29th IEEE International Conference on Data Engineering (ICDE), pp. 805–816. ACM, Brisban (2013)
10. Zheng, C., Shen, G., Li, S., Shenker, S.: Distributed segment tree: support of range query and cover query over DHT. In: 5th International workshop on Peer-To-Peer Systems (IPTPS), Santa Barbara (2006)
11. Chang, F., Dean, J., Ghemawat, S., et al.: Bigtable: a distributed storage system for structured data. In: Proceedings of Operating Systems Design and Implementation (OSDI). USENIX Association, Seattle, pp. 205–218 (2006)
12. Apache HBase. http://hbase.apache.org/
13. Cooper, B.F., Ramakrishnan, R., Srivastava, U., Silberstein, A. et al.: PNUTS: Yahoo!s hosted data serving platform. In: Proceedings of VLDB Endowment. ACM, Auckland, pp. 1277–1288 (2008)
14. Lakshman, A., Malik, P.: Cassandra: a decentralized structured storage system. ACM Oper. Syst. Rev. (SIGOPS) **44**(2), 35–40 (2010)
15. DeCandia, G., Hastorun, D., Jampani, M. et al.: Dynamo: Amazons highly available key-value store. In: Proceedings of the 21st ACM Symposium on Operating Systems Principles (SOSP), pp. 205–220. ACM, Stevenson (2007)

Filtering Uncertain XML Documents by Threshold XPEs

Bo Ning[(✉)], Yu Wang, Ansheng Deng, Yi Li, and Yawen Zheng

Dalian Maritime University, Dalian, China
ningbo@dlmu.edu.cn

Abstract. Uncertainty can be expressed naturally by XML format, however the existing SDI technologies for XML document cannot deal with uncertain data. In this paper, a probabilistic index architecture is proposed for indexing the XPath expressions which are described by the users, and a algorithm for filtering probabilistic XML document is proposed. Experiments are conducted to verify the feasibility and effectiveness of our proposed index and algorithm. The result shows that the novel method is efficient and can meet users' requirements better.

1 Introduction

Document distribution technique is to solve the large scale growth of network data, to reduce the utilization rate of effective information, then is distributed to users with the taken information source document filtering. Many classic systems Many classic systems select appropriate documents for distribution to users by means of keyword matching, predicate contrasting and attribute weighted value, such as Siena [1], Gryphon [2] and Elvin [3] systems. But a large number of unrelated documents are also returned to the user, the reason is that the document structure does not meet the user requirements or the document description submitted by the user is not detailed enough [4]. Then XML (Extensible Markup Language) appears as the standard of data exchange in the network, then, document filtering technique, which is proposed based on XML document, solve the problem that the document structure does not meet the user requirements. Meanwhile, the XPath expression [5,6] appears as a technology for positioning XML document's cable element, then the document filtering system based on XPath expression comes into being.

Many of the existing classic document filtering systems include XTrie [7] algorithm based on string matching, XFilter [8] and YFilter [9] algorithm based on finite state machine, NIndex [10] algorithm based on relation table and XPush [11] algorithm based on stack structure. XTrie algorithm based on string matching decomposes the user query, build a query index tree and index table, and it makes document filtering completed by matching the document's string in the information source. The XFilter algorithm based on the finite state machine uses the complex index structure and the improved finite state machine, and achieves the effective filtering of the document through the state transition of the finite

S. Song and Y. Tong (Eds.): WAIM 2016 Workshops, LNCS 9998, pp. 292–302, 2016.
DOI: 10.1007/978-3-319-47121-1_25

state machine (FSM). XFilte cannot achieve multi-query document filtering for only processing a single query, then the YFilter algorithm is proposed, and it can achieve multi-query document filtering through the finite state machine. NIndex algorithm stores multiple queries on the PXPETree, and it can achieve document filtering through the model of query relational table. The XPush algorithm achieves the document filtering based on the Twig model. Well-developed XML document filtering technology is to determine the data, with the continuous growth of the number of uncertain data, the problem of document distribution for uncertain data is eager to be resolved, so it leads to the research of the XML with uncertain data, the document filtering algorithm in this paper.

The contribution of our proposed algorithm can be concluded on the following aspects. Firstly, we establish the two level index structure and index users' query information and correlation, secondly, we use SAX event driven method to analyze XML document containing uncertain data and complete the specific query matching process. At last, we propose an efficient document filtering algorithm for uncertain data to meet the user's probabilistic query requirements.

2   Selective Distribution for Uncertain Data

2.1   Uncertain Data and Probabilistic XML Document

Many of the data in the document information are uncertain. Since the query and storage of uncertain data make the original data get complex, so it is as far as possible to avoid uncertain data. But in many special fields, it is impossible to avoid uncertain data, so that the uncertain data attracts a lot of researchers's attention. Uncertain data is usually expressed in the form of a probability value, moreover, it attaches a probability value with expression data (probability is used to express uncertainty). Using two-dimensional form alone to store the probability value information will cause the waste of space, so in this paper we employee the XML format to express the uncertain data. In the document tree, a probability value can be directly attached to the document tree as an attribute.

Probabilistic XML can naturally describe the semi-structured data with uncertain information. Probabilistic document tree is a kind of data model, which is used to describe the uncertain information. The probability attribute can be set in probabilistic XML document tree. There are two kinds of nodes in the XML tree, one is the ordinary node which represents element node in document tree, and the other is the distribution node which represents the probabilistic distribution among the uncertain elements.

2.2   XPath Expression of Probabilistic XML Document

With XML document and user query, it is not only to meet the query structure of the user query expression tree, but also to meet the query probabilistic constraint, so that we can return this document to the user. The expression of query is through XPath expression to determine whether to meet the ancestor descendant relationship. When the user submits a query request, it is authorized to raise the threshold requirement to the accuracy of the results, when the document meets the requirements of the structure and content of the user as well as the threshold requirement, it will return query results. User threshold requirements can be divided into two types: one is path query threshold, which need to be given by users to set the threshold of a single path. The other is the overall threshold query, since the threshold probabilistic value of the query, we estimate the whole query tree of the XPath in this paper, and select specific query results to return to users.

2.3   Document Distribution of Probabilistic XML

Document distribution is also called selective dissemination of information (Selective Dissemination of Information, SDI), SDI is a mechanism which is based on the needs of users to select the information for the user. There are two input parts in the SDI system: one is the user's submitted documents; The other is the information document. User documentation is to describe the user's query and user's information, and the information document in the SDI system will be converted into XML document tree.

According to uncertain data, probabilistic XML document distribution is to build a probabilistic index tree structure, to grant the user to determine the threshold value of the definition of the authority, to achieve the results of the user's choice of the purpose. Document distribution as long as is the study of the uncertain data contains XML document, will be submitted to the user of the threshold value division, through the document containing the uncertain information matching the user requirements, at the end the documents are returned back to the users. The overall probability value can be calculated in XML document parsing process, and the user's threshold will enter the document filtering engine with the user's submission information. Probability value is used as a constraint to determine whether the current document is consistent with the user's requirements in the process of filtering.

3   Probabilistic XPath Query Index: PXtrie

3.1   Probabilistic XML Document Filtering

XPath query expression is decomposed into small strings, through its relevance to construct probabilistic PXtrie index tree T, while constructing the index table corresponding to the PXtrie index table ST. PXtrie index structure can be completed after the document filtering. First, the probability XML, the document analysis, when the content of the analysis to the PXtrie index tree in the matching of information, to determine whether to meet the current information corresponding to the requirements, when the above two requirements are satisfied, the probability of the sub string to determine whether to meet the user threshold requirements. If there is a need for the user of the document, then the

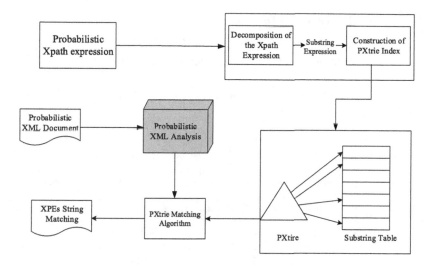

Fig. 1. Filtering probabilistic XML

user corresponds to the ST table returns, after the completion of the document filtering, through the success matching of the query string to determine, the document returned to the successful matching of the user, the specific matching process is shown in Fig. 1.

3.2 Query Decomposition

Probability index PXtrie XPath query is based on the probability of string matching document distribution structure, it uses an XPath expression of multi-user query request to build an index. Given the efficiency and the widespread availability of XTrie index, add on XTrie structure probability value attribute implementation document filtering algorithm based on PXtrie index structure. In order to describe the PXtrie index structure, this paper makes a detailed description of the PXtrie index structure. For more than one user query into a sub string query decomposition, and then integrated out the index structure of the whole. User query decomposition can effectively improve the efficiency of document search and match, and reduce the waste of storage space. And threshold of user information through the construction of secondary indexes, the probability index subtree achieve complete matching query processing. Under the PXtrie index structure, the input content includes the user submits the documents information; An information document information. The content of the output is a collection of users that successfully match the current XML document. User queries using XPath query expressions to express, and then decomposed into several boy series, based on user sub string splicing implementation of eliminating the same user query information section. In order to maintain the ancestor descendant relation and the characteristic of the child string, the PXtrie index table (ST table) corresponding to the PXtrie index tree is constructed.

$$P_1 = //a/b/*/c/d \quad \langle ab,cd \rangle$$
$$P_2 = /c/d/*/a/b[*/b/c]//d \quad \langle cd,ab,bc,d \rangle$$
$$P_3 = //d//d//*/a/b[c/d]/*/c/d \quad \langle d,ab,abcd,cd \rangle$$

Fig. 2. Examples of query decomposition

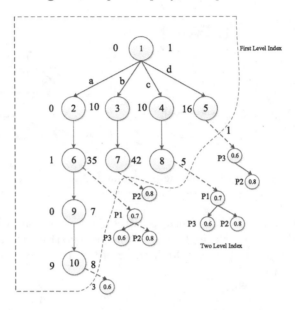

Fig. 3. Examples of XPEs tree

The construction principle of PXtrie index tree: user the XPath expression of the decomposed expression can contain only the child axis and descendant axis. The detailed decomposition is shown in Fig. 2, where P_1, P_2, P_3 are XPath expressions from users with different requirements respectively.

3.3   PXtrie Index Structure

PXtrie index tree T not only labeled user query subtree integration results but also contains the user's probability threshold. In Fig. 2 the above queries are shown, and we assuming that threshold requirements are respectively 0.7, 0.8 and 0.6. Figure 3 is constructed from the probabilistic PXtrie index tree T. Figure 4 is a probabilistic PXtrie index table ST, where the P denote user XPath expression. The S denote sub strings after the user query expression is decomposed. The *ParentRow* denote father sub string of S (If the current sub string is the root then the sub string of father is 0). The *Rel* Level denote the relative length of sub string. The *Rank* denote sub string is the which child of the sub string in parent. The *NumChild* denote sub string there are several children. The *Next* denote point to the next same sub string for different users.

	Parent Row	Rel Level	Rank	Num Child	Next	TV
1	0	2, ∞	1	1	4	0.7
2	1	3, 3	1	0	3	0.7
3	0	2, 2	1	1	10	0.8
4	3	3, 3	1	2	8	0.8
5	4	3, 3	1	0	0	0.8
6	4	1, ∞	2	0	7	0.8
7	0	1, ∞	1	1	0	0.6
8	7	3, ∞	1	2	0	0.6
9	8	2, 2	1	0	0	0.6
10	8	3, 3	2	0	0	0.6

Fig. 4. Index structure

The TV denote query threshold required by the user that represents the current sub string.

Definition 1 *The sub string pointer.* *Sub string pointer is PXtrie index tree T and PXtrie index table ST the connection, we set sub string pointer P (Node) in the PXtrie structure, user decomposition of sub string will by sub string pointer P (Node) to point to the ST sequence table.*

When p (node) is equal to zero indicates that the current position of the sub string is not enough to constitute one of the user sub string. When sub string is nonzero, the tag value representative of the corresponding sub string on table row number, the left side of the internal nodes in T.

Definition 2 *The maximum suffix pointer.* *The document node does not con-form to the current string requirements, then jump to the most likely to be in accordance with sub string up. The maximum suffix pointer Q (Node) is constructed to mark the next most likely to match the success of the sub string, the right side of the internal nodes in T.*

Definition 3 *Precision counter.* *When the document and the user query matching, mark of determine whether the information document is in line with the user's requirements. Precision counter B (L, I), which: on behalf of the current sub string, Z represents the current sub string where the XML document level.*

Initialization B (I, L) set 0, the user query to match the success of B (I, 1) for 1, sub- string of child sub string when the match is successful in a B (I, 1) plus 1, B (I, 1) is equal to the number of children sub string plus 1 to complete the number of the current user to complete the overall query.

Definition 4 *Successful match flag.* *Set up the successful match mark C (P, L) for the complete query of the user, P denote to match users of success, L denote sub string in XML levels of position. Initialize C (p, l) for 0, when value of C (p, l) is equal to 1, denote the current user query has been successful.*

4   Probabilistic XML Document Filtering Algorithm

When document filtering, the PXtrie search algorithm is designed for find the needs of users. Algorithm input user query to build the PXtrie index tree T, PXtrie index table ST and wait to match the XML document, only when the document contains the user query sub string, document can be returned to the user. When PXtrie search algorithm complete, the sub string sequence number will be returned. When all the user's query sub string are satisfied, the user can be placed in the return collection of the document. The search algorithm is shown in Algorithm 1.

In Algorithm 1 PXtrie query algorithm input three parts, including PXtrie index tree, PXtrie index table and probability XML document, the output of the algorithm is table set C. Firstly we initialize the result set C as is empty, the current pointer refers to the node N in the T set, as it is the root node, and then it parse the XML the document. When the tag is encountered, we increase the current document level by 1, and in the index structure, we search the node of current tag in T. If the current root node does not have t tag, the current root node pointer does not move, and continue to parse the next start tag in the document, from top to bottom, until the path tag from the root node label T is found. If the T tag in the document is parsed, Node[i] points to the current location, and the next start of the tag position N also points to the N'. The probability value of current node sub string can be calculated and determined. Finally, if a complete sub string be found, then jump to the matching function. If N' points to the root node, indicates that the current child string does not exist in the PXtrie tree.

5   Experimental Results and Analysis

5.1   Experimental Platform

Through the analysis of artificial data, in this paper, we have carried out experiments and tests on the probabilistic document distribution algorithm. In MyEclipse compiler environment, we use JDK 7 to achieve all of the algorithm, all tests in PC of 62 bit Windows 7 operating system with a 2.10 CPU GHz, 2 GB memory, 500 GB.

5.2   The Influence of the Document Width

In the test of the effect of the document width on the running time of the algorithm, we can test the running time of different users in the algorithm by

---

Algorithm 1. SEARCH (D, ST, T)

Data: XML document D, PXtrie index table ST and PXtrie tree T;
Result: The user documents XPath expression and the XML document D the successful matching collection C.

1 **begin**
2 | Initialization C is empty;
3 | Set the Node[i] as the root node of the PXtrie tree T, i = 0 to Lmax;
4 | Initialization L=0;
5 | Initialize N as the root node of the PXtrie tree T;
6 | Initialize prob(t) is 1;
7 | Initialize B (I, l) precision counter for 0.C (P, l) successfully match the flag for 0;
8 | **if** *the start tag t of D has been parsed* **then**
9 | | L=L+1;
10 | | N'=N;;
11 | **while** *the tag t is not found and N is not a root node* **do**
12 | | N=Q(N);
13 | | **if** *In T there is a path label t from N to N'* **then**
14 | | | Node[l]=N';
15 | | | N=N';
16 | **while** *N' is not the root node* **do**
17 | | **if** *P(N')¿0* **then**
18 | | | C=CUMATCH(STP(N')LB);
19 | | | N'=Q(N);
20 | | **else**
21 | | | **if** *In the search of D has encountered the end tag* **then**
22 | | | | B (I, L) of nodes in length within of the whole query sub string is set to 0;
23 | | | | root node of Node[i]=T;
24 | | | | L=L-1;
25 | | | | N=Node[l];
26 | /* return C */
27 **end**

---

controlling the size of the document, and we can draw the folded line shown in Fig. 5 by comparison. When the current document depth is 10 layers, the size of the document is 10 Mb,20 Mb,30 Mb,20 Mb, 50 Mb. And by the analysis of Fig. 5, when the document width increases, the running time increases linearly, and the increase of the number of users will result in the increase of the running time.

5.3 The Influence of the Document Depth

There is the influence of the depth of the document on the algorithm time in Fig. 6, where the control of query document is 10 Mb, and experiments are carried out at the layer of 5, 6, 7, 8, 9, and 10, respectively. In order to make the case

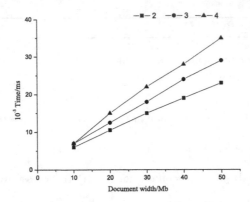

Fig. 5. Examples of query decomposition

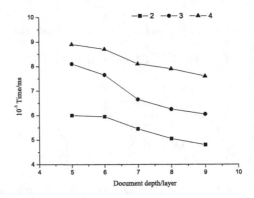

Fig. 6. Examples of query decomposition

comparable, we compare with different number of users, and from the experimental data of Fig. 6, the running time decreases with the increase of the document depth and increases with the increase of user's number, and the increase of document depth makes the running time longer. The algorithm implements the operation of redundant query of users, whenever the operation is successful, it will stop the user's search. When the document depth is increased, the possibility of finding users on the same branch in-creases, and there is the improvement of the time efficiency for the decrease of residual search for the number of users.

5.4   The Influence of the Number of Users

The number of users is tested on the time of algorithm in Fig. 7, and the test is carried out on a 10 layer document with 10 Mb. Running time increases when the number of users is increased, but when the number of users always increases, there is a logarithmic growth for time. It indicates that when the number of users increases to a certain number, its influence on the time is stable. In order to

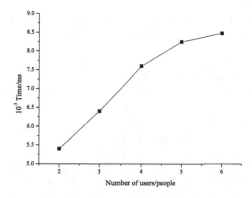

Fig. 7. Examples of query decomposition

improve the operational efficiency of this algorithm, the processing optimization of next step will be processed by the algorithm of parallel query.

6   Conclusions

In this paper, we propose a document filtering algorithm based on PXtrie index structure for uncertain data, constructing probabilistic index structure is a convenient way to filter information for multi-user request. Filtering algorithm realizes document filtering of the threshold query requirements for user by querying, matching, updating operation on the XML document with uncertain information. We verify the availability and effectiveness of the probabilistic structural document filtering algorithm by an example, and solve the problem of document distribution with uncertain data. Finally, we conclude that the increase of the number of users will increase the running time by the comparative analysis. Thus, with the improvement of the document filtering efficiency with uncertain data as the center, we will reduce the number of users in the PXtrie index tree to go on the next research in this paper.

Acknowledgement. This research was supported by the National Natural Science Foundation of China (Grant No. 61202083, 61272171), the Liaoning Province project (Grant No. 12014055), the Fundamental Research Funds for the Central Universities of China (Grant No. 3132016034).

References

1. Aguilera, M.K., Storm, R.E., Strurman, D.C., et al.: Matching events in a content-based subscription system. In: Proceedings of the 18th Annual ACM Symposium on Principles of Distributed Computing, pp. 53–61. ACM, New York (1999)
2. Carzaniga, A., Rosenblum, D.S., Wolf, A.L.: Design, evaluation of a wide-area event notification service. ACM Trans. Comput. Syst. (TOCS) **19**(3), 332–383 (2001)

3. Segall, B., Arnold, D., Boot, J., et al.: Content based routing with elvin4. In: Proceedings of AUUG2K, pp. 890–901. IEEE Computer Society, Los Alamitos (2000)
4. Chan, C.Y., Felber, P., Garofalakis, M., et al.: Efficient filtering of XML documents with XPath expressions. VLDB J. **11**(4), 354–379 (2002)
5. Al-Khalifa, S., Srivastava, D., Jagadish, H.V., Koudas, N., Patel, J.M., Wu, Y.: Structural joins: a primitive for efficient XML query pattern matching. In: Proceedings of the ICDE, pp. 141–152 (2002)
6. Bruno, N., Koudas, N., Srivastava, D.: Holistic twig joins: optimal XML pattern matching. In: Proceedings of SIGMOD, pp. 310–321 (2002)
7. Chan, C.Y., Felber, P., Garofalakis, M., et al.: Efficient filtering of XML documents with XPath expressions. In: Proceedings of the 18th International Conference on Data Engineering (ICDE 2002), pp. 235–244. IEEE Computer Society, Los Alamitos (2002)
8. Diao, Y., Fischer, P., Franklin, M.J., et al.: Yfilter: efficient and scalable filtering of XML documents. In: Proceedings of the 18th International Conference on Data Engineering, pp. 341–342. IEEEE Computer Society, Los Alamitos (2002)
9. Ning, B., Liu, C.: XML filtering with XPath expressions containing parent and ancestor axes. Inf. Sci. **2**(10), 41–54 (2012)
10. Gou, G., Chirkova, R.: Efficiently querying large XML data repositories: a survey. IEEE Trans. Knowl. Data Eng. **19**(10), 1381–1403 (2007)
11. Gupta, A.K., Suciu, D.: Stream processing of XPath queries with predicates. In: Proceedings of the 2003 ACM SIGMOD International Conference on Management of Data, pp. 419–430. ACM, New York (2003)

Storing and Querying Semi-structured Spatio-Temporal Data in HBase

Chong Zhang[1,2](✉), Xiaoying Chen[1,2], Xiaosheng Feng[1,2], and Bin Ge[1,2]

[1] Science and Technology on Information Systems Engineering Laboratory,
National University of Defense Technology, Changsha 410073, China
leocheung8286@yahoo.com, chenxiaoying1991@yahoo.com, xsfeng@nudt.edu.cn,
gebin1978@gmail.com
[2] Collaborative Innovation Center of Geospatial Technology, Wuhan, China

Abstract. With the development of remote sensing, positioning and other technology, a large amount of spatio-temporal data require effective management. In the current research status, a lot of works have focused on how to effectively use HBase to store and quickly find structured spatio-temporal data. However, some spatio-temporal data exists in the semi-structured documents, such as metadata that describes the remote sensing products, under such context, the query is changed to spatio-temporal query + semi-structured query (XPath), which is less studies in previous works. In this paper, we focus on how to efficiently and economically achieve semi-structured spatio-temporal data storage and query in HBase. Firstly, the formal description of the problem is presented. Secondly, we propose HSSST storage model using a semi-structured approach TwigStack. On this basis, semi-structured spatio-temporal range query and kNN queries are carried out. Experiments are conducted on real dataset, comparing with MongoDB which need higher hardware configuration, the results show that in moderate configuration of machines, the performance of semi-structured spatio-temporal query algorithms are superior to MongoDB, thus it has advantage in real application.

Keywords: Spatio-temporal · Semi-structured · HBase · Range query · KNN Query

1 Introduction

With remote sensing, telecommunications and other technologies' development, huge amount of spatio-temporal data are collected and exploited in various applications, which is a challenge to database community. Most previous works focus on structured spatio-temporal data, i.e., spatio-temporal objects are formatted in record $< location, time, other\_attributes >$, however, for retrieving remote sensing data, the case is different, i.e., the objects in remote sensing can't be formatted in structured record, for instance, a satellite image. To retrieve easily, users usually use other textual data to describe the original data in remote

This work is supported by NSF of China grant 61303062 and 71331008.

S. Song and Y. Tong (Eds.): WAIM 2016 Workshops, LNCS 9998, pp. 303–314, 2016.
DOI: 10.1007/978-3-319-47121-1_26

```
<Remote Sensing>
  <Type>Satellite Image</Type>
  <Sensor>
    <Type>CCD Camera</Type>
    <Resolution>50</Resolution>
    <Spectrum>
      <Upper>0.927μm</upper>
      <Lower>0.9μm</Lower>
    </Spectrum>
  </Sensor>
  <Geo Range>
    <Long>89.223799, 90.156234</Long>
    <Lat>28.123457, 32.73485</Lat>
  </Geo Range>
  <Time Range>1347898654, 1358907896</Time Range>
  <Path>/path/1125.tif</Path>
</Remote Sensing>
```

Fig. 1. An example of spatio-temporal semi-structured data

sensing application, which is called *meta data*. Further, for convenient, the meta data is usually in semi-structured format, e.g., XML or JSON, which is flexible to express. Thus, the retrieval work is transfered to retrieve spatio-temporal semi-structured document (or object, each meta data file can be viewed as an object), i.e., the problem is changed, it is not to simply query on spatio-temporal data, it is spatio-temporal query + semi-structured query (such as XPath).

Figure 1 shows a remote sensing meta data sample, users can query remote sensing objects by declaring spatio-temporal predicates and XPath predicate. For instance, a query could be, given a spatial area $R = (c, r)$, where c is centroid, r is radius, find meta data documents which are satisfied with XPath query /remote sensing [type = "Satellite Image" and //type = "CCD Camera"] within R during recent two weeks, note that the XPath here is twig query.

A straight forward solution is to use MongoDB [1] to store each meta data document as Mongo's document, and build spatial index of MongoDB, and then use Mongo's query language to carry out the query execution. However, MongoDB needs high-performance configure hardware to support efficient retrieval, which would cost much in real application. In this paper, we argue that HBase is a better solution to accomplish the query task, which is leveraged by the distribution of machines.

To achieve our goal, we first study on how to store spatio-temporal and semi-structured information into HBase, together, we propose HBase Semi-Structured Spatio-Temporal (HSSST) model to realize our idea, and then we present range query and kNN query algorithms to support the two retrieval operations. We conduct experiment on real remote sensing dataset and the results show that HSSST outperforms MongoDB, and is capable for real applications. Our contributions are summarized as follows:

- We propose semi-structured spatio-temporal query type, which is useful for remote sensing application.
- We propose HSSST model to support storage and index of semi-structured spatio-temporal data.
- We design range query and kNN query algorithms.

The rest of this paper is organized as follows. Section 2 reviews related works. Section 3 formally defines the problem and prerequisites. Section 4 presents HSSST structure. In Sect. 5, algorithms for range and kNN queries are presented. And we experimentally evaluate HSSST compared with MongoDB in Sect. 6. Finally, Sect. 7 concludes the paper with directions for future works.

2   Related Works

MongoDB [1] is a scalable, high-performance, open source, document-oriented database, classified as a NoSQL database. MongoDB can directly support spatial data storage and indexing classes meet appropriate functional requirements, such as: GeoJSON may be geospatial (2d) index on the spatial coordinate data stored in the document on the (Legacy Coordinate Pairs) index, then calculates Geohash value (Geohash geocoding is based on latitude and longitude). Other NoSQL database does not directly support spatial data coding method, however, it can be extended by Geohash method, in MongoDB, the dimension of time and space dimensions combined with its own query is not supported.

Selecting a storage method of semi-structured temporal data should also consider its conversion into other forms of data to application performance, OpenAIRE [7] research data source metadata into Linked Open Data (LOD) method to explore, compared HBase, CSV, XML into RDF conversion performance of the three methods, the results showed the highest conversion efficiency map of HBase.

Nishimura et al. [5] address multidimensional queries for PaaS by proposing MD-HBase. It uses k-d-trees and quad-trees to partition space and adopts Z-curve to convert multidimensional data to a single dimension, and supports multi-dimensional range and nearest neighbor queries, which leverages a multi-dimensional index structure layered over HBase. However, MD-HBase builds index in the meta table, which does not index inner structure of regions, so that scan operations are carried out to find results, which reduces its efficiency.

Hsu et al. [4] propose a novel Key formulation scheme based on R^+-tree, called KR^+-tree, and based on it, spatial query algorithm of kNN query and range query are designed. Moreover, the proposed key formulation schemes are implemented on HBase and Cassandra. With the experiment on real spatial data, it demonstrates that KR^+-tree outperforms MD-HBase. KR^+-tree is able to balance the number of false-positive and the number of sub-queries so that it improves the efficiency of range query and kNN query a lot. This work designs the index according to the features found in experiments on HBase and Cassandra. However, it still does not consider the inner structure of HBase.

Han et al. [3] propose HGrid data model for HBase. HGrid data model is based on a hybrid index structure, combining a quad-tree and a regular grid as primary and secondary indices, supports efficient performance for range and kNN queries. This paper also formulates a set of guidelines on how to organize data for geo-spatial applications in HBase. This model does not outperform all its competitors in terms of query response time. However, it requires less space than the corresponding quad-tree and regular-grid indices.

HBaseSpatial, a scalable spatial data storage based on HBase, proposed by Zhang et al. [8]. Compared with MongoDB and MySQL, experimental results show it can effectively enhance the query efficiency of big spatial data and provide a good solution for storage. But this model does not compare with other distributed index method.

3   Problem Definition and Preliminaries

In this section, we formally define the problem and query types, and then introduce a classic XPath query algorithm, TwigStack.

3.1   Problem Definition

Generally, spatio-temporal semi-structured data (object or document) could be formally modeled as $< sid, (x_l, y_l, x_u, y_u), [t_s, t_e], T >$, where sid is identifier of semi-structured data (object or document), (x_l, y_l, x_u, y_u) is the spatial range window referenced or decribed by document or object sid, $[t_s, t_e]$ is the valid period for (x_l, y_l, x_u, y_u) referenced or decribed by document or object sid, and T is the semi-structured (e.g., XML or JSON) data (no spatio-temporal data is contained) of document or object sid.

We formally define two spatio-temporal queries in the context of semi-structured data:

(1) Range query $(t_{qs}, t_{qe}, c, r, xq)$. Given a time period $[t_{qs}, t_{qe}]$, a spatial range $R_q = (c, r)$, where c is the centroid, r is radius, and an XPath query predicate xq, range query aims to find all spatio-temporal semi-structrued objects (or documents) which satisfy the conditions, $(x_l, y_l, x_u, y_u) \cap R_q \neq \phi$, $[t_s, t_e] \cap [t_{qs}, t_{qe}] \neq \phi$, and $Predicate(T, xq) = true$.

(2) kNN query $(t_{qs}, t_{qe}, q, k, xq)$. Given a time period $[t_{qs}, t_{qe}]$, a spatial point q, an integer k, and an XPath query predicate xq, kNN query aims to find k spatio-temporal semi-structrued objects (or documents) which are nearest neighbors to q accordingly, and $[t_s, t_e] \cap [t_{qs}, t_{qe}] \neq \phi$, and $Predicate(T, xq) = true$.

3.2   TwigStack

TwigStack algorithm is an effective method to solve XML structural queries, especially queries against the twig join. It uses *region code* to encode each node

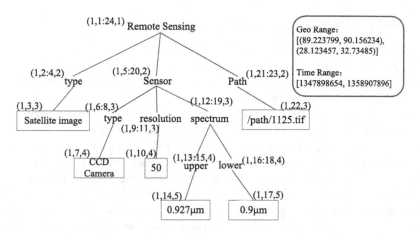

Fig. 2. An example of region code

in XML document, and then leverage stack with relationships in the code to accomplish twig query on XMLs. In particular, for a non-leaf node n_i in XML document xd, the region code is in format (xid, $enter\_code$: $exit\_code$, l), where xid is the identifier of xd, and $enter\_code$ and $exit\_code$ are sequence numbers for arriving and leaving n_i by a preorder traversal of the XML tree, respectively, and l is n_i' level. Then for a leaf node n_l in xd, the code is in format (xid, $enter\_code$, l), and the notations are similar to non-leaf node, except there is no exit code for a leaf node.

For example, Fig. 2 illustrates the tree format and region code for Fig. 1. Assuming that the XML document identifier is 1, then region code for root *Remote Sensing* is (1, 1: 24, 1), where number 24 is the sequence number after visiting all nodes in XML. After encoding, given any two codes, the structural relationship between the corresponding nodes can be determined, for instance, such as *spectrum* (1, 12: 19, 3) and *upper* (1, 13: 15, 4), since 12 <13 <15 <19, so *spectrum* is the ancestor of *upper*, further, the difference between two nodes' levels is 1, so we can defer that *spectrum* is the parent node of *upper*. On the contrary, there is no structural relationship between *spectrum* (1, 12: 19, 3) and *50* (1, 10, 4), due to that 10 is not covered by (12, 19).

Here, we use functions to represent operations in TwigStack.

(1) *encodeNode(n)*: encode node n with region code;
(2) *decom(xq)*: decompose XPath query xq into paths;
(3) *queryPath(p)*: search the semi-structure data by path p;
(4) *mergePath(R)*: merge temporary result set R into final result.

4   SSSH Model

In this section, we present HBase Semi-Structured Spatio-Temporal (HSSST) model. First, we address how to store and index spatio-temporal data in HBase, and then we add semi-structure information into the model.

4.1   Spatio-Temporal Storage and Index

Usually, an HBase cluster is composed of at least one administrative server, called *Master*, and several other servers holding data, called *RegionServers*. Physically, a table in HBase is horizontally partitioned along rows into several regions, each of which is maintained by exactly one *RegionServer*. In *RegionServer*, data are organized into list-like structure called *StoreFile*, where each entry is pre-configured with the same fixed size (usually 64 KB) and the size of a certain number of entries is equal to that of the block of the underlying storage system. The catalog table *meta* stores the relation {[table name]:[start row key]:[region id]:[region server]}, thus given a row key, the corresponding *RegionServer* can be found, and then the *RegionServer* searches the *value* locally on *StoreFile*.

Base on the above description, we can use the *meta*'s function to index data. We use Hilbert curve to partition the whole space as the initial granularity. According to the design rationale of HBase, the prefix of row key should be different so that the overhead of inserting data could be distributed over *RegionServers*. And such design is able to satisfy this demand. Hilbert curve is a kind of space filling curve which maps multi-dimensional space into one-dimensional space. In particular, the whole space is partitioned into equal-size cells and then a curve is passed through each cell for only once in term of some sequence, so that every cell is assigned a sequence number. Different space filling curves are distinguished by different sequencing methods. Due to information loss in the transformation, different space filling curves are evaluated by the criteria, locality preservation, meaning that how much the change of proximities is from original space to one-dimensional space. Hilbert curve is proved to be the best locality preserved space filling curve [2]. With Hilbert curve, any object in the original space is transformed into $[0, 2^{2\lambda} - 1]$ space, where λ is called the order of Hilbert curve. Figure 3 shows four Hilbert curves in two-dimensional space with $\lambda = 1, 2, 3$ and 4.

Based on above descriptions, we use Hilbert cell value as row key in the meta table to index spatio-temporal data as first level, thus, each record can be placed into the corresponding region according to Hilbert value of spatial part of the record. In particular, the following mapping structure is built in the meta table (for simplicity, table name is omitted): {[start Hilbert cell, end Hilbert

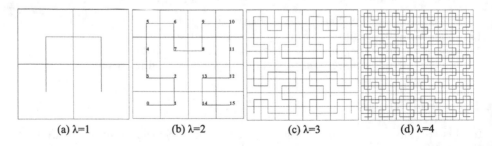

(a) λ=1   (b) λ=2   (c) λ=3   (d) λ=4

Fig. 3. Hilbert curves

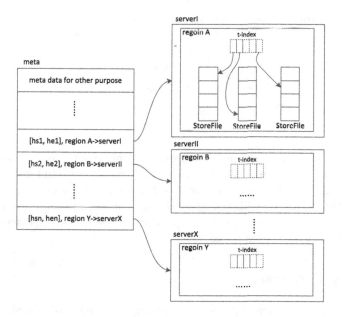

Fig. 4. Overview of STEHIX

cell]:[region id]:[region server]}. Initially, assuming there are N regions across M *RegionServer*s, we can uniformly assign Hilbert cells to these regions, for instance, the first entry could be $\{[0, ((2^{2\lambda} - 1)/N) - 1] : regionA : serverI\}$, and the second $\{[((2^{2\lambda} - 1)/N), (2 * (2^{2\lambda} - 1)/N) - 1] : regionB : serverII\}$.

For retrieving local data efficiently, we design *t-index* to index *StoreFile* by temporal dimension. In particular, *t-index* is a list-like in-memory structure, each entry of which points to a list of addresses referring to key-value data in the *StoreFile*. For building *t-index*, we use a period T to bound the length of the list of *t-index*, and such consideration is based on the fact that there may be some cycle for the spatial change of objects. The period T is divided into several segments, each of which is corresponding to an entry in *t-index*. Each entry points to a list of addresses referring to key-value data in *StoreFile*, whose temporal component modulo T lies in the segment. Figure 4 shows an overview of spatio-temporal storage and indexing architecture.

4.2 Semi-structure Data Storage

We encode semi-structure information with region code and combine the code with spatio-temporal dimensions into HBase. In particular, firstly, we use function *encodeNode()* to encode each node in the tree, e.g., tn_1 is encoded into ec_1, and we add tn_1 into HBase as column qualifier, and ec_1 as the value for the corresponding cell.

Table 1. Logical view for semi-structured data storage

	hssst				
	space	time	tn_1	tn_2	...
rk_1	(x_l, y_l, x_u, y_u)	(t_s, t_e)	ec_1	ec_2	...
rk_2	...	...	...	...	...

Table 1 illustrates an example of storing semi-structured data into HBase combined with spatio-temporal dimensions, where rk_1, rk_2 and etc. are Hilbert values indexed by *meta* structure, and *hssst* is used as column family, *space* and *time* these two column qualifiers are spatial dimension and temporal dimension, respectively, and the following is the tree nodes in the semi-structured data, and their cells' values are region codes, accordingly.

5   Query Processing

In this section, we present the query processing for range query and kNN query for spatio-temporal semi-structured data in HBase.

5.1   Range Query

For a range query $(t_{qs}, t_{qe}, c, r, xq)$, basic work flow is described as follows, first, using Hilbert curve, a set of one-dimensional intervals I_q is calculated by intersecting spatial predicate (c, r) with Hilbert cells, then according to mapping relation in *meta* structure, the involved *RegionServer*s are informed to search the corresponding regions locally, utilized by *t-index*. After locating the corresponding *StoreFile*, the entry of the *StoreFile* is fetched and TwigStack algorithm is used to inspect whether the document or object is satisfied the XPath query.

Algorithm 1 describes range query processing for HSSST. In line 1, spatial range (c, r) is regarded as parameter to retrieve the intersected Hilbert cell set I_q, and the temporal predicate is converted into $[0, T]$ interval in line 2. In line 3, function $findRegions()$ finds the involved regions which intersect with I_q. From line 4 to 14, each corresponding *t-index* is inspected to retrieve *tempEntrySet* of *StoreFile*, and then xq is decomposed into paths (line 6), and for each path p, XPath query execution is carried out to find candidate $xFile$ (line 8), if $xFile$ is satisfied with query predicates, it is added into $xFileSet$ (line 9 and 10), in line 13, function $mergePath()$ is invoked to form the final results.

5.2 kNN Query

For a kNN query $(t_{qs}, t_{qe}, q, k, xq)$, basic idea is, proximity objects of point q are constantly, incrementally retrieved until k results are found. In particular, first, Hilbert cell h containing point q is located, then the corresponding *t-index* is utilized to retrieve all records lie in h, meanwhile, neighbor cells of h are also

Algorithm 1. Range Query Processing

Input:

   $(t_{qs}, t_{qe}, c, r, xq)$

Output:

   $xFileSet$ //result list

1: $I_q = toIntervals(c, r)$
2: $key_{qs} = t_{qs} \bmod T$, $\ key_{qe} = t_{qe} \bmod T$
3: $Regions = findRegions(I_q)$
   /*the following processing is executed separately in each $region$*/
4: **for** each $region \in Regions$ **do**
5:    $tempEntrySet = t\text{-}index.searchIndex(key_{qs}, key_{qe})$
6:    $P = decom(xq)$
7:    **for** each $p \in P$ **do**
8:       $xFile = queryPath(p, tempEntrySet)$
9:       **if** isSatisfied($xFile, t_{qs}, t_{qe}, c, r$) **then**
10:          $xFileSet \leftarrow xFile$
11:       **end if**
12:    **end for**
13:    $xFileSet = mergePath(xFileSet)$
14: **end for**
15: **return** $xFileSet$

retrieved, and these records and Hilbert cells are all enqueued into a priority queue where priority metric is the distance from q to record or Hilbert cell. Then top element is constantly dequeued and inspected by TwigStack algorithm, either being added to result list or being followed to retrieve neighbor cells to be enqueued, until k results are found.

Algorithm 2 presents kNN query processing. The first line initializes a priority queue PQ where each element is ordered by the distance from q to the element. The element can be Hilbert cell or semi-structured document or object (we use document in the algorithm), and if it is a Hilbert cell, the distance is $MINDIST$ [6], other wise, the distance is the Euclidean distance from q to geo-location of the document. In line 2, the Hilbert cell containing q is gained, and is enqueued in line 3. From line 4, the procedure constantly retrieves top element e from PQ (line 5) and processes it, in particular, if e is a Hilbert cell (line 6), find the corresponding region rg from the meta table (line 7), and then the corresponding $t\text{-}index$ is searched to retrieve all the documents satisfying temporal predicate (line 8), which are enqueued into PQ (line 9 to 11), after that, the neighbor cells of e are obtained and enqueued into PQ (line 12 to 15); other wise, i.e., if e is a document (line 16), and further e is inspected whether satisfying xq predicate (line 17), if it does, which means e is a result, e is added into $Qlist$ (line 18), and the above procedure is looped until the size of $Qlist$ reaches k (line 19 to 21).

Algorithm 2. kNN Query Processing

Input:

$(t_{qs}, t_{qe}, q, k, xq)$

Output:

$Qlist$ //result list

1: $PQ=null$ //initial a priority queue
2: $h=coorToCell(q)$
3: $PQ.enqueue(h, MINDIST(q, h))$
4: **while** $PQ \neq \phi$ **do**
5:      $e=PQ.dequeue()$
6:      **if** e is typeof cell **then**
7:          $rg=findRegions(e)$
8:          $RS=rg.findDocuments(e, (t_{qs}, t_{qe}))$
9:          **for** each $doc \in RS$ **do**
10:             $PQ.enqueue(doc, dist(q, doc))$
11:         **end for**
12:         $CellSet=getNeighborCells(e.center)$
13:         **for** each $cell \in CellSet$ **do**
14:             $PQ.enqueue(cell, MINDIST(q, cell))$
15:         **end for**
16:     **else if** e is typeof document **then**
17:         **if** $isSatisfied(e, xq)$ **then**
18:             $Qlist \leftarrow e$
19:             **if** $Qlist.size()=k$ **then**
20:                 **return** $Qlist$
21:             **end if**
22:         **end if**
23:     **end if**
24: **end while**

---

6   Experimental Evaluation

We implement the algorithms and run them on real remote sensing dataset, which consists of meteorologic data, oceanic data, imagery data and etc., and contains about 1 million spatio-temporal semi-structured meta-data (document).

Our algorithms are implemented in Hadoop 2.5.1 and HBase 0.98.6, and run on a cluster with size varied from 3 to 11, in which each node is equipped with Intel(R) Core(TM) i3 CPU @ 2.40 GHz, 2 GB main memory, and 500 GB storage, and operating system is CentOS release 6.5 64 bit, and network bandwidth is 10 Mbps. For comparison, we choose MongoDB, which is a document-oriented database, it requires that most of the index is loaded in memory, thus accelerating retrieval speed, however, such requirement increase the cost for hardware, and in real application, it would cost more money than HBase case. And the setting for MongoDB is Intel Xeon 3.60 GHz * 4 CPU, 32 G memory. According to MongoDB design, each semi-structured documents are stored in MongoDB as a document, and we use its own index structure to handle spatio-temporal query associated with XPath query.

6.1 Range Query

For evaluating range query, we first vary query *selectivity* which is defined as the product of spatial range, temporal range and XPath's selectivity over the whole spatio-temporal range. A large *selectivity* means to select more documents. When *selectivity* is increased, we can see from the results (Fig. 5(a)), the responding time is also increased, which can be explained that a larger selectivity involves more documents to be inspected and thus costs more time. For detail, the performances of HSSST and MongoDB are similar, this is because that, the indexes of two are tree-like structures, while the configuration for MongoDB is more higher than HSSST, thus we can see our algorithm is efficient and effective.

And then, we vary HBase cluster number, Fig. 5(b) shows the results. Amazingly, our HSSST outperforms MongoDB when the number of cluster nodes is increased.

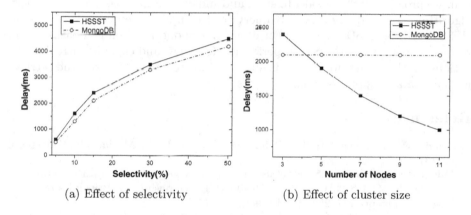

(a) Effect of selectivity   (b) Effect of cluster size

Fig. 5. Experimental results for range queries

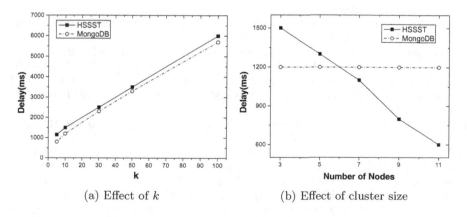

(a) Effect of k (b) Effect of cluster size

Fig. 6. Experimental results for kNN queries

6.2   kNN Query

Next, we run kNN query to evaluate our algorithm. Figure 6 shows the results. When k is increased, both responding time are increased, which can be explained that a larger k would cause more cells and documents to be enqueued and checked. Similarly, their performances are close to each other. And then we very the cluster number for HBase, and the results prove that our method is capable and efficient.

7   Conclusion

With further application of semi-structured data, more and more people will pay attention to retrieve semi-structured spatio-temporal objects. This paper originally proposes query problem on semi-structured spatio-temporal data in HBase, which is a new challenge to database community. To address this problem, we propose HSSST model to store and index data in HBase, and two query algorithms, range query and kNN query to accomplish the query. We compare HSSST with MongoDB, a document-orient no-sql database, on real dataset, and results show that our HSSST outperforms MongoDB and capable for real applications. In the future, we plan to utilize this idea to efficiently store and retrieve graph data and apply to social networks.

References

1. Chodorow, K.: MongoDB: The Definitive Guide. O'Reilly Media, Inc., Sebastopol (2013)
2. Faloutsos, C., Roseman, S.: Fractals for secondary key retrieval. In: Proceedings of the Eighth ACM SIGACT-SIGMOD-SIGART Symposium on Principles of Database Systems, pp. 247–252. ACM (1989)
3. Han, D., Stroulia, E.: Hgrid: a data model for large geospatial data sets in hbase. In: 2013 IEEE Sixth International Conference on Cloud Computing (CLOUD), pp. 910–917. IEEE (2013)
4. Hsu, Y.T., Pan, Y.C., Wei, L.Y., Peng, W.C., Lee, W.C.: Key formulation schemes for spatial index in cloud data managements. In: 2012 IEEE 13th International Conference on Mobile Data Management (MDM), pp. 21–26. IEEE (2012)
5. Nishimura, S., Das, S., Agrawal, D., Abbadi, A.E.: Md-hbase: a scalable multi-dimensional data infrastructure for location aware services. In: 2011 12th IEEE International Conference on Mobile Data Management (MDM), vol. 1, pp. 7–16. IEEE (2011)
6. Roussopoulos, N., Kelley, S., Vincent, F.: Nearest neighbor queries. In: ACM Sigmod Record, vol. 24, pp. 71–79. ACM (1995)
7. Vahdati, S., Karim, F., Huang, J.-Y., Lange, C.: Mapping large scale research metadata to linked data: a performance comparison of HBase, CSV and XML. In: Garoufallou, E., Hartley, R.J., Gaitanou, P. (eds.) MTSR 2015. CCIS, vol. 544, pp. 261–273. Springer, Heidelberg (2015). doi:10.1007/978-3-319-24129-6_23
8. Zhang, N., Zheng, G., Chen, H., Chen, J., Chen, X.: Hbasespatial: a scalable spatial data storage based on hbase. In: 2014 IEEE 13th International Conference on Trust, Security and Privacy in Computing and Communications (TrustCom), pp. 644–651. IEEE (2014)

Efficient Approximation of Well-Designed SPARQL Queries

Zhenyu Song[1,2], Zhiyong Feng[1,2], Xiaowang Zhang[1,2]([✉]), Xin Wang[1,2],
and Guozheng Rao[1,2]

[1] School of Computer Science and Technology, Tianjin University,
Tianjin 300350, China
[2] Tianjin Key Laboratory of Cognitive Computing and Application,
Tianjin 300350, China
xiaowangzhang@tju.edu.cn

Abstract. Query response time often influences user experience in the real world. However, the time of answering a SPARQL query with its all exact solutions in large scale RDF dataset possibly exceeds users' tolerable waiting time, especially when it contains the OPT operations since the OPT operation is the least conventional operator in SPARQL. So it becomes essential to make a trade-off between the query response time and the accuracy of their solutions. That is, partial answers can be provided for users to reduce the response query time within their tolerable waiting time. In this paper, based on the depth of the OPT operation occurring in a query, we propose an approach to obtain its all approximate queries with less depth of the OPT operation. Although queries are approximated in this method, it remains the "non-optional" query patterns from users. This paper mainly discusses those queries with well-designed patterns since the OPT operation in a well-designed pattern is really "optional". We remove "optional" triple patterns with less depth of the OPT operation and then obtain approximate queries with different depths of the OPT operation. Furthermore, we evaluate the approximate query efficiency and solutions precision with the degree of approximation. It shows that users can keep the balance between query efficiency and solutions precision by changing the degree of approximation.

Keywords: RDF · SPARQL · Well-designed patterns · Approximate queries

1 Introduction

Currently, there is renewed interest in the classical topic of graph databases [13]. Much of this interest has been sparked by SPARQL: the query language for RDF. Resource Description Framework (RDF) [7] is the standard data model in the Semantic Web. RDF describes the relationship of entities or resources using directed label graph. RDF has a broad range of applications in the Semantic Web, social network, bio-informatics, geographical data, etc [15]. An example in Table 1 has been given to describe the entities of *Jon Smith* and *Liz Ben*.

© Springer International Publishing AG 2016
S. Song and Y. Tong (Eds.): WAIM 2016 Workshops, LNCS 9998, pp. 315–327, 2016.
DOI: 10.1007/978-3-319-47121-1_27

Table 1. professor.rdf

Jon Smith	workFor	Semantic University
Jon Smith	teachOf	Liz Ben
Jon Smith	rdf:type	professor
Liz Ben	rdf:type	master
Liz Ben	advisor	Jon Smith
Liz Ben	takesCourse	Ontology

For example, in the first line, it describes that the person *Jon smith* works for *Semantic University*. SPARQL [11] recommended by W3C has become the standard language for querying RDF data since 2008 by inheriting classical relational languages such as SQL.

In the process of information retrieval, users' tolerable waiting time is limited [9]. Users also might have tolerable waiting time for querying RDF data. For a SPARQL query, if it contains the OPT operation, it will take much time to query optional pattern in SPARQL since OPT is the least conventional operator in AND, OPT, FILTER, SELECT and UNION [14]. It has been shown in [10] that the complexity of SPARQL query evaluation raises from PTIME-membership for the conjunctive fragment to PSPACE-completeness when OPT operation is considered. So it is important to make a trade-off between query response time and accuracy of solutions, which is a traditional topic in databases [1]. Since it is hard to obtain all exact solutions of a SPARQL query in a fixed time, a natural idea to reduce the response time of SPARQL query is by removing some "optional" parts of this query (i.e., occurrences of the OPT operator). Moreover, we still expect to preserve its "non-optional" part of this query. For instance, consider a pattern Q as follows:

$$Q = ((?x, rdf\text{:}type, professor) \ \text{OPT} \ ((?x, workFor, ?y) \ \text{OPT} \ (?x, teachOf, ?z))).$$

Here $(?x, rdf : type, professor)$ is a "non-optional" pattern in this query while both $(?x, workFor, ?y)$ and $(?x, teachOf, ?z))$ are "optional" patterns. Based on this natural idea, there are three possible new patterns with less OPT operators as follows:

- $Q_1 = (?x, rdf : type, professor)$;
- $Q_2 = ((?x, rdf : type, professor) \ \text{OPT} \ (?x, workFor, ?y))$;
- $Q_3 = ((?x, rdf : type, professor) \ \text{OPT} \ (?x, teachOf, ?z))$.

Clearly, we can find that Q_1 and Q_2 are ideal candidates which contain less optional patterns with protecting "non-optional" patterns while Q_3 is not since $(?x, teachOf, ?z)$ directly depends on $(?x, workFor, ?y)$.

In 2015, Barceló, Pichler, and Skritek [2] proposed the notion of approximation (for short, BPS's *approximation*) to characterize "*partial answer*", that is, an answer can be extended to a "*maximal answer*" (i.e., exact answer) of a

SPARQL query represented in well-designed pattern trees [8]. In this sense, the evaluation problems of Q_1 and Q_2 are taken as the partial evaluation problems of Q. However, we investigated that the BPS's approximation did not provide a fine-grained classification between Q_1 and Q_2. For users, they can't judge which one will lead to less query response time within tolerable waiting time.

In this paper, based on the depth of OPT operation occurring in a query, we propose an approach to obtain its all approximate queries with less depth of the OPT operation. We mainly consider the fragment of UNION-free well-designed SPARQL patterns where the OPT operator is a really "optional" operation [10] in characterizing a weak monotonicity [6]. Besides, the UNION-free well-designed SPARQL fragment is indeed maximal among all fragments of LSQ [4] - a linked dataset describing SPARQL queries extracted from the logs of public SPARQL endpoints in our real world [12]. For simplification, we directly call well-designed patterns instead of UNION-free well-designed SPARQL patterns. The main contributions of this paper can be summarized as follows:

- Firstly, we provide the conception of OPT-depth. For a well-designed pattern in *OPT normal form*, its OPT-depth can describe the depth of OPT operation occurring in this pattern. Our approximation method is proposed based on the *OPT normal form* via OPT-depth.
- Secondly, we treat a well-designed pattern in *OPT normal form* as a well-designed tree, whose inner nodes are labeled by OPT operation. We apply our approximation method by removing "optional" subtrees of a well-designed tree.
- Finally, through comparison with the non-approximate queries on LUBM dataset, the approximate queries lead to better performance.

The rest of this paper is organized as follows: Sect. 2 briefly introduces the SPARQL and conception of well-designed patterns. Section 3 defines the k-approximation queries. Section 4 presents the well-designed tree to capture k-approximation queries and Sect. 5 evaluates experimental results. Finally, Sect. 6 summarizes the paper.

2 Preliminaries

In this section, we introduce RDF and SPARQL patterns and well-designed patterns [10].

2.1 RDF

Let I, B and L be infinite sets of *IRIs*, *blank nodes* and *literals*, respectively. These three sets are pairwise disjoint. We denote the union $I \cup B \cup L$ by U, and elements of $I \cup L$ will be referred to as *constants*.

A triple $(s, p, o) \in (I \cup B) \times I \times (I \cup B \cup L)$ is called an *RDF triple*. An *RDF graph* is a finite set of RDF triples.

2.2   SPARQL Patterns

Assume furthermore an infinite set V of *variables*, disjoint from U. The convention is to write variables starting with the character '?'.

SPARQL *patterns* are inductively defined as follows.

- Any triple from $(I \cup L \cup V) \times (I \cup V) \times (I \cup L \cup V)$ is a pattern (called a *triple pattern*). A Basic Graph Pattern (BGP) is a set of triple patterns.
- If P_1 and P_2 are patterns, then so are the following: P_1 UNION P_2, P_1 AND P_2 and P_1 OPT P_2.
- If P is a pattern and S is a finite set of variables then SELECT$_S(P)$ is a pattern.
- If P is a pattern and C is a constraint (defined next), then P FILTER C is a pattern; we call C the *filter condition*. Here, a *constraint* is a boolean combination of *atomic constraints*.

2.3   Well-Designed Patterns

The notion of well-designed patterns is introduced to characterize the *weak monotonicity* [10].

A UNION-free pattern P is well-designed if the followings hold:

- P is safe, that is, each subpattern of the form Q FILTER C of P holds the condition: $var(C) \subseteq var(Q)$.
- for every subpattern $P' = (P_1 \text{OPT} P_2)$ of P and for every variable ?x occurring in P, the following condition holds: If ?x occurs both inside P_2 and outside P', then it also occurs in P_1.

For instance, the pattern Q in Sect. 1 is a well-designed pattern. However, consider the pattern $(((?x, p, ?y) \text{ OPT } (?y, q, ?z)) \text{ OPT } (?x, r, ?z))$, it is not a well-designed pattern since ?z occurs in both $(?y, q, ?z)$ and $(?x, r, ?z)$ but ?z does not occur in $(?x, p, ?y)$.

Note that the OPT operation provides really optional left-outer join due to the weak monotonicity [10], which is an important property to characterize the satisfiability of SPARQL [16]. For instance, consider the pattern Q in Sect. 1, $(?x, workFor, ?y)$ and $(?x, teachOf, ?z)$ are freely optional.

3   Approximate Queries

In this section, we introduce our approximation method in the *OPT normal form*.

3.1   OPT Normal Form

A UNION-free pattern P is in *OPT normal form* [10] if P meets one of the following two conditions:

- P is constructed by using only the AND and FILTER operators;
- $P = (P_1 \text{ OPT } P_2)$ where P_1 and P_2 patterns are in OPT normal form.

For instance, the pattern Q stated in Sect. 1 is in OPT normal form. However, consider the pattern $(((?x, p, ?y) \text{ OPT } (?x, q, ?z)) \text{ AND } (?x, r, ?z))$ is not in OPT normal form.

Note that all patterns in OPT normal form have the following form[1]:

$$P_0 \text{ OPT } P_1 \text{ OPT} \ldots \text{OPT } P_m; \tag{1}$$

where P_0 is an OPT-free pattern, that is, P_0 contains only AND and FILTER operations (called AF-pattern). In this sense, we use $BGP(P)$ to denote P_0 and $\mathcal{O}(P)$ to denote $\{P_1, \ldots, P_m\}$, i.e., the collection of optional patterns occurring in P.

Proposition 1 [10, Theorem 4.11]. *For every UNION-free well-designed pattern P, there exists a pattern Q in OPT normal form such that P and Q are equivalent.*

In the proof of Proposition 1, we apply three rewriting rules based on the following equations: let P, Q, R be patterns and C a constraint,

- $(P \text{ OPT } R) \text{ FILTER } C \equiv (P \text{ FILTER } C) \text{ OPT } R;$
- $(P \text{ OPT } R) \text{ AND } Q \equiv (P \text{ AND } Q) \text{ OPT } R;$
- $P \text{ AND } (Q \text{ OPT } R) \equiv (P \text{ AND } Q) \text{ OPT } R.$

Since each UNION-free well-designed pattern is equivalent to a pattern in OPT normal form by Proposition 1, we mainly consider all well-designed patterns in OPT normal form in the following.

To further observe some features of patterns in OPT normal form, we consider a complicated pattern P, where the OPT operation is deeply nested, as follows:

$$P = (t_1 \text{ OPT } (t_2 \text{ OPT } t_3)) \text{ OPT } (t_4 \text{ OPT } t_5). \tag{2}$$

Note that, in P, t_1 is non-optional while t_2, t_3, t_4 and t_5 are optional. Furthermore, if we consider the subpattern $(t_2 \text{ OPT } t_3), t_2$ is non-optional while t_3 is still optional. Analogously, if we consider the subpattern $(t_4 \text{ OPT } t_5), t_4$ is non-optional while t_5 is still optional. Now, if we observe the figure of P shown in Fig. 1, t_2 and t_4 are on top of t_3 and t_4, respectively.

3.2   OPT-depth in OPT Normal Form

To characterize the different levels of optional patterns, we define $OPT\text{-}depth$ of patterns in OPT normal form.

---

[1] We abbreviate $((P_0 \text{ OPT } P_1) \text{ OPT} \ldots \text{OPT } P_m)$ as $P_0 \text{ OPT } P_1 \text{ OPT} \ldots \text{OPT } P_m$.

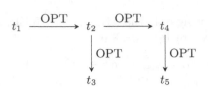

Fig. 1. The figure of OPT normal form

Definition 1 (OPT-depth). Let P be a pattern in OPT normal form. We use $dep(P)$ to denote its *OPT-depth* as follows:

- $dep(P) = 0$ if P is an AF-pattern;
- $dep(P) = \max\{dep(P_1), \ldots, dep(P_m)\} + 1$ if $\mathcal{O}(P) = \{P_1, \ldots, P_m\}$.

For instance, the OPT-depth of the pattern Q stated in Sect. 1 and the pattern P in Eq. (2) is 2.

3.3 Approximate Queries

To define our approximate queries, we introduce an important notion called *reduction* [10].

We say that a pattern P' is a *reduction* of a pattern P, if P' can be obtained from P by replacing subpattern $(P_1 \text{ OPT } P_2)$ with P_1, that is, P' is obtained by deleting some optional parts of P. The reflexive and transitive closure of the reduction relation is denoted by $\trianglelefteq$. In this sense, for a pattern, its reductions can be taken as "inexact" patterns, which can be obtained by reducing the OPT operation. For instance, in Sect. 1, Q_1 and Q_2 are reductions of Q.

Inspired from the notion of reduction, we introduce our *k-approximate patterns*.

Definition 2 (k-approximation). Let P be a pattern in OPT normal form $(P_0 \text{ OPT } P_1 \text{ OPT} \ldots \text{OPT } P_m)$ and k be a natural number. The *k-approximate pattern* of P (written as $P^{(k)}$) can be obtained in the following inductive way:

- $P^{(k)} = BGP(P)$ if $k = 0$;
- $P^{(k)} = P_0 \text{ OPT } P_1^{(k-1)} \text{ OPT} \ldots \text{OPT } P_m^{(k-1)}$ if $1 \leq k \leq dep(P) - 1$;
- $P^{(k)} = P$ if $k \geq dep(P)$.

Intuitively, approximate patterns are subpatterns obtained by reducing their OPT-depths. For a well-designed query Q and its k-approximation query $Q^{(k)}$, $Q^{(k)}$ is more closed to Q with higher value of k. In this sense, our approximation generalizes reduction [10] in a fine-grained way. Since there exists the unique OPT-depth for each OPT in OPT normal form, we have the following proposition:

Proposition 2. *Let P be a pattern in OPT normal form and k be a natural number. $P^{(k)}$ exists and $P^{(k)}$ is unique.*

For instance, in Sect. 1, $Q^{(0)} = Q_1$ and $Q^{(1)} = Q_2$. In Eq. (2), $P^{(0)} = t_1$ and $P^{(1)} = ((t_1 \text{ OPT } t_2) \text{ OPT } t_4)$. $Q^{(0)}$ and $Q^{(1)}$ are the reductions of Q. Analogously, $P^{(0)}$ and $P^{(1)}$ are the reductions of P.

4 K-Approximation Computation

In this section, we propose a method to compute all approximate patterns based on a redesigned parse tree called *well-designed tree*.

Now, we introduce the notion of *well-designed tree*.

Definition 3 (well-designed tree). Let P be a well-designed pattern in OPT normal form. A well-designed tree T based on P is a redesigned parse tree, which can be defined as follows:

- All inner nodes in T are labeled by OPT operations and leaf nodes are labeled by AF-patterns.
- For each subpattern $(P_1 \text{ OPT } P_2)$ of P, the well-designed tree T_1 of P_1 and the well-designed tree T_2 of P_2 have the same parent node.

For instance, given a pattern P^2 in OPT normal form,

$$P = (((((t_1 \text{ AND } t_3) \text{ FILTER } C) \text{ OPT}_2 \ t_2) \text{ OPT}_1$$
$$((t_4 \text{ OPT}_4 \ t_5) \text{ OPT}_5 (t_6 \text{ OPT}_6 \ t_7))).$$

We write $((t_1 \text{ AND } t_3) \text{ FILTER } C)$ as p_0 for short, which is the non-optional part of P. The well-designed tree T is shown in Fig. 2. Some pruning strategies can be applied to the well-designed tree to achieve k-approximation. After removing optional subtrees from the well-designed tree, we get a k-approximation spanning tree (KST for short) which is also a well-designed tree. We denote a k-approximation spanning tree from well-designed tree T as $KST_T^{(k)}$. In order to obtain $KST_T^{(k)}$, we define a special traversal method for the well-designed tree based on the conception of OPT-depth, called *Left-Deep Level Traversal*. Before defining *Left-Deep Level Traversal*, we provide a partial traversal approach called *Leftmost Traversal*.

For a well-designed tree, *Leftmost Traversal* of this tree is by only traversing the left subtree after visiting root node. For instance, consider T in Fig. 2, the leftmost traversal of T is denoted by $LT(T) = \{\text{OPT}_1, \text{OPT}_2, p_0\}$. *Left-Deep Level Traversal* of the well-designed tree is proposed as follows:

Definition 4 (left-deep level traversal). Let T be a well-designed tree. Left-Deep Level Traversal denoted by $LD(T)$ is composed of levels. $level(i)$ can be obtained by leftmost traversing each node's right children node (called candidate) in $level(i-1)$. Especially, $level(0) = LT(T)$.

---

[2] We give each OPT operator a subscript to differentiate them so that readers understand clearly.

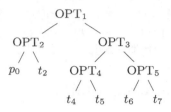

Fig. 2. Well-designed tree

For each subtree t in the well-designed tree, the leftmost leaf node written as $LM(t)$ is the non-optional part of t. For instance, for the well-designed tree T in Fig. 2, $LM(T) = \{p_0\}$. We construct $KST_T^{(k)}$ by removing the subtrees below $level(k-1)$ from T. Particularly, $KST_T^{(0)}$ can be built by returning $LM(T)$.

In the process of building $KST_T^{(k)}$, firstly we compute each node's candidate in $level(k-1)$. Secondly we obtain the $LM(n)$ for each OPT node n in $level(k-1)$. Finally $KST_T^{(k)}$ can be constructed by replacing the leftmost nodes with corresponding OPT nodes in T. We obtain the k-approximation query through traversing on $KST_T^{(k)}$. The process of building $KST_T^{(k)}$ is described in Algorithm 1.

Algorithm 1. K-approximation Spanning Tree

Input: Well-designed tree T from pattern P, Leftmost list $leftmost$, k-approximation
    with k
1: Initialize Candidate $candidate$ with $T, i \leftarrow 0$
Output: K-approximation Spanning Tree
2: **if** $k = 0$ **then return** $LM(T)$
3: **else if** $k \geq dep(P)$ **then return** T
4: **else**
5:     **while** $i \neq k$ **do**
6:         $level(i) \leftarrow LT(candidate)$
7:         $candidate \leftarrow GetCandidate(level(i))$
8:         **for** each $node$ in $candidate$ **do**
9:             **if** $node$ is OPT **then**
10:                $leftmost \leftarrow LM(node)$
11:             **end if**
12:         **end for**
13:     **end while**
14:     Replace nodes in $leftmost$ with corresponding OPT nodes in T. **return** T
15: **end if**

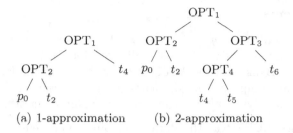

(a) 1-approximation   (b) 2-approximation

Fig. 3. Approximation spanning tree (a) 1-approximation, (b) 2-approximation

Example 1. Consider the well-designed tree T in Fig. 2 from pattern P. The $LD(T)$[3] with candidates and leftmost list can be described as follows:

Level	Traversal list	Candidates	Leftmost list
0	OPT_1, OPT_2, p_0	OPT_3, t_2	$t_4, \times$
1	OPT_3, OPT_4, t_4, t_2	OPT_5, t_5	$t_6, \times$
2	OPT_5, t_6, t_5	t_7	$\times$
3	t_7		

In $KST_T^{(0)}$, p_0 is set as the root node without any child node. If we want to obtain $KST_T^{(1)}$, we can replace t_4 with OPT_3 in T based on $level(0)$. Analogously, $KST_T^{(2)}$ can be obtained by replacing t_6 with OPT_5 in T based on $level(1)$. Since $dep(P) = 3$, $KST_T^{(3)}$ is regarded as T itself. Both $KST_T^{(1)}$ and $KST_T^{(2)}$ are shown in Fig. 3.

$P^{(1)}$ and $P^{(2)}$ are shown as follows:

$$P^{(1)} = ((((t_1 \ AND \ t_3) \ FILTER \ C) \ OPT_2 \ t_2) \ OPT_1 \ t_4),$$

$$P^{(2)} = ((((t_1 \ AND \ t_3) \ FILTER \ C) \ OPT_2 \ t_2) \ OPT_1 \ ((t_4 \ OPT_4 \ t_5) \ OPT_5 \ t_6)).$$

5 Experiments and Evaluations

This section presents our experiments. The purpose of the experiments is to evaluate (1) the performance improvement of approximate well-designed SPARQL queries, and (2) the appropriate k to reduce the users' waiting time for solutions.

---

[3] We use $\times$ to denote that for each non-OPT node n in candidates, there exists no corresponding $LM(n)$ in leftmost list.

5.1 Experiments

Implementations and running environment. All experiments were carried out on a machine running Linux, which has one CPU with four cores of 2.40 GHz, 32 GB memory and 500 GB disk storage. All of the algorithms were implemented in Java with Eclipse as our compiler. Jena [5] (Jena-3.0.1) and Sesame [3] (Sesame-4.1.1) are used as the underlying query engines of approximate queries.

Dataset. We used LUBM[4] as the dataset in our experiments to look for the relationship between approximate query efficiency and k. LUBM, which features an ontology for the university domain, is a standard benchmark to evaluate the performance of Semantic Web repositories, In our experiments, we used LUBM1, LUBM5 and LUBM10 as query datasets.

SPARQL queries. The queries over LUBM were designed as three forms in Appendix A. Obviously, OPT nesting in Q_3 is the most complex among three forms. Furthermore, we built AND and FILTER operations in each query. All of query patterns have k ranging from 0 to 4. Specially, since $dep(Q_1)$ is 1, we regard k-approximate query as Q_2 itself when $k > 1$.

Solutions precision. After approximating well-designed SPARQL queries, different k leads to different solution precision which can be reflected by the amount of variables in queries and the amount of solutions. Here we show the amount of variables in k-approximation queries in Table 2 and the amount of queries solutions over LUBM10 dataset in Table 3.

5.2 Efficiency of Approximate Queries

The variation tendencies of query response time shown in Figs. 4 and 5 are similar. Query efficiency is promoted with lower response time when k is decreasing

Table 2. Amount of variables after approximation

k	k = 0	k = 1	k = 2	k = 3	k =4
Q_1	2	6	6	6	6
Q_2	2	3	4	5	6
Q_3	2	5	11	15	16

Table 3. Amount of solutions after approximation over LUBM10

k	k = 0	k = 1	k = 2	k = 3	k = 4
Q_1	1602	4797	4797	4797	4797
Q_2	1602	1602	1602	4797	4797
Q_3	1602	38478438	228797712	228797712	228797712

---

[4] http://swat.cse.lehigh.edu/projects/lubm/.

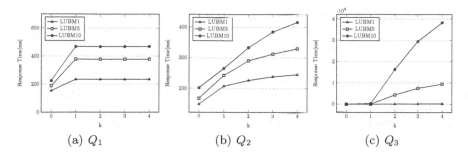

Fig. 4. K-approximation on Jena (a) Q_1, (b) Q_2, (c) Q_3

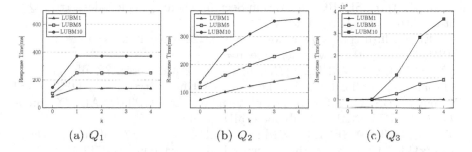

Fig. 5. K-approximation on Sesame (a) Q_1, (b) Q_2, (c) Q_3

(approximation degree becomes larger). Furthermore, there has been a significant increase in query efficiency when the dataset scale grows up. For instance, we observe Q_3, which corresponds to a full well-designed tree. When the dataset is LUBM10, its query response time is more than an hour implemented by Jena and Sesame without any approximation ($Q_3^{(4)}$). Approximate queries can efficiently reduce the query response time and users' waiting time. We assume that Q_3 on LUBM10 comes from users, we can conclude that solutions of $Q_3^{(3)}$ with 15 variables save 25 % query response time than solutions of $Q_3^{(4)}$ with 16 variables. Furthermore, there is no difference between the amount of solutions in $Q_3^{(3)}$ and $Q_3^{(4)}$. In this scene, it can be approximated as $Q_3^{(3)}$ with inexact solutions to improve user experience.

6   Conclusion

In this paper, we have presented the approximation of well-designed SPARQL patterns in OPT normal form based on the depth of OPT operation. Theoretically, our proposal k-approximation generalizes reductions of patterns in a fine-grained way. The k-approximation provides rich and various approximate queries to answer user's query within a fixed time. Approximate queries always remain the non-optional query pattern from users. Our experimental results show that our approximation on the depth of OPT operation is reasonable and useful.

In the future, we are going to handle other non-well-designed patterns and deal with more operations such as UNION. Besides, we will extend the approximation method to obtain other approximation queries.

Acknowledgments. This work is supported by the program of the National Key Research and Development Program of China under 2016YFB1000603 and the National Natural Science Foundation of China (NSFC) under 61502336, 61373035. Xiaowang Zhang is supported by the Tianjin Thousand Young Talents Program.

Appendix A SPARQL Queries

Here are the three SPARQL queries in Sect. 5, where Q_1 is left-deep tree, Q_2 is right-deep tree and Q_3 is full tree.

PREFIX rdf:$\langle$http://www.w3.org/1999/02/22-rdf-syntax-ns#$\rangle$

PREFIX ub: $\langle$http://www.lehigh.edu#$\rangle$

Q_1: SELECT * WHERE {{?X rdf:type ub:FullProfessor.?X ub:mastersDegreeFrom ?x1} FILTER (?x1!="$\langle$http://www.University0.edu$\rangle$") OPTIONAL {?X ub:name ?x2} OPTIONAL {?X ub:telephone ?x3} OPTIONAL {?X ub:teacherOf ?x4} OPTIONAL {?X ub:doctoralDegreeFrom ?x5} }

Q_2: SELECT * WHERE { {?X rdf:type ub:FullProfessor.?X ub:mastersDegreeFrom ?x1} FILTER (?x1!="$\langle$http://www.University0.edu$\rangle$") OPTIONAL { {?X ub:name ?x2} OPTIONAL { {?X ub:telephone ?x3} OPTIONAL {{?X ub:teacherOf ?x4} OPTIONAL {?X ub:doctoralDegreeFrom ?x5}}}} }

Q_3: SELECT * WHERE { {{{{{?X rdf:type ub:FullProfessor.?X ub:mastersDegreeFrom ?x1} FILTER (?x1!="$\langle$http://www.University0.edu$\rangle$")} OPTIONAL {?X ub:worksFor $\langle$http://www.Department0.University0.edu$\rangle$}} OPTIONAL {{?X ub:name ?x2} OPTIONAL {?X ub:emailAddress ?x3}}} OPTIONAL {{{?X ub:telephone ?x4} OPTIONAL {?X ub:teacherOf ?x5}} OPTIONAL { {?X ub:doctoralDegreeFrom ?x6} OPTIONAL {?X ub:researchInterest ?x7}}}} OPTIONAL { { { {?Y rdf:type ub:GraduateStudent} OPTIONAL {?Y ub:name ?y1}} OPTIONAL {{?Y ub:takesCourse ?y2} OPTIONAL {?Y ub:telephone ?y3}}}OPTIONAL {{{?Y ub:emailAddress ?y4} OPTIONAL {?Y ub:memberOf ?y5}} OPTIONAL {{?Y ub:advisor ?y6} OPTIONAL {?Y ub:und- ergraduateDegreeFrom ?y7}}}} }

References

1. Abiteboul, S., Richard, H., Vianu, V.: Foundations of Databases. Addison Wesley, Reading (1995)
2. Barcelo, P., Pichler, R., Skritek, S.: Efficient evaluation and approximation of well-designed pattern trees. In: Proceedings of PODS 2015, pp. 131–144. ACM (2015)
3. Broekstra, J., Kampman, A., Harmelen, F.: Sesame: a generic architecture for storing and querying RDF and RDF schema. In: Horrocks, I., Hendler, J. (eds.) ISWC 2002. LNCS, vol. 2342, pp. 54–68. Springer, Heidelberg (2002). doi:10.1007/3-540-48005-6_7

4. Han, X., Feng, Z., Zhang, X., Wang, X., Rao, G., Jiang, S.: On the statistical analysis of practical SPARQL queries. In: Proceedings of WebDB (2016). http://dx.doi.org/10.1145/2932194.2932196
5. Carroll, J.J., Dickinson, I., Dollin, C., Reynolds, D., Seaborne, A., Wilkinson, K.: Jena: implementing the semantic web recommendations. In: Proceedings of WWW 2004, pp. 74–83 (2004)
6. Kaminski, M., Kostylev, E.V.: Beyond well-designed SPARQL. In: Proceedings of ICDT 2016, pp. 5:1–5:18 (2016)
7. Klyne, G., Jeremy, C.J., McBride, B.: Resource description framework (RDF): concepts and abstract syntax. W3C Recommendation (2004)
8. Letelier, A., Prez, J., Pichler, R., Skritek, S.: Static analysis, optimization of semantic web queries. Proc. PODS **38**(4), 84–87 (2012)
9. Nah, F.H.: A study on tolerable waiting time: How long are web users willing to wait? Behav. Inf. Technol. **23**(3), 153–163 (2003)
10. Prez, J., Arenas, M., Gutierrez, C.: Semantics and complexity of SPARQL. ACM Trans. Database Syst. **34**(3), 30–43 (2009)
11. Prud'Hommeaux, E., Seaborne, A.: SPARQL query language for RDF. W3C Recommendation (2008)
12. Saleem, M., Ali, M.I., Hogan, A., Mehmood, Q., Ngomo, A.C.N.: LSQ: the linked SPARQL queries dataset. In: Arenas, M., et al. (eds.) ISWC 2015. LNCS, vol. 9367, pp. 261–269. Springer, Heidelberg (2015). doi:10.1007/978-3-319-25010-6_15
13. Wood, P.: Query languages for graph databases. SIGMOD Rec. **41**(1), 50–60 (2012)
14. Zhang, X., Van den Bussche, J.: On the primitivity of operators in SPARQL. Inf. Process. Lett. **114**(9), 480–485 (2014)
15. Zhang, X., Van den Bussche, J.: On the power of SPARQL in expressing navigational queries. Comput. J. **58**(11), 2841–2851 (2016)
16. Zhang, X., Van den Bussche, J., Picalausa, F.: On the satisfiability problem for SPARQL patterns. J. Artif. Intell. Res. **56**, 403–428 (2016)

Author Index

Printed in the United States
By Bookmasters